Ecohydraulics: An Integrated Approach

Ecohydraulics: An Integrated Approach

Edited by Paul Stanley

SYRAWOOD
PUBLISHING HOUSE

New York

Published by Syrawood Publishing House,
750 Third Avenue, 9th Floor,
New York, NY 10017, USA
www.syrawoodpublishinghouse.com

Ecohydraulics: An Integrated Approach
Edited by Paul Stanley

International Standard Book Number: 978-1-68286-528-6 (Hardback)

Cataloging-in-Publication Data

Ecohydraulics : an integrated approach / edited by Paul Stanley.
    p. cm.
Includes bibliographical references and index.
ISBN 978-1-68286-528-6
1. Ecohydrology. 2. Hydrology. I. Stanley, Paul.
QH541.15.E19 E26 2018
577.6--dc23

TABLE OF CONTENTS

Permissions

List of Contributors

Index

PREFACE

The study of water resources as well as water distribution and regulation is known as hydrology. Water is a scarce resource and there are unique processes involved in the restoration and movement of water on Earth. The preservation of ecological conditions that replenish sources of water is crucial for biodiversity as well as environmental processes. Watersheds, basins and sand bars are unique formations that occur through the various mechanisms that water sources go through. The various advancements in ecohydrology are glanced at and their applications as well as ramifications are looked at herein. With state-of-the-art inputs by acclaimed experts of this field, this book targets students and professionals. For all those who are interested in hydrology, this book can prove to be an essential guide.

After months of intensive research and writing, this book is the end result of all who devoted their time and efforts in the initiation and progress of this book. It will surely be a source of reference in enhancing the required knowledge of the new developments in the area. During the course of developing this book, certain measures such as accuracy, authenticity and research focused analytical studies were given preference in order to produce a comprehensive book in the area of study.

This book would not have been possible without the efforts of the authors and the publisher. I extend my sincere thanks to them. Secondly, I express my gratitude to my family and well-wishers. And most importantly, I thank my students for constantly expressing their willingness and curiosity in enhancing their knowledge in the field, which encourages me to take up further research projects for the advancement of the area.

Editor

The role of oxygen in degradation of Hydrocarbons in sediments of an Estuary in Nigeria

Dike Henry Ogbuagu[1, *], Ebenezer Temitope Adebayo[2], Emmanuel Ikechukwu Iwuchukwu[3]

[1]Department of Environmental Technology, Federal University of Technology, PMB 1526, Owerri, Nigeria
[2]Department of Fisheries and Aquaculture Technology, Federal University of Technology, Owerri, Nigeria
[3]Department of Agricultural Engineering, Federal Polytechnic, Nekede, Owerri, Nigeria

Email address:
henrydike2002@yahoo.com (Ogbuagu D. H.), adebayotemitope.et@gmail.com (Adebayo E. T.),
iwuchukwuemmanuel14@gmail.com (Iwuchukwu E. I.)

Abstract: The rates of degradation of Aliphatic (AHs) and Polynuclear Aromatic Hydrocarbons (PAHs) in sediments of the Bonny Estuary were investigated. Degradation was enhanced by aerobic condition in the order of treatments 10000ppm>1000ppm>100ppm (AHs), 100ppm>1000ppm>10000ppm (PAHs) than in the anaerobic condition which was in the order 10000ppm>100ppm>1000ppm (AHs), 100ppm>1000ppm>10000ppm (PAHs). Rates of degradation differed over the harvest days [$F_{(10.90)}>F_{crit(4.17)}$] and between the oxygen conditions [$F_{(10.92)}>F_{crit(4.17)}$] ($P<0.05$). Degradation enhancements by oxygenation were by 9.85, 63.64 and 62.66% (AHs) and 2.07, 3.15 and 3.95% (PAHs) on Days 14, 28 and 42 respectively. Rate of degradation was dose-dependent and slower especially of the PAHs.

Keywords: Degradation, Hydrocarbons, Bonny Estuary, Marine Sediments, Anoxia

1. Introduction

Sediments on the sea floor are formed by particulate matter that settles out of the water column, and may consist of anything from coarse gravel and sand to clay and organic ooze. In natural aquatic ecosystems, persistent organic pollutants like the toxic metals and hydrocarbons occur in low concentrations mainly due to weathering of soils and their associated bedrocks, as well as geological dispositions. However, due to human activities in recent times, there has been an unprecedented increase in the level of these pollutants, and the occurrence of such species in excess of natural load has become a problem of increasing concern not only to environmentalists but also to health practitioners.

For example, of more than 120 polluted sediment sites investigated in harbours and fjords in Norway between 1993-1996 alone, about 90 were heavily polluted with contaminants such as PCBs, PAHs, tributyl-triphenyl tin compounds, mercury, lead and cadmium [1]. Skei [2] and Karimov [3] have also made related observations. Contaminated sediments are most commonly found near population centres and harbours [4-9] like the Bonny Estuary

in the Niger Delta, where there are many sources of pollution of various kinds. The Bonny Estuary has industrial installations, including the oil, gas, and ancillary operations facilities located proximal to the shores, and so, contamination of sediments could be a major environmental problem. According to Ogbuagu et al. [10, 11], pollutants in sediments can spread to the surrounding water, re-suspend when sediments are disturbed, or be absorbed by benthic organisms through bioaccumulation. Worse still, because of certain mechanisms, contaminated sediments may continue to release hazardous chemicals to the surroundings long after the land-based sources of the pollution have been eliminated. Consequently, contaminated sediments can have serious effects on living organisms [10, 12-16] and the entire ecosystems [17], including higher mortality, reduced growth rates and impaired reproductive processes in marine organisms. Because many contaminants accumulate in food chains, they can also affect human health through trophic relationships.

Sediments of marine ecosystems are usually depleted of dissolved oxygen; a condition that is generally found in areas that have restricted water exchange. In most cases, oxygen is prevented from reaching the deeper levels by a physical barrier such as mud [18] as well as by a pronounced density stratification in which, for instance, heavier hyper-saline waters rest at the bottom of a basin. Anoxia, among other effects could tremendously slow down the rate of biodegradation of pollutants, especially of the already persistent ones such as crude oil and petroleum pollutants. With increasing inputs of such pollutants into the Bonny Estuary by the many oil, gas, shipping, and ancillary companies, there is need to investigate the fate of such POP as petroleum hydrocarbons in a simulated anoxic and oxygenated conditions as to infer natural removal of pollutant loadings over time.

2. Materials and Methods

2.1. Study Area

Bonny Island, with hot and humid weather conditions experiences heavy rains for about 9 months in the year; with average precipitation of 3800mm, ambient temperature of 33°C during the day, and humidities of almost 100%. The coastal settlement is located between latitudes 4° 22′ and 4° 32′N and longitudes 7° 12′ and 7° 18′E (Fig. 1). The Bonny Estuary is the main passage for merchant shipping to the inland Port of Onne and for tankers and merchant shipping going to Port Harcourt and the Nigerian National Petroleum Company refined products terminal at Okirika in the Niger Delta region of Nigeria. The estuary receives effluents and other wastes from a number of industries and ancillary companies on the Island. The Bonny Oil and Gas Terminal (BOGT), the Nigeria Liquefied Natural Gas complex (NLNG), and other service companies are all located close to the estuary.

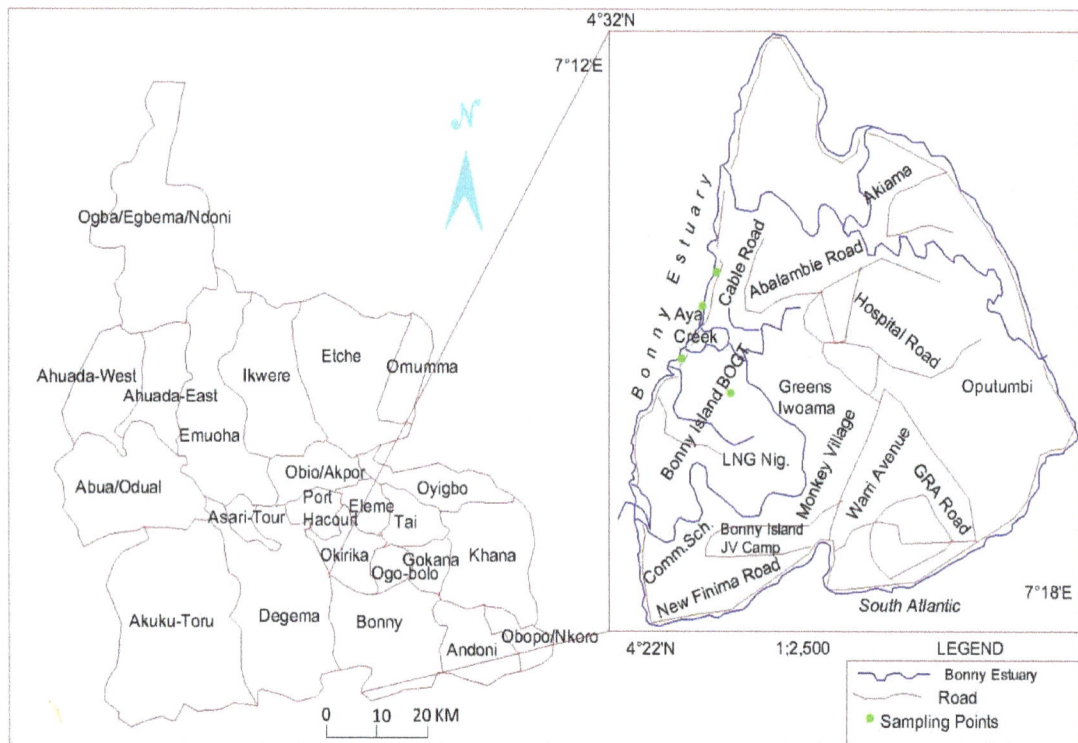

Fig 1. *Map of Rivers State showing Bonny Estuary, west of the Bonny Local Government Area*

2.2. Sediment Sample Collection

Sediments were collected from seabed at two points (Coordinates 4° 25' 698" Northing, 7° 09' 325" Easting and 4° 26' 546" Northing, 7° 09' 475" Easting for Upstream and Downstream respectively) that were relatively pristine, between the NLNG and The Shell Petroleum Development Company Limited (SPDC) BOGT facility in the Estuary. A pre-grab sampler was deployed from the deck of a vessel and successive grab samples from about 20meters depth of the river were emptied into 5 liters containers completely lined internally with aluminum foil. Labeled samples were transported to the Laboratory and preserved in the refrigerator until they were needed.

2.3. Laboratory Procedures

2.3.1. Test Sample

The Bonny Light crude oil test sample was obtained from the SPDC BOGT facility and preserved in the refrigerator until when needed.

2.3.2. Treatment of Sediment Samples

The sediments were sieved with their habitat water through 2mm, 1mm and 0.5mm sieves to remove large stones, mollusks, crustaceans and organic debris, and sieved sediments allowed to settle over night before being treated with various concentrations of crude oil.

Successive one kilogram of sieved sediments were

weighed and mixed with pre-determined quantities of crude oil (with a total organic content of 85.4%) in such a manner as to achieve the required concentrations of 100ppm, 1000ppm and 10000ppm. These concentrations were based on the dry weight of the sediment. The crude oil concentrations and sediments were homogenized by mixing thoroughly with a laboratory desk top mixer for about 10 minutes in increasing order of concentration. The mixture container was washed and properly cleaned prior to mixing each new treatment.

The treated and homogenized sediments were dispensed into separately labeled glass petri dishes, making sure that there were no air bubbles trapped or overflow of sediment. Each concentration was prepared for four harvest day in both anoxic and aerobic conditions.

Samples for anoxic condition monitoring were packed inside anaerobic jar while those for the aerobic condition monitoring were kept under the laboratory condition. The anaerobic jar was kept in an incubator at the temperature of 18°C. Sediments were then harvested on Days 0, 14, 28, and 42 for all the concentrations (100ppm, 1000ppm and 10000ppm).

2.3.3. Determination of Hydrocarbons

Five grams of harvested sediment sample was weighed into a beaker and 10mL of analytical grade hexane added to it. The mixture was shaken for 5minutes, filtered, and filtrates were used for the determination of hydrocarbons using the flame ionization detector Gas Chromatography (GC). This involved setting up the GC conditions (oven temperature, detector temperature, gas flow rates) and injecting the filtrates into the GC for run prompting. For every hydrocarbon parameter measured, standards and blanks were prepared and used for calibration of the instrument. The calibrated chromatograms were used to quantify the concentration of hydrocarbons in the extracted samples.

2.4. Statistical Analysis

Bivariate and descriptive statistical analyses as provided by the SPSS v.17.0 and the MS Excel 2007 softwares were used to analyze data. Variation plots were used to represent degradation levels of the hydrocarbons in sediment samples over the 42 days of experiment; the one-way ANOVA was used to test for homogeneity in mean levels of degradation over the experimental period, and post-hoc structure of group means was detected with means plots at $P<0.05$.

3. Results

3.1. Degradation of Hydrocarbons

Of the aliphatic hydrocarbons (AHs), the 100, 1000 and 10000ppm crude oil treatments degraded from 98.25, 980.28 and 9850.25ppm at the commencement of the experiment, through 86.45, 758.45 and 7035.74ppm on Day 14, 65.38, 608.15 and 2278.49ppm on Day 28 to 48.25, 485.35 and 1544.80ppm on Day 42 in the aerobic setup. This translates

to percentage degradation rates of 12.01, 22.63 and 28.57% on Day 14, 33.46, 37.96 and 76.86% on Day 28 and 50.89, 50.49 and 83.31% on Day 42 respectively (Fig. 2).

In the anaerobic setup, degradation in the 100, 1000 and 10000ppm treatment concentrations proceeded from 98.25, 980.28 and 9850.25ppm on Day 0, through 98.25, 880.35 and 7763.37ppm on Day 14, 75.88, 784.25 and 7257.85ppm on Day 28 to 29.87, 28.78 and 51.28ppm on Day 42. This amounts to percentage degradations of 0.00, 10.19 and 21.19% on Day 14, 22.77, 19.99 and 27.83% on Day 28 and 29.87, 28.78 and 51.28% on Day 42 (Fig. 3).

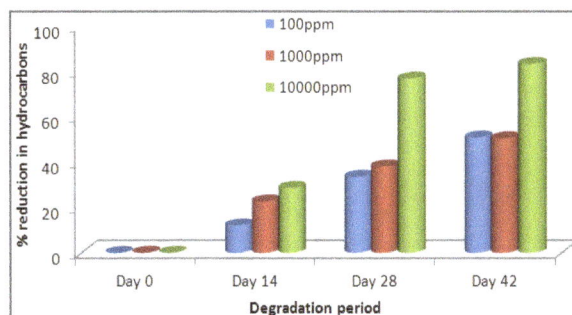

Fig 2. Mean residual concentrations of aliphatic hydrocarbons after 42 days under aerobic condition

Of the PAHs, degradation in the aerobic setups proceeded from 4.80, 5.45 and 10.84ppm at the commencement of the experiment through 4.35, 5.30 and 10.25ppm on Day 14, 4.18, 5.08 and 10.12ppm on Day 28 to 3.98, 4.95 and 10.05ppm at the end of the experiment. This gives the percentage degradation rates of 9.38, 2.75 and 1.19% on Day 14, 12.92, 6.79 and 6.64% on Day 28 and 17.08, 9.17 and 7.29% at the end of the experiment (Fig. 4).

However, in the anaerobic setups, degradation in the PAHs components of the three crude oil treatments proceeded from 4.80, 5.45 and 10.84ppm at the commencement of experiment through 4.40, 5.40 and 10.50ppm on Day 14, 4.32, 5.25 and 10.45ppm on Day 28 to 4.28, 5.10 and 10.38ppm on Day 42. This translates to percentage degradations of 8.33, 0.92 and 3.14% on Day 14, 10.00, 3.67 and 3.59% on Day 28 to 10.83, 6.42 and 4.24% at the end of the experiment (Fig. 5).

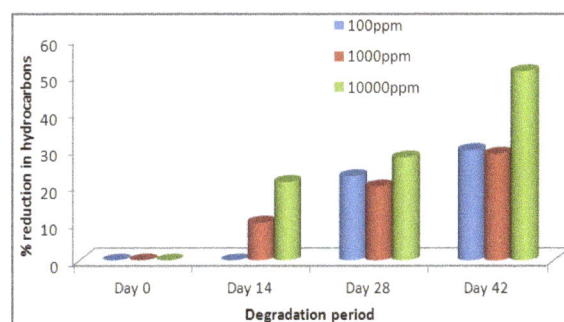

Fig 3. Mean residual concentrations of aliphatic hydrocarbons after 42 days under anaerobic condition

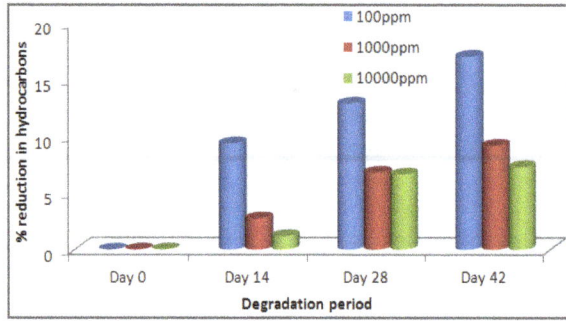

Fig 4. Mean residual concentrations of polynuclear aromatic hydrocarbons after 42 days under aerobic condition

A test of homogeneity in mean variance revealed significant difference in rates of degradation of combined hydrocarbons on each of Days 14, 28 and 42 [$F_{(10.90)}>F_{crit(4.17)}$] at $P<0.05$.

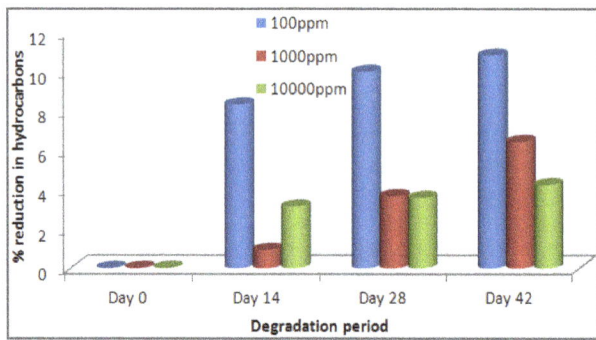

Fig 5. Mean residual concentrations of polynuclear aromatic hydrocarbons after 42 days under anaerobic condition

A post-hoc structure of group means that utilized Day 0, the commencement of the experiment as predictor variable revealed that on Day 14 (Fig. 6), Day 28 (Fig. 7) and Day 42 (Fig. 8), the aliphatic hydrocarbon treatment (3642.93 ppm concentration) contributed the heterogeneity most.

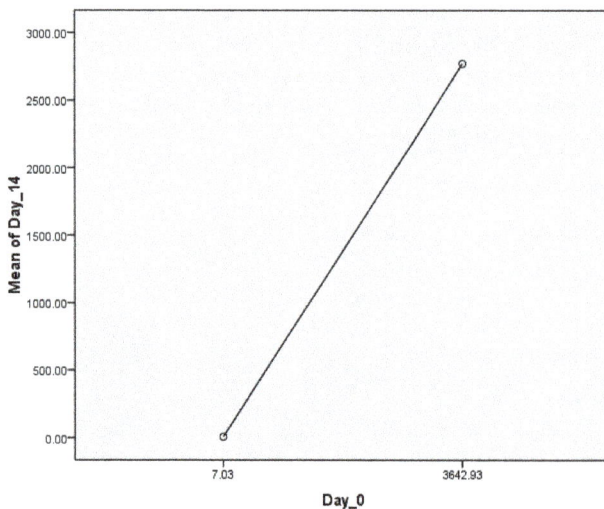

Fig 6. Means plot in rate of degradation of hydrocarbons between Day 0 and Day 14

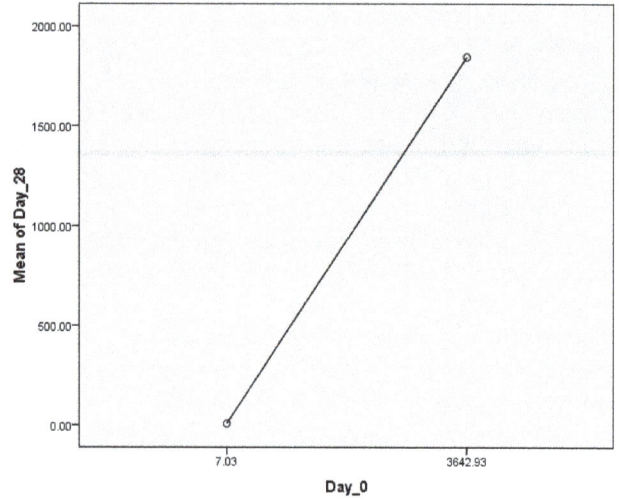

Fig 7. Means plot in rate of degradation of hydrocarbons between Day 0 and Day 28

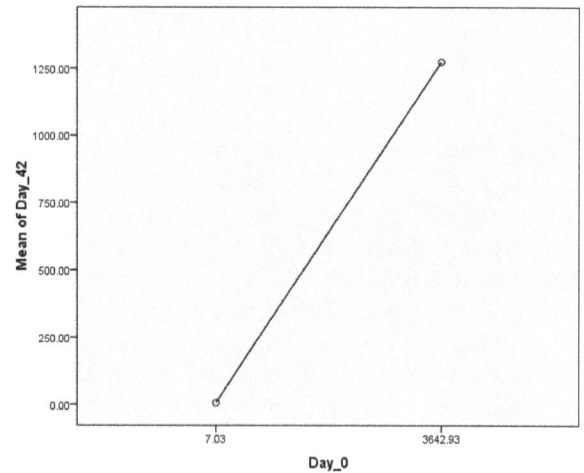

Fig 8. Means plot in rate of degradation of hydrocarbons between Day 0 and Day 42

3.2. Comparison of Rates of Degradation under Aerobic and Anaerobic Conditions

Mean degradations in pooled concentrations of the AHs and PAHs over the 42 days experimental period under aerobic and anaerobic conditions are presented in Table 1. In the aerobic setup, the aliphatics degraded from mean concentration of 3642.93 ppm on Day 0, through 2626.88 ppm on Day 14, 984.01 ppm on Day 28 to 692.80 ppm on Day 42. However, in the anaerobic setup, they degraded through 2913.99 ppm on Day 14, 2705.99 ppm on Day 28 to 1855.34 ppm on Day 42.

Table 1. Mean concentrations of hydrocarbons in aerobic and anaerobic degradation conditions

Hydrocarbons(ppm)	Day 0	Day 14	Day 28	Day 42
Aliphatics (aerobic)	3642.93	2626.88	984.01	692.80
Aliphatics (anaerobic)	3642.93	2913.99	2705.99	1855.34
PAHs (aerobic)	7.03	6.63	6.46	6.33
PAHs (anaerobic)	7.03	6.77	6.67	6.59

Of the PAHs, mean concentrations degraded from 7.03ppm at the commencement of experiment, through 6.63, 6.46 to 6.33ppm in the aerobic setup, and 7.77, 6.67 to 6.59ppm in the anaerobic setup on Days 14, 28 and 42 respectively.

The One-way ANOVA test revealed significant heterogeneity in mean concentrations of the petroleum hydrocarbons treatments under aerobic and anaerobic degradation conditions [$F_{(10.92)} > F_{crit(4.17)}$] at $P < 0.05$. Oxidation therefore enhanced the degradation of the AHs by 9.85% on Day 14, 63.64% on Day 28 and 62.66% on Day 42 (Fig. 9). However, in the PAHs, oxygenation enhanced degradation by 2.07% on Day 14, 3.15% on Day 28 and 3.95% on Day 42.

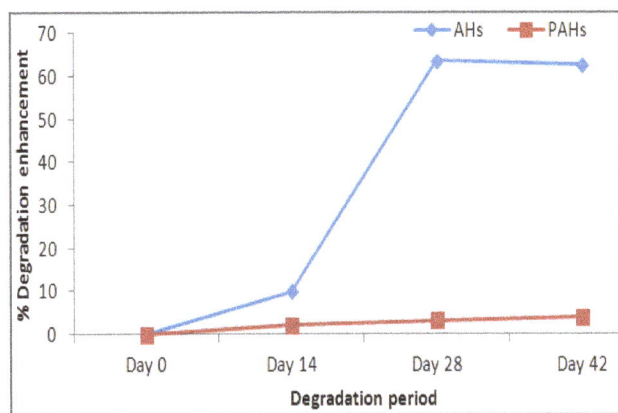

Fig 9. Percentage oxidative enhancement of degradation of the aliphatic (AHs) and polynuclear aromatic hydrocarbons (PAHs)

4. Discussion

The current study clearly indicates that the polynuclear aromatic hydrocarbons are more recalcitrant to biodegradation than the aliphatics, especially in anoxic conditions, as exemplified in sediments of the marine environment under simulation. The persistence of petroleum hydrocarbons, as members of the Persistent Organic Pollutants (POPs) group in sediments of Estuaries had also been noted by Unlu et al. [19] in the Ambarli Port area of the Sea of Marmara in Turkey, Chen and Chen [20] in sediments of Kaohsiung Harbor in Taiwan and Mostafa et al. [21] in sediments of the western harbour of Alexandria in Egypt.

Generally, degradation rates were slow in the 100 and 1000ppm aliphatic hydrocarbons treatments and even slower in all the concentrations of the polynuclear aromatic hydrocarbons. Of the aliphatic hydrocarbons in aerobic and anaerobic conditions, the 100000 ppm concentration treatment degraded fastest, followed by the 1000 and 100 ppm treatments respectively. This indicates a dose-related degradation rate of the hydrocarbons. The effects of persistent pollutants on aquatic fauna, especially in the benthic regions of have been elucidated by Ogbuagu et al. [10]. However, of the polynuclear aromatic hydrocarbons under both aerobic and anaerobic conditions, the 100 ppm treatments degraded fastest, followed by the 10000ppm concentrations. This indicates an inverse dose-related pattern; a direct opposite of the degradation patterns of the aliphatic

hydrocarbons category.

Degradation rates clearly differed as the period progressed from the 14th, through 28th and to the 42nd day of experimentation. Maximum degradations were recorded at the end of the experiments and this implies that duration or time is a determinant factor in biodegradation of pollutants in aquatic environments. However, this marked difference in degradation rates appeared to be mostly associated with the initial concentrations of the aliphatic hydrocarbons.

Oxygenation (aerobiosis) clearly aided degradation of both the aliphatic and polynuclear aromatic hydrocarbons, as confirmed by the ANOVA test. This underpins the role of oxygen in the enhancement of biodegradation process by aerobic microorganisms, and further underscores the role of oxygen in sediment pollutants detoxification and purification of water bodies.

5. Summary

Degradation rates of the petroleum hydrocarbons was dose-dependent and that of the polynuclear aromatic hydrocarbons was very slow in marine sediments. Rates of degradation differed markedly during the experimental period and was enhanced by aerobiosis.

6. Conclusion

This study revealed that petroleum hydrocarbons were persistent in the environment and that their degradations in marine sediments were enhanced by the presence of oxygen.

Acknowledgements

We are grateful to Transcontinental Petrotech Nigeria Limited for granting us participation in this study.

References

[1] Norwegian Environment Agency (NEA), Polluted marine sediments, (2009). Retrieved 25/07/2013 from http://www.environment.no/Topics/Marine-areas/Hazardous-chemicals-in-coastal-waters/Polluted-marine-sediments/

[2] Skei J (2009), The entrapment of pollutants in Norwegian Fjord sediments- A beneficial situation for the North Sea, In: Holocene marine sedimentation in the North Sea Basin, S.D. Nio, R.T.E. Shüttenhelm & Tj. C.E. Van Weering (Eds.), Blackwell Publishing Ltd., Oxford, UK. doi: 10.1002/9781444303759.ch32.

[3] Karimov M (2007), Heavy metals in the bottom sediments of the Western Norwegian fjords, MSc Thesis, Geology Department, University of Tartu, pp.122.

[4] Ahnert A and Borowski C (2000), Environmental risk assessment of anthropogenic activity in the deep sea. Jr. Aquat. Ecosyst. Stress & Recov.; 7(4): 299. Available at http://web.ebscohost.com/ehost/pdf?vid=5&hid=2&sid=4b3a3 0cd-c7ec-4838-ba3c-48ce12f26813%40sessionmgr12. Accessed 11 October 2013.

[5]   Daoji L and Dag D (2004), Ocean pollution from land-based sources: East China Sea. AMBIO– A Jr. of the Human Envt., 33(1/2): 107-113.

[6]   Kenneth RW (2006), Plague of Plastic Chokes the Seas. Los Angeles Times Archived from http://www.pulitzer.org/year/2007/explanatory-reporting/works/oceans04.html. Accessed 2008-04-01.

[7]   Milkov AV (2004), Global estimates of hydrate-bound gas in marine sediments: how much is really out there?. Earth Sci .Rev., 66(3–4): 183-197. Bibcode:2004ESRv...66..183M. doi:10.1016/j.earscirev. Accessed 2 November 2003.

[8]   Rios LM, Moore C and Jones PR (2007), Persistent organic pollutants carried by Synthetic polymers in the ocean environment. Marine Pollut. Bull., 54(8): 1230–1237. doi:10.1016/j.marpolbul.2007.03.022. PMID 17532349.

[9]   Stull JK (1989), Contaminants in sediments near a major marine outfall: history, effects and future. OCEANS '89 Proceedings 2: 481-484.

[10] Ogbuagu DH, Njoku JD and Ayoade AA (2011a), Trends in macrobenthal biotypes of Imo River in a Nigerian Delta region. Jr. Biodiv. Envir. Sci. (JBES), 1(4): 22-28. Available at http://www.innspub.net. Accessed 11 October 2013.

[11] Ogbuagu DH, Okoli CG, Emereibeole EI, Anyanwu IC, Onuoha O, Ubah NO, Ndugbu CO, Okoroama ON, Okafor A, Ewa E, Ossai R and Ukah F (2011b), Trace metals accumulation in biofilms of the upper and middle reaches of Otamiri river in Owerri, Nigeria. Jr. Biodiv. Envir. Sci. (JBES), 1(3): 19-26. Available at http://www.innspub.net . Accessed 10 October 2013.

[12] Kroeger T (2012), Dollars and Sense: Economic Benefits and Impacts from two Oyster Reef Restoration Projects in the Northern Gulf of Mexico. TNC Report. Retrieved from http://www.nature.org/ourinitiatives/regions/northamerica/oyster-restoration-study-kroeger.pdf.

[13] Moore C, Moore SL, Leecaster MK and Weisberg SB (2001), A Comparison of Plastic and Plankton in the North Pacific Central Gyre. Marine Pollut. Bullet., 42(12): 1297–1300. doi:10.1016/S0025-326X(01)00114-X. PMID 11827116.

[14] Newell RIE, Cornwell JC and Owens MS (2002), Influence of simulated bivalve biodepositon and microphytobenthos on sediment nitrogen dynamics, a laboratory study. Limnol. and Oceanogr., 47:1367-1379.

[15] Orr JC, Fabry VJ, Aumont O, Bopp L, Doney SC et al. (2005), Anthropogenic ocean acidification over the twenty-first century and its impact on calcifying organisms. Nature, 437 (7059): 681–686. Bibcode:2005Natur.437.681O. doi:10.1038/nature04095. ISSN 0028-0836. PMID 16193043.

[16] Tanabe S, Watanabe M, Minh TB, Kunisue T, Nakanishi S, Ono H and Tanaka H (2004), PCDDs, PCDFs, and coplanar PCBs in albatross from the North Pacific and Southern Oceans: Levels, patterns, and toxicological implications. Envir. Sci. & Technol., 38 (2): 403–413. doi:10.1021/es034966x.

[17] Abowei JFN and Sikoki FD (2005), Water Pollution Management and Control. Doubletrust Publication Company, Port Harcourt, 236pp.

[18] Gerlach (1974), Marine Pollution, Springer, Berlin.

[19] Unlu S, Sari E, Apak R, Balci N and Koldemir B (2013), Distribution and source identification of polycyclic aromatic hydrocarbons in sediments of the Ambarli Port area, Sea of Marmara, Turkey. Global Adv. Res. Jr. Envir. Sci. Toxicol. (GARJEST), 2(6):144-151. Available online at http://garj.org/garjest/index.htm

[20] Chen CW and Chen CF (2011), Distribution, origin, and potential toxicological significance of polycyclic aromatic hydrocarbons (PAHs) in sediments of Kaohsiung Harbor, Taiwan. Mar. Pollut. Bull., 63: 417-423.

[21] Mostafa AR, Barakat AO, Qian Y and Wade TL (2003), Composition, distribution and sources of polycyclic aromatic hydrocarbons in sediments of the western harbour of Alexandria, Egypt. Jr. Soils Sed., 3: 173–179.

Desalination using algae ponds under nature Egyptian conditions

El Nadi M. H. A.[1], El Sergany F. A. G. H.[2], El Hosseiny O. M.[3]

[1]Public Works Dept., Faculty of Eng., ASU, Cairo, Egypt
[2]Civil Eng. Dept., HIT, 10th of Ramadan City, Egypt
[3]Civil Eng. Dept., Future Univ., Cairo, Egypt

Email address:
mhnweg@yahoo.com (E. Nadi M. H. A.), fatenelsergany@gmail.com (E. Sergany F. A. G. H.), Cloudvii@gmail.com (E. Hosseiny O. M.)

Abstract: This study aims to study the suitability of using algae for water desalination as a new conceptual technique using algae ponds under the nature circumstances. A pilot unit consisted of three serial basins each divided to three parallel basins was erected outdoor in open area to be under natural climatic conditions. The algae were prepared in the lab using BG11 media to encourage its growth under saline water then put in the basins with 0.4 lit/path/run for algae rate and 0.1 lit/path/run for BG11 rate. The basins fed by several concentrations from saline water from 2000 ppm till 40000 ppm with several runs to get the TDS removal efficiency. The experimental work shows that the removal efficiency for TDS in the first basin varied between 61& 80.5% and became after second basin between 71& 93% which increased by third basin to 83 to 97.7%. These variations were due to inlet TDS concentration and the climatic conditions (Temperature & sunlight period). The results achieved *high removal efficiency for TDS* and the results in the highest inlet concentration were inside the permissible limits for drinking water after the third basin.

Keywords: Water Desalination, Innovative Techniques for Desalination, Biological Desalination, Desalination by Algae

1. Introduction

Several microorganisms are applied to remove salts from water. Some artificial bacteria, which are still under research, were applied for water desalination 3 years ago. Also, some algae spices from 5 years ago started to be applied under research manner to desalinate sea water with successful promising results.

Algae are simple plants without roots, stems, and leaves contain chlorophyll. They are a heterogeneous group of organisms and vary in size from microscopic unicellular forms to sea weeds of many feet in length with many different shapes. Algae kinds grow wherever in water either salinity water in seas or fresh water in rivers and lakes and under different weathers from snow weather to hot climatic, wherever they could synthesize their components and food by photo-synthesis [1].

Scendesmus species growth was successfully obtained in saline water as it absorbs salts and makes use of them in its metabolism, while *Chlorella* and *Scenedesmus* are described as the most active Algae in stabilization ponds, because they are extremely, and commonly exhibit extremely wide range of salt tolerance in their habitat. [2]

Micro-algae have a high capacity for inorganic nutrient uptake and can be used in mass culture in outdoor solar bioreactors. Unicellular green algae such as *Chlorella* and *Scendesmus* have been widely used in wastewater treatment because the often colonize the ponds naturally, and they have fast growth rates and high nutrient uptake capabilities. However, one of the major drawbacks of using micro-algae in wastewater treatment is the harvesting of biomass. [3]

Green algae was used for treatment of industrial wastewater from natural gas production fields of 1500 m^3 daily flow, which contains salinity up to 25000 ppm and oil content up to 100 ppm. The effluent could be re-used in irrigation of crops with a flow enough for about 60 fed./day. While the produced algae 1.5 ton/day that could be used in medical industry, pigments for industry, functional food, bio-fertilizers, and animal or fish fodders [4].

Another algae application in stabilization ponds for industrial wastewater treatment of high salinity and oil content was made through a pilot plant which was built in the site of N/D field in Abu-Mady north of Egypt. Several runs were made to ensure the results. The influent loads were varied between 35000 & 10900 ppm for TDS, 250 & 65 ppm for Oil, 212 & 59ppm for BOD & 344 & 88ppm for COD. The effluent concentrations after 7 days retention time in the pond were ranged between 2750 & 2120 ppm for TDS, 2 & 0.5ppm for Oil, 7,4 ppm for BOD & 16 & 12 ppm for COD. These removal efficiencies proved the suitability of algae application for saline industrial wastewater treatment [5].

Desalination based on the use of algae in the removal of salts from saline water, and water production for use in different purposes is a new concept and it has been used and tested in previous researches in industrial waste water treatment, where using the algae reduced the cost to the minimum while maintained the efficiency with no reduction. The achieved results were promising and good in the desalination of sea water and is successful, continuing to reach the removal efficiency up to 95% till the rates are relatively affordable for possible use in different purposes, opening the door to a new direction may succeed in solving the problem of water desalination with cost reduction to the minimum possible [6].

A study was made to assess the removal efficiencies of different nutrients in saline water by the means of the green algae *Scendesmus species*. Total dissolved solids, Sodium, Chloride, and Phosphate removal efficiencies measured in the output of the reactor after 14 days reached around 97%. While the removal efficiencies of both Nitrate and Sulfate reached around 93% and all the parameters in the effluent were inside the permissible limits for potable water except the presence of some enzymes due to algae decay. This shows the success of this system to produce suitable potable water from sea water. [7]

In March 2011, another study was made. The objective of this study was to assess the removal efficiencies of different nutrients in saline water by the means of the green algae *Scendesmus species* through a continuous flow treatment system. The saline water from Red Sea, Ismailia Governorate (about 40000 ppm) was used. Algae were added to two consecutive reactors in which a 7 days retention time was applied in each to prevent the enzymes formations in the basins. The experiment in the continuous flow was repeated for 4 runs and the average values were taken into consideration. Analysis of growth media was daily determined. Total dissolved solids, Sodium, Chloride, and Phosphate removal efficiencies measured in the output of the second reactor reached around 97%. While the removal efficiencies of both Nitrate and Sulfate reached around 93% [8, 9].

El Hosseiny, O.M., et al,[10] in their study proved the suitability of applying algae for biological desalination under nature Egyptian conditions with high removal ratios for TDS depending on the inlet water salinity and both of the sunlight period and the temperature.

2. Materials & Methods

A new system for water desalination was applied by using Algae for a natural, environment friendly and economic

desalination. The *Scendesmus* algae species was the choice for operation process, thanks to its natural feature of growing very well in almost any mineral medium. The application was successfully worked under suitable conditions inside the lab as illustrated by El Nadi [5][6][9], El Sergany [5][6], Saad [7] & Badawy [8]. In this study the system is built outside the lab to meet the normal nature conditions.

The pilot plant located on the roof of the Material Lab of the Ain Shams university engineering faculty specified for open air pilots. The pilot plant consists of three storage tanks, three basins each divided to three equal paths and equipped with manual algae separators units. Figure (1) illustrates the pilot photo

Fig (1). Photo of the applied Pilot for algae Ponds

The pilot plant was operated from February 2013 till June 2013 to obtain the system suitability. It consisted of three runs, each with three different TDS concentrations for the three parts of each basin. Several concentrations of saline water 2000, 4000, 10000, 15000, 20000, 25000, 30000, 35000 & 40000 ppm were applied three in each run. The feeding media for algae BG-11 solution was added to give algae enough nutrition with rate 0.1 lit/path/run. *Scendesmus* algae were added by rate 0.4 lit/path/run to treat the saline water for a retention time of 7 days in first basin. Water after separating algae in the first basin was then transferred to the second basin. *Scendesmus* algae in second basin did the second stage of treatment for another 7 days. Then the work repeated in the third basin. Desalinated water was collected in the effluent after the third basin. Table (1) shows the chemical composition of BG-11 nutrient solutions.

Table (1). Chemical composition of BG-11 nutrient solutions [6]

Chemical	BG-11(g/l)
$NaNO_3$	1.5
$K_2HPO_4 3H_2O$	0.04
Na_2CO_3	0.02
$MgSO_4.7H_2O$	0.075
$CaCl_2.2H_2O$	0.036
EDTA-Na_2	0.001
$Fe(NH_3)_2$Citrate	0.006
Citric acid	0.006

Samples were taken each day from each part of each basin, as well as the effluent tank. The sample volume was 50 ml. Water samples were routinely collected at 9:00 am each morning and analyzed to investigate water quality during the examination period. The measured parameters were as Total dissolved solids (TDS), Air Temperature, Humidity, Sunlight period & Algae growth rate.

3. Results & Discussions

The first run started with small salts concentrations. The feeding tanks had concentrations of 2000, 4000 and 10000 mg/l. The run was applied for 21 days through the three

basins for the three parallel lines.

The study measured the TDS concentrations at each basin effluent in addition to the climatic conditions during the run operation days. The results of the measured data during First, second & third runs are shown in tables (2), (3) & (4)

The pilot was found successful in removing TDS concentrations from the medium using Algae with BG11 media and under Natural variable conditions. However, the Removal efficiencies average differed according to inlet TDS concentration. Table (5) and figure (2) illustrated the variations in removal ratios due to the inlet TDS concentration.

Table (2). *Measured Parameters during First Run*

Basin	Date	Air Climatic Conditions				TDS mg/l		
		Temperature °C		Humidity %	Sun Light Period (hr)	Left Basin	Middle Basin	Right Basin
		Day	Night					
	1/3/13	21	11	54	12.00	2000	4000	10000
	2/3/13	28	12	42	12.02	1020	3200	7920
	3/3/13	29	17	22	12.05	920	2920	6130
Basin 1	4/3/13	21	15	57	12.07	890	2610	5080
	5/3/13	19	12	60	12.10	860	2370	4000
	6/3/13	20	11	53	12.12	820	2020	3333
	7/3/13	23	9	44	12.15	780	1730	2990
	8/3/13	27	12	34	12.18	780	1730	2990
	9/3/13	23	14	63	12.20	730	1460	2420
	10/3/13	27	13	46	12.22	690	1190	2000
Basin 2	11/3/13	30	17	30	12.25	650	980	1709
	12/3/13	34	22	18	12.28	620	910	1410
	13/3/13	28	17	51	12.30	600	840	1209
	14/3/13	36	16	36	12.32	580	790	1170
	15/3/13	22	17	27	12.35	580	790	1170
	16/3/13	20	15	53	12.37	510	750	950
	17/3/13	14	14	48	12.40	450	710	850
Basin 3	18/3/13	22	12	52	12.43	400	680	770
	19/3/13	24	11	57	12.46	380	650	730
	20/3/13	31	12	57	12.50	360	630	700
	21/3/13	25	12	47	12.54	340	610	680

Table (3). *Measured Parameters during Second Run*

Basin	Date	Air Climatic Conditions				TDS mg/l		
		Temperature °C		Humidity %	Sun Light Period (hr)	Left Basin	Middle Basin	Right Basin
		Day	Night					
	1/4/13	28	21	29	12.57	20000	30000	40000
	2/4/13	28	16	34	13.00	13840	27030	34060
	3/4/13	32	19	28	13.04	9260	20200	29050
Basin 1	4/4/13	23	16	53	13.07	7370	13270	22980
	5/4/13	24	15	52	13.10	6130	8960	14460
	6/4/13	29	13	40	13.14	5280	7340	9320
	7/4/13	35	19	22	13.17	4050	6220	7830
	8/4/13	26	18	2	13.20	4050	6220	7830
	9/4/13	24	16	8	13.23	3240	5130	6120
	10/4/13	23	14	10	13.27	2890	4240	4980
Basin 2	11/4/13	23	14	8	13.30	1940	3480	4020
	12/4/13	24	13	9	13.33	1330	3170	3250
	13/4/13	24	13	10	13.36	950	2760	3080
	14/4/13	28	14	11	13.39	890	2590	2840
	15/4/13	29	14	48	13.42	890	2590	2840
	16/4/13	23	15	54	13.45	850	2260	2320
	17/4/13	22	15	53	13.47	810	1920	1980
Basin 3	18/4/13	23	14	52	13.49	770	1580	1609
	19/4/13	23	14	49	13.51	730	1290	1330
	20/4/13	22	14	44	13.53	690	1090	1150
	21/4/13	21	14	51	13.55	640	900	920

Table (4). *Measured Parameters during Third Run*

Basin	Date	Air Climatic Conditions				TDS mg/l		
		Temperature °C		Humidity %	Sun Light Period (hr)	Left Basin	Middle Basin	Right Basin
		Day	Night					
	1/5/13	35	18	35	14.10	25000	35000	40000
	2/5/13	32	19	43	14.12	15160	30090	33480
	3/5/13	32	19	51	14.14	10470	24540	28270
Basin 1	4/5/13	36	19	46	14.16	7920	15480	22120
	5/5/13	34	19	39	14.18	6300	10760	13540
	6/5/13	33	19	42	14.20	4880	8130	8720
	7/5/13	33	20	51	14.22	3940	7250	7630
	8/5/13	31	20	56	14.24	3940	7250	7630
	9/5/13	31	20	52	14.26	3330	5820	6010
	10/5/13	33	21	48	14.28	2690	4210	4580
Basin 2	11/5/13	30	20	41	14.3	2480	3370	3730
	12/5/13	27	19	53	14.32	2130	2740	3070
	13/5/13	34	21	34	14.34	1840	2560	2750
	14/5/13	27	18	42	14.36	1620	2280	2520
	15/5/13	27	19	42	14.38	1620	2280	2520
	16/5/13	28	18	52	14.4	1370	1950	2240
	17/5/13	32	20	55	14.42	1010	1670	1960
Basin 3	18/5/13	34	20	41	14.43	860	1410	1600
	19/5/13	38	22	36	14.45	810	1130	1310
	20/5/13	33	21	28	14.46	750	860	1170
	21/5/13	33	21	50	14.48	690	830	880

Table (5). *TDS Removal Ratio*

Time (days)	Average Removal Efficiency %		
	Inlet TDS Above 20000 ppm	Inlet TDS Between 20000 and 1000 ppm	Inlet TDS Below 1000 ppm
1	0.0	0.0	0.0
2	20.9	17.6	7.0
3	38.5	31.9	13.3
4	55.0	43.4	18.8
5	68.9	52.1	22.7
6	76.9	58.1	25.9
7	80.6	62.3	29.5

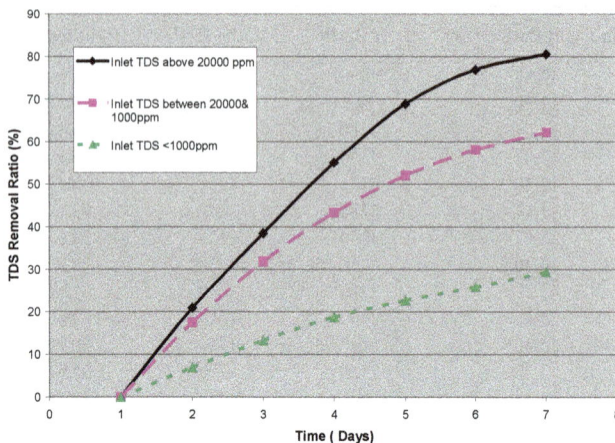

Figure (2). *TDS Removal Ratio with time for different inlet TDS concentrations*

From previous Table & figure, it is obvious that the removal efficiency increases with the increase of TDS. In other words, the removal efficiency average is directly proportional to the starting TDS concentration.

The results are showing the same trend from previous work made inside lab [6, 7, 8 & 9]. There is some deviation though which is occurring because of the effect of the nature condition of the day and night periods which was not applied in laboratory work.

In El Nadi et al [9] work, TDS average removal efficiency by 81.3% was achieved in 7 days in laboratory condition for any inlet TDS concentration. But in this study, three categories had been seen due to the inlet TDS concentration, the first for TDS more than 20000 ppm that achieved maximum removal ratio 77% in 7 days. The second for inlet TDS between 20000 and 1000 ppm and obtained maximum removal ratio of 64.7% and the third for inlet TDS less than 1000ppm that reached maximum removal efficiency 28.1%.

These variations could be explained for the effect of Temperature variations and the effect of sun light period differences in the nature circumstances.

This was also noticed on the continuous flow ponds system that the removal efficiency was very high in the first basin then decreased in the second one and be very slow changed in the third basin as shown in figure (3).

In figure (3), the shown TDS values resemble the average TDS values present in the three basins in series when starting with TDS of 40000 ppm. The behavior of Algae appears to be very effective in the first basin, and it decreases after in the second before turning to nearly constant in the third basin.

For the continuous flow pond system it was noticed that with inlet TDS 40000ppm, the overall efficiency of the 3 basins reached 97.7%, while the efficiency of the first two basins alone reached 93% while first basin alone achieved 80.6%.

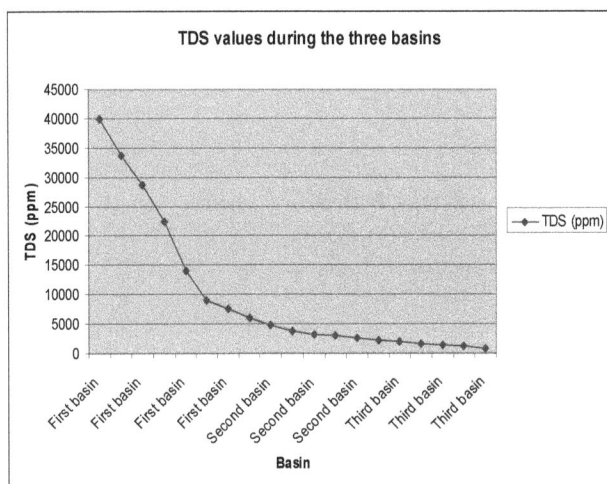

Figure (3). TDS values all through the three basins

4. Conclusions

Generally, results encourage the use of green algae for desalination and put an easy equation for the system design. The study had shown the following specific conclusions:-

1 The continuous flow algae ponds in the field under nature variable conditions achieved high removal for TDS from sea water.
2 *Scendesmus species* growth was successfully obtained in saline water as it uptakes salts and make use of them in its metabolism.
3 The effectiveness of the system increase as the starting total dissolved solids content increases. It was with high removal efficiency when TDS was above 20000 ppm. It was with moderate removal efficiency when TDS was between 20000 and 1000 ppm. And it was with low removal efficiency when TDS was below 1000 ppm.
4 The system efficiency affected mainly with the inlet TDS concentration.
5 For more removal for salinity second or third reactor could be applied with successful results.
6 The effect of the nature conditions of temperature,

sunlight period was noticeable by about 5-15% lower in removal efficiency compared with the previous work made inside the laboratory
7 The advantages of the proposed continuous flow system are low cost, easy construction with low operation & maintenance cost, and low energy requirements.
8 The system gives low pollution to surrounding environment, with maximum benefit of by-products, besides it solves the problem of getting rid of the waste products.

References

[1] Pelczar, M.J. and Reid , R.D., *"The Algae"*,: Mc-graw - Hill Book Company, N.Y. , USA 1965.

[2] Gimmler. *"The Metabolic Response of the Halotolerant Green Alga Dunalella parva to Hypertonic Shocks"*. Gottingen : Ber. Deutsch.Bot. Ges. Bd., Vol. 94, 1981.

[3] G, Laliberte, et al., *"Effect of Phosphorus Addition on Nutrient Removal from Wastewater with the Cyanobacterium Phprmidium bohneri"* . s.l. : Bioresource Technol, Vol. 59. 227-233,1997.

[4] Ibrahim, M.S., *"Reuse and Recycle of Wastewater in Natural Gas Industry."*, PhD Thesis, Inst. Of Environmental Studies & Researchs, ASU, Cairo, Egypt , 2005.

[5] El Nadi, M. H., El Sergany, F.A. R & Ibrahim, M.S.M., *"Use of Algae for Wastewater Treatment In Natural Gas Industry."*, ASU, Faculty of Engineering Scientific Bulliton, Vol.1.,1687-1695, Cairo, Egypt, September 2008.

[6] El Nadi, M. H., El Sergany, F.A. GH. R *"Water Desalination by Algea"*, .ASU Journal of Civil Engineering Vol. 2., pp 105-114, September 2010.

[7] El Nadi, M.H. A., Waheb, I.S. A. , Saad, S. A.H.A.,*"USING CONTINUOUS FLOW ALGAE PONDS FOR WATER DESALINATION "*, El Azhar Univ., Faculty of Eng., CERM of civil Eng., vol. 33, No. 4, December, 2011.

[8] Badawy M.A., El Nadi, M.H. & Nasr, N.A.H.,*" BIOLOGICAL DESALINATION TECHNIQUE BY ALGAE APPLICATION"*, Ain Shams Univ., Institute for Environmental Studies and Research, Journal of Environmental Science, vol. 17, No. 4, December, 2011.

[9] El Nadi, M.H., Nasr, N.A.H., El Hosseiny, O.M. & Badawy M.A., *"ALGAE APPLICATION FOR BIOLOGICAL DESALINATION."*, 2nd International Conference & Exhibition, Sustainable water supply & sanitation, (SWSSC2012), holding company for water and wastewater, Cairo, Egypt, December. 2012.

[10] El Hosseiny, O.M., El Nadi, M.H., Fergala, M.A.A. & El Khouly, M.S.M., *"Water Desalination by Continuous Flow Algae Ponds under nature conditions."*, El Azhar Univ., Faculty of Eng., CERM of civil Eng., vol. 35, No. 4, October 2013.

3

Is the Mediterranean Sea Circulation in a Steady State

Abed El Rahman Hassoun[1, 2, 3, *], **Véronique Guglielmi**[2, 3], **Elissar Gemayel**[1, 2, 3], **Catherine Goyet**[2, 3], **Marie Abboud-Abi Saab**[1], **Michele Giani**[4], **Patrizia Ziveri**[5], **Gianmarco Ingrosso**[4], **Franck Touratier**[2, 3]

[1]National Council for Scientific Research, National Center for Marine Sciences, Batroun, Lebanon
[2]IMAGES_ESPACE-DEV, Université de Perpignan Via Domitia, Perpignan Cedex, France
[3]ESPACE-DEV, UG UA UR UM IRD, Maison de la télédétection, Montpellier Cedex, France
[4]OGS (Istituto Nazionale di Oceanografia e di Geofisica Sperimentale), Oceanography Section, Via A. Piccard, Trieste, Italy
[5]ICREA - Institute of Environmental Science and Technology (ICTA), Universitat Autònoma de Barcelona, Ed. Z, ICTA-ICP, Carrer de les columnes, E- 08193 Bellaterra, Barcelona, Spain

Email address:

abedhassoun@gmail.com (A. E. R. Hassoun), veronique.guglielmi@univ-perp.fr (V. Guglielmi), elissargemayel@hotmail.com (E. Gemayel), cgoyet@univ-perp.fr (C. Goyet), mabisaab@cnrs.edu.lb (M. Abboud-Abi Saab), mgiani@inogs.it (M. Giani), patrizia.ziveri@uab.cat (P. Ziveri), gingrosso@ogs.trieste.it (G. Ingrosso), touratier@univ-perp.fr (F. Touratier)

Abstract: Most global ocean models are based on the assumption of a "steady state" ocean. Here, we investigate the validation of this hypothesis for the anthropized Mediterranean Sea. In order to do so, we calculated the mixing coefficients of the water masses detected in this sea via an optimum multiparameter analysis referred to as the MIX approach, using data from the BOUM (2008) and MedSeA (2013) cruises. The comparison of the mixing coefficients of each water mass, between 2008 and 2013, indicates that some of their proportions have significantly changed. Surface water mass proportions did not change significantly (Δ 0.05-0.1), while intermediate and deep water mass mixing coefficients of both Eastern and Western basins were significantly modified ($\sim\Delta$ 0.35). This study clearly shows that the Mediterranean seawater is not in a "steady state".

Keywords: Mixing Coefficient, MIX Approach, Water Masses, Seawater Circulation, Mediterranean Sea, Climate Change

1. Introduction

The Mediterranean Sea is a mid-latitude semi-enclosed sea. It witnesses two well-defined seasons, wet-cold winter and dry-warm summer, with short period of transition between them [1] [2]. An excess of evaporation over precipitation can be quantified by calculating the annual mean budget "evaporation minus precipitation" (E-P) over the whole Mediterranean Sea (350-750 mm yr^{-1}) [1]. Thus, Atlantic Water (AW) inflows at the surface and Mediterranean Water outflows along the bottom. The Atlantic water entering into Gibraltar in the surface layer, after travelling to the easternmost Levantine Sub-basin, is transformed into one of the saltiest seawater masses through air-sea heat and moisture fluxes. This relatively cold and salty water, which crosses the entire Mediterranean Sea in the opposite direction of the surface Atlantic Water, finally exits from the Strait of Gibraltar at the mid-depths. In wintertime, a significant negative heat budget [1] [3] causes a buoyancy loss, initiating deep and/or intermediate dense water formation. These events are frequent over the shelf areas and in the offshore regions, both in the Western and the Eastern basins [4] [5]. The Mediterranean Sea also exchanges water with the Black Sea through the Turkish Strait System (the Dardanelles, the Marmara Sea and the Bosphorus Strait) and receives significant amounts of freshwater from river discharge. In addition, this sea is always provided by groundwater discharges and sewage, which are likely, an important source of freshwater, nutrients, trace metals, alkalinity and other elements to the Mediterranean System [6] [7]. The Mediterranean Sea is composed mainly of two nearly equal size basins, the Western and the Eastern ones, connected by the relatively shallow Strait of Sicily (sill depth ~ 500 m). In each basin a number of sub-basins are characterized by different water masses circulations, chemical and biological features and rough bottom topography. Note that we reserve the term "sea" for the Mediterranean Sea and "basin" for the Eastern and Western parts of it; any smaller entity is a sub-basin.

The Mediterranean Sea is a particular system, characterized by a complex thermohaline, wind, and water flux-driven multi-scale circulation with interactive variability [8]. This land-locked sea is exporting warm and salty intermediate waters to the North Atlantic Ocean by the narrow and shallow Strait of Gibraltar [width ~ 13 km ; sill depth ~ 300 m] affecting the global thermohaline conveyor belt [9] [10]. Through pathways to the Atlantic Polar regions or through indirect mixing processes, the salty Mediterranean water preconditions the deep convection cells of the Polar Atlantic. There the North Atlantic Deep Water is formed and successively spreads throughout the world ocean constituting the core of the global thermohaline circulation [11] [12]. Thus, the salty water of Mediterranean origin may affect water formation processes and variability and even the stability of the global thermohaline equilibrium state. Hence, the study of the Mediterranean water masses (their formation, spreading, mixing, and impact on general circulation) is essential for a better understanding of the ocean circulation and variability. Moreover, the existence of a thermohaline cell (the Eastern Mediterranean "Conveyor Belt"), multiple scales of motion defining the general circulation (basin/sub-basin/mesoscale), and deep water mass formation processes, make this sea a "test basin" for general circulation studies.

Aiming to study the hydrographic situation in the Mediterranean Sea, many researchers have used an optimum multiparameter analysis ([13] in the Adriatic Sub-basin, and [14] in the Eastern Mediterranean Sub-basin; [15] in the Western Mediterranean Sub-basin; [16] in the entire Mediterranean Sea). However, a study of the evolution of water mass mixing coefficients in this semi-enclosed sea is needed to better understand the hydrographic system in this sea.

Reference [17] introduced the MIX approach based upon an optimum multiparameter mixing analysis. The use of this approach is particularly recommended for regional studies where the distribution of water masses can be clearly defined (this is the case of the Mediterranean Sea). In the present paper, our main objective is to evaluate the variation (if any) of the water mass circulation in the Mediterranean Sea based on their mixing coefficients calculated via the MIX approach using data collected in 2008 and 2013.

2. Methodology

2.1. Study Area

The BOUM cruise (Biogeochemistry from the Oligotrophic to the Ultra oligotrophic Mediterranean Sea [18]) was conducted during summer 2008, from 20 June to 22 July, on board the R/V L'Atalante [19]. It consists of a longitudinal transect (more than 3000 km long from the Levantine Sub-basin to the Northwestern Mediterranean Sea) of 27 short-term stations and 3 long-term stations referred as A, B, and C (Fig.1).

The MedSeA (Mediterranean Sea Acidification in a Changing Climate; [20] [21]) cruise occurred during spring 2013, on board of the Spanish R/V Angeles Alvariño, from 2 May to 2 June. The full cruise track (more than 8000 km long) consisted of two longitudinal legs where 23 stations along the Mediterranean Sea were sampled throughout the water column. During the first leg, samples were collected from Atlantic waters off Cadiz, Spain to the Levantine Sub-basin in the Eastern Mediterranean basin (3879 km long, 15 stations, 279 sampled points, maximum sampled depth = 3720 m). The second leg was conducted in the Northern part of the Mediterranean from Heraklion, Crete, Greece in the Eastern Mediterranean basin to Barcelona, Spain in the North Western Mediterranean basin (3232.5 km long, 8 stations, 183 sampled points, maximum sampled depth = 3000 m ; Fig.2).

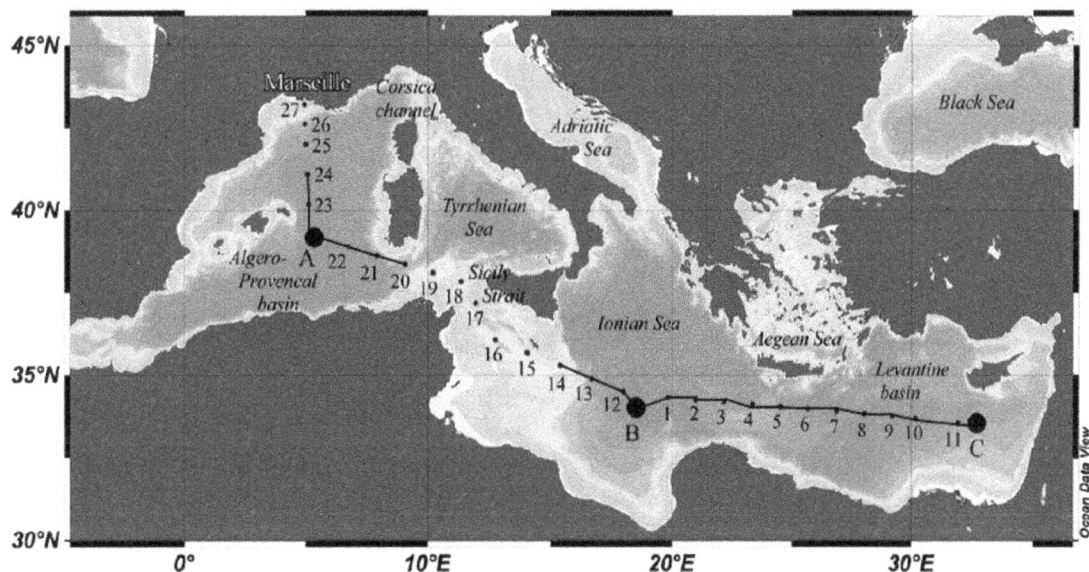

Figure 1. Map of the 2008 BOUM cruise in the Mediterranean Sea. Short-term stations are indicated by numbers (from 1 to 27); the three long-term stations are referred as A, B, and C (Touratier et al., 2012). The thick black lines referred to the sections used in this study.

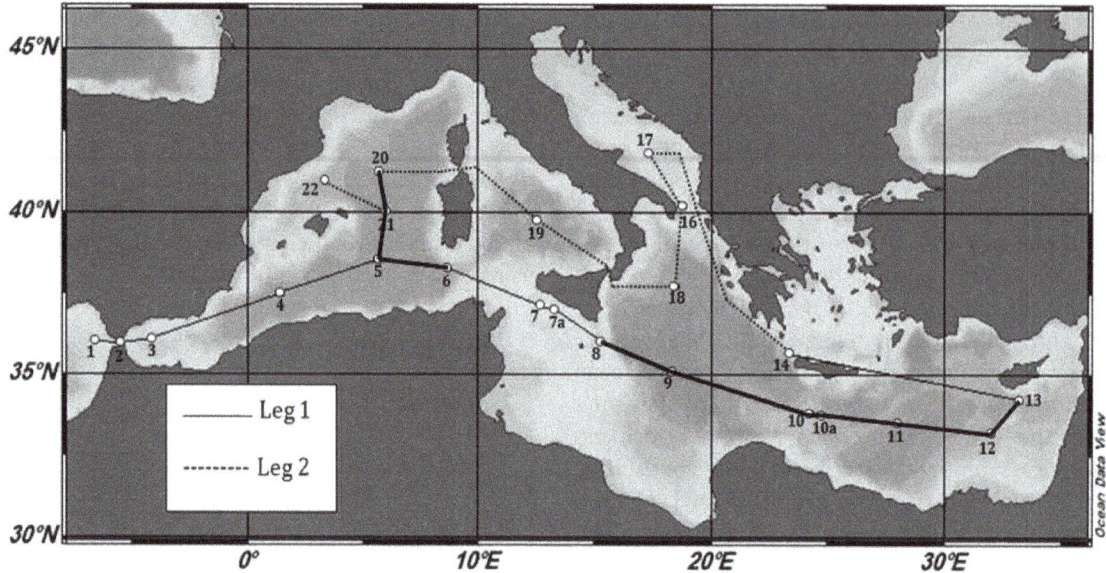

Figure 2. Map of the 2013 MedSeA cruise in the Mediterranean Sea. The numbers from 1 to 22 correspond to the sampled stations. The thick black lines referred to the sections used in this study.

2.2. Sampling and Measurements

In the present study, we use the following properties: potential temperature (θ ; °C), salinity (S), dissolved oxygen (O_2 ; μmol kg^{-1}), nitrates (NO_3 ; μmol kg^{-1}), and phosphates (PO_4 ; μmol kg^{-1}).

For the 2008 BOUM cruise, the profiles for θ, S, and O_2 were obtained using a Sea-Bird Electronics 911 PLUS CTD system. Each CTD cast was associated with a carousel of 24 Niskin bottles to collect seawater samples used to perform the analysis of the other chemical and biological properties. Concerning the nutrients (NO_3 and PO_4), the description of the methods used for the analysis are explained by [22] and [23].

For the 2013 MedSeA cruise, hydrologic properties [S and T (°C)] were measured *in situ* using a Sea-Bird Electronics CTD system (SBE 911plus) associated with a General Oceanic rosette sampler, equipped with twenty four 12 L Niskin bottles. The precision of measurements is ± 0.001 °C for T and ± 0.0003 for S. Water samples for dissolved oxygen determination were collected in calibrated BOD 60 ml bottles. Oxygen concentrations were measured using a Winkler iodometric titration [24] with a Mettler-Toledo. DL-21 potentiometric titrator with a Pt ring redox electrode for the determination of the equivalence point [25]. The analytical precision and accuracy are ± 1.5 μmol kg^{-1}. For the nitrates (NO_3) and phosphates (PO_4), water samples were pre-filtered over glass fiber filters (Whatman GF/F) with a pore size of 0.7 microns, immediately after sampling. Then, they were stored at -20°C in polyethylene bottles until their analysis. The samples were thawed and analyzed colorimetrically with a Bran + Luebbe 3 autoanalyzer according to [26], at OGS laboratory in Trieste, Italy.

2.3. Modeling: The MIX Approach

Aiming to quantify the contribution of the different water sources to the collected data, we used the MIX approach

developed by [17]. This approach uses a multi-parameter analysis. Optimum multiparameter methods [27] [28] are based on the assumption that observed water properties at a hydrographic station are the result of mixing among two or more "source waters". They have been used to evaluate water mass properties in both the Southern Ocean [29] [30] and other regions [28] [31].

In the present study, four conservative tracers (S, θ, NO, and PO) are used in the MIX approach. Reference [32] defined the conservative tracers NO and PO as follows:

$$NO = O_2 + R_{ON}NO_3 \qquad (1)$$

$$PO = O_2 + R_{OP}PO_4 \qquad (2)$$

where NO and PO are composite tracers of the non conservative tracers O_2 (dissolved oxygen concentration in μmol kg^{-1}) and NO_3 (nitrate concentration in μmol kg^{-1}), and PO_4 (phosphate concentration in μmol kg^{-1}), respectively. According to the equation of [33] which describes the average photosynthesis and aerobic respiration in the interior of the ocean, the composite tracers NO and PO were built using the fact that the consumption of oxygen is balanced by the production of nutrients during the processes of respiration and decomposition. The two constants R_{ON} and R_{OP} are the ratios of the stoichiometric coefficients (ψ) involved in the Redfield equation ($R_{ON} = \psi O_2/\psi NO_3$; $R_{OP} = \psi O_2/\psi PO_4$).

The general conservation equation for a conservative tracer (Ω) is given by :

$$\Omega = \sum_{j=1}^{n} K_j \Omega_j \qquad (3)$$

where K_j represents the contribution (also called "mixing coefficient") of a water source j, n is the number of water sources in the system, and Ω_j is the value of the conservative tracer for the water source j.

For each seawater sample, where the conservative tracers Ω are either measured (S and θ) or calculated (NO and PO ; see equations 1 and 2), the contributions K_j are estimated after resolving the system of equations with the following constraints :

$$\sum_{j=1}^{n} K_j = 1 \text{ (mass conservation)} \quad (4)$$

$$\forall_j, \ 0 \leq K_j \leq 1 \quad (5)$$

It is a constrained linear least-squares problem, with one equality constraint and n inequality constraints. To solve it, we use a medium-scale optimization algorithm (with MATLAB) similar to that described by [34]. The difference in range between tracers leads to an ill-conditioned problem: the linear equation solution involves the inversion of the matrix containing the water sources characteristics, but this matrix is really badly scaled. So, prior to the inversion, rows and columns of the matrix are normalized.

The stability of the results (linked to the correct placement of sources) is then tested by adding thirty perturbations to the values of Ω and the Ω_j which provide at each point a mean solution for each K_j [17]. These perturbations are independent Gaussian random variables with zero mean and given standard deviation. We entered the standard deviations for each tracer with a given value for the measurement points and another (larger) value for sources. These thirty Gaussian perturbations allowed us not only to test the stability of the results but also to take into account the differences between the errors of measurement based on the tracers.

2.4. Determination of the Water Sources

The Appropriate definition of water source properties is crucial to achieving physically meaningful results from the MIX approach. The different water sources j were typically identified using the conservative tracers S and θ, their corresponding θ/S diagrams (Fig. 3 and 6) and referring to the literature. In order to lower the number of water sources, the Mediterranean Sea was tested as two independent systems : the Western basin (5° E $\leq$ longitude $\leq$ 9.5°E) and the Eastern basin (longitude $\geq$ 15°E). Consequently, the relatively shallow stations located in the Sicily Strait between 9.5°E and 15°E were ignored when the MIX approach was applied.

Table 1. Physical and chemical properties of water sources used by the MIX approach in the Western Mediterranean basin for BOUM 2008 and MedSeA 2013.

Water Sources	Depth	T (°C)	S	θ (°C)	NO$_3$ (μmol kg^{-1})	PO$_4$ (μmol kg^{-1})	O$_2$ (μmol kg^{-1})
BOUM 2008							
AW	50	17.21	37.269	17.20	0.00	0.002	240.8
EOW	50	15.12	38.163	15.12	0.05	0.032	259.0
LIW	496	14.15	38.853	14.07	5.69	0.192	181.8
TDW	2839	13.31	38.472	12.86	8.61	0.391	196.2
WMDW	1483	13.08	38.459	12.86	8.78	0.383	190.0
MedSeA 2013							
AW	50	16.17	36.607	16.16	0.63	0.058	234.8
EOW	50	15.12	38.163	15.12	0.05	0.032	259.0
LIW	496	14.15	38.853	14.07	5.69	0.192	181.8
TDW	2839	13.31	38.472	12.86	8.61	0.391	196.2
WMDW	1483	13.08	38.459	12.86	8.78	0.383	190.0

AW, Atlantic Water;
EOW, Effluent Outflow Water;
LIW, Levantine Intermediate Water;
TDW, Tyrrhenian Deep Water;
WMDW, Western Mediterranean Deep Water.

For the Western basin, we used five water sources (n = 5). Whereas, six ones were used for the Eastern basin (n = 6). The physical and chemical properties of these water sources, for both BOUM 2008 and MedSeA 2013 cruises, are shown in Tables 1 and 2 for the Western and the Eastern Mediterranean basins respectively. These water sources were identified based on the data of the two cruises and on the availability of the input parameters (S, θ, O$_2$, NO$_3$, PO$_4$) manipulated by MIX approach to calculate the mixing coefficients. The main water sources are indicated on the corresponding θ/S diagrams of the Western and Eastern basins (Fig. 3 and 6). When the water sources of the MedSeA cruise have similar θ/S characteristics compared to the BOUM ones, we decided to take into consideration the characteristics of the BOUM water masses in our mixing coefficients calculation to maintain the same ratios previously calculated for the Mediterranean Sea based on the 2008 BOUM cruise.

Table 2. Physical and chemical properties of water sources used by the MIX approach in the Eastern Mediterranean basin for BOUM 2008 and MedSeA 2013.

Water Sources	Depth	T (°C)	S	θ (°C)	NO$_3$ (μmol kg^{-1})	PO$_4$ (μmol kg^{-1})	O$_2$ (μmol kg^{-1})
BOUM 2008							
MAW	73	15.73	37.618	15.72	0.00	0.001	239.7
EOW	174	15.59	38.436	15.56	2.37	0.102	205.9
LIW	248	16.25	39.163	16.21	0.96	0.013	220.6
CIW	173	15.17	39.045	15.14	3.34	0.090	204.9

Water Sources	Depth	T (°C)	S	θ (°C)	NO$_3$ (µmol kg^{-1})	PO$_4$ (µmol kg^{-1})	O$_2$ (µmol kg^{-1})
EMDW-Adr	2965	13.86	38.710	13.38	4.91	0.142	195.4
EMDW-Ag	2222	14.00	38.789	13.64	4.84	0.167	189.3
MedSeA 2013							
MAW	50	17.72	38.873	17.72	0.01	0.029	241.3
EOW	174	15.59	38.436	15.56	2.37	0.102	205.9
LIW	248	16.25	39.163	16.21	0.96	0.013	220.6
CIW	173	15.17	39.045	15.14	3.34	0.090	204.9
EMDW$_{Adr}$	950	13.18	38.702	13.04	1.65	0.087	219.7
EMDW$_{Ag}$	2222	14.00	38.789	13.64	4.83	0.167	189.3

MAW, Modified Atlantic Water ;
EOW, Effluent Outflow Water ;
LIW, Levantine Intermediate Water ;
CIW, Cretan Intermediate Water ;
EMDW$_{Adr}$, Eastern Mediterranean Water-Adriatic origin ;
EMDW$_{Ag}$, Eastern Mediterranean Deep Water-Aegean origin.

2.5. Determination of the Redfield Ratios R_{ON} and R_{OP}

Typical values given by [33] are R_{ON} = 8.6 and R_{OP} = 138 (molar ratios). However, these ratios cannot be representative for the Mediterranean waters since [33] never used observations from this region. In this paper, we use the R_{ON} and R_{OP} ratios previously estimated, specifically for the Mediterranean Sea by [35] using the 2008 BOUM database. The best results for the regressions between the nutrients and O$_2$ were obtained after considering two different layers for both the Western and the Eastern basins : the surface/intermediate layer (from 50 up to 750 m), and the deep layer (from 750 m to the bottom). We hypothesize that these layers are related to the traditional Mediterranean circulation scheme which is described by two cells [36] : the well ventilated surface cell which contains water masses like MAW, WIW, and LIW ; and the deep cells which contains the WMDW and TDW in the Western basin, and the EMDW$_{Adr}$ and EMDW$_{Ag}$ in the Eastern basin.

3. Results and Discussion

In order to assess the goodness of the solutions and to check possible errors due to the water-source description and to uncertainties associated with the measurements, the residual vector was calculated for each of the four conservative tracers. This vector corresponds to the difference between the predicted and the measured data. For BOUM 2008 and MedSeA 2013, the MIX method performs well, producing unbiased small residuals (always less than 10%) for all measures.

Hereafter, we discuss the evolution of the main water sources mixing coefficients for the Western and Eastern Mediterranean basins.

3.1. Western Basin

3.1.1. Atlantic Water, AW

The mixing coefficients of the surface water masses did not change significantly since 2008 till 2013 (Fig.5). The Mediterranean Sea has an active water exchange with the Northern Atlantic through the Strait of Gibraltar. The surface Atlantic flow entering the Mediterranean Sea does not only fill the water deficit of 1 m per year, but it also replaces the Mediterranean deep outflow, which represents a loss of 20 m of water per year for the whole Mediterranean Sea [37]. As a consequence of the excess of evaporation over precipitation (~ 0.62 – 1.16 m year^{-1}) [38], heating and various physical phenomena (gyres, eddies,...), the proportion of the AW decreases progressively because its characteristics change while propagating Eastward to be referred as Modified Atlantic Water (MAW). This trend was comparable during the two oceanographic surveys (BOUM 2008 and MedSeA 2013).

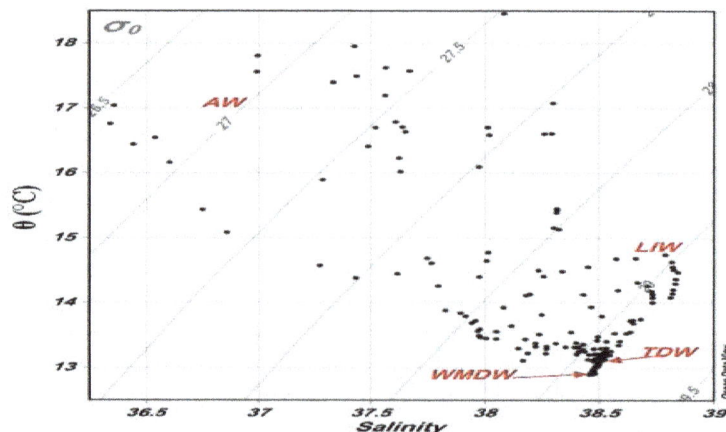

Figure 3. *θ/S diagram illustrating the main water masses detected in the Western basin during the MedSeA cruise in May 2013 (AW: Atlantic Water, LIW: Levantine Intermediate Water, TDW: Tyrrhenian Deep Water, WMDW: Western Mediterranean Deep Water).*

3.1.2. Levantine Intermediate Water, LIW

The amount of this water mass has increased during the study period (2008-2013). For example, its proportion increased from 0.55 in 2008 to 0.9 in 2013 within the intermediate layers (~ 300 m) of the Sardinia Strait (Fig.5). The LIW contributes predominately to the non-returning efflux, mixed with both EMDW and WMDW in the Strait of Gibraltar, and to the outflow into the Atlantic Ocean [2] [8]

[39]. Our results show recent changes in the Mediterranean outflow. These increasing proportions could be caused by the excessive evaporation and the decrease in precipitation and freshwater supplies to the Eastern Mediterranean Sea [40] which can be direct consequences of the unequivocal global warming [41]. Decreasing trends of river discharges in the Eastern Mediterranean Sea [42] has been observed, as in the Adriatic Sub-basin [43].

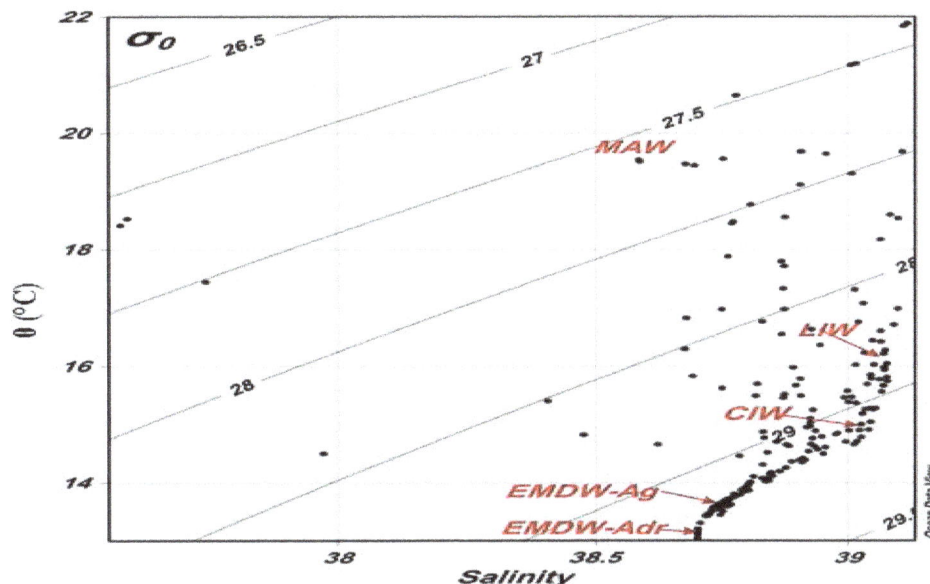

Figure 4. θ/S diagram illustrating the main water masses detected in the Eastern basin during the MedSeA cruise in May 2013 (MAW: Modified Atlantic Water, LIW: Levantine Intermediate Water, CIW: Cretan Intermediate Water, EMDW-Ag: Eastern Mediterranean Deep Water originated from the Aegean Sub-basin, EMDW-Adr: Eastern Mediterranean Deep Water originated from the Adriatic Sub-basin).

3.1.3. Tyrrhenian Deep Water, TDW

The proportion of this water mass has clearly decreased below 1000 m in the Western basin. Its amount during 2013 become less abundant in the Southern sub-basins compared to the one recorded for 2008 (Fig.5). This water, which is a product of the mixing between Eastern (LIW and EMDW) and western waters (WMDW), fills the Tyrrhenian Sub-basin down to the bottom [44]. Thus, this decrease could be attributed to the modification in the EMDW formation, since the Adriatic Sub-basin seems to retrieve its role as a main contributor in the deep water mass formation in the Eastern basin instead of the Aegean during the EMT (Eastern Mediterranean Transient ; data of the MedSeA cruise 2013). However, [45] argue that the TDW might result from a dense water formation process occurring within the Tyrrhenian itself, in the East of the Bonifacio Strait. This means that the decrease in the mixing proportion could also be attributed to climatic changes affecting temperature and salinity, thus the deep water formation.

Nevertheless, the mixing coefficients proportions of the TDW have slightly increased from 0.05 in 2008 to 0.2 in 2013, exclusively in the North of the Western basin between 0 and 500 m. This fact could be firstly related to the harsh winter convective mixing in this wind-driven area [46]. Secondly, this smooth proportions increase could be also connected to

the mixing caused by heavy freshwater inputs from the Northern rivers during May (time of the snow melting) [47].

3.1.4. Western Mediterranean Deep Water, WMDW

The mixing coefficients schemes of this water mass (Fig.5) show that its amount decreased slightly below 1000 m, between 2008 and 2013, particularly in the South of the Western basin. The WMDW remains well distributed in the Western basin with higher proportions in the North of this basin, where it is originated (Fig.5). The observed decrease could be due to the significant warming and salinification, widely mentioned in the literature [48] [49] [50] [51] [52]. However, at a station in the south of the Tyrrhenian Sub-basin sampled in September 1999 during the PROSOPE cruise [53], the total alkalinity in deep water was higher than in deep water at the Dyfamed site [54]. If it is confirmed that the Tyrrhenian deep water contributes to the deep water in the Liguro-Provençal Sub-basin, this observation could explain the modifications in the WMDW proportions which could also be attributed to the changes in the TDW linked to the Eastern deep water formation. Based on 23 years eddy-permitting reanalysis, [55] argued that the largest water mass formation event of the past 23 years occurred in the Western Mediterranean basin in 2005-2006. This event was presumably preconditioned by the EMT which modified the characteristics of the LIW crossing the Sicily Strait.

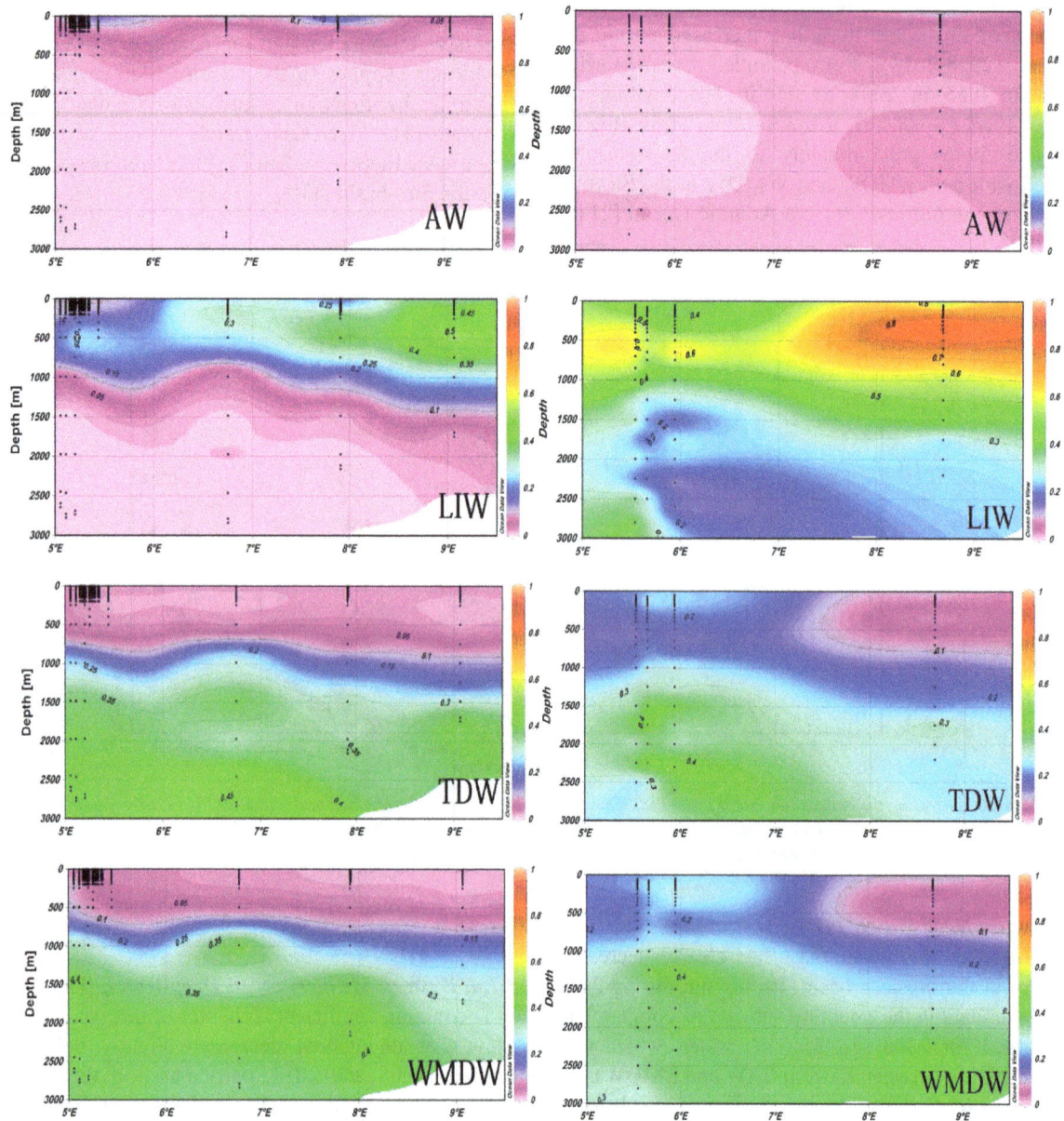

Figure 5. Profiles of the mixing coefficients of water masses in the Western Mediterranean basin during the 2013 MedSeA cruise (right column) and the 2008 BOUM cruise (left column

3.2. Eastern Basin

However, similarly to the TDW, a slight rise of the mixing coefficients has been registered within the layer above 1000 m in the Northern part of the Western basin. This boost could also be attached to the vertical mixing related to the tough meteorological winter conditions in the Gulf of Lions, as well as in Balearic and Liguro-Provençal Sub-basins [46] [56].

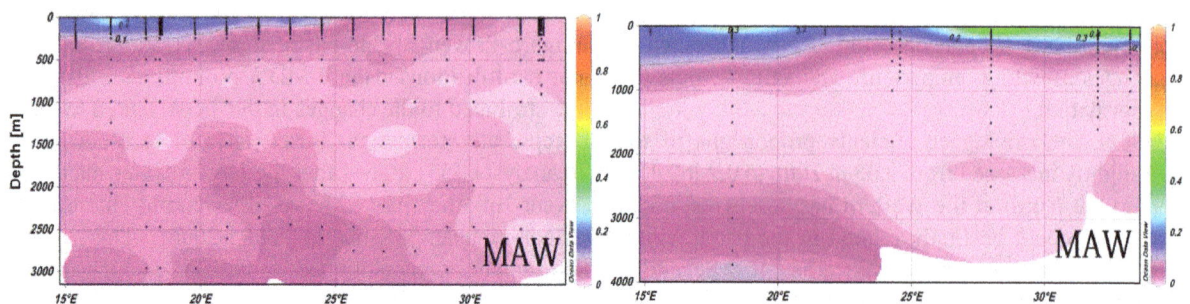

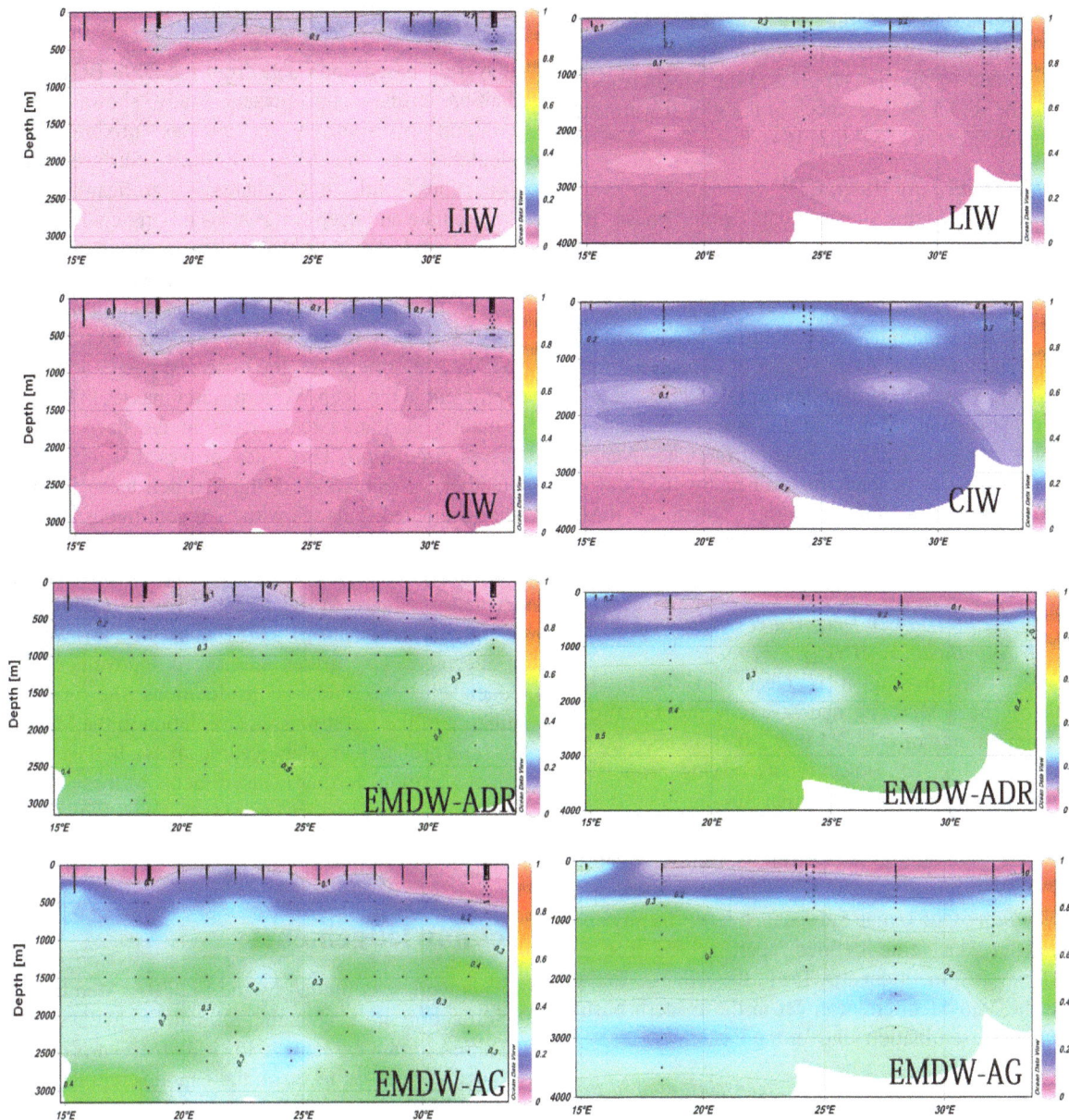

Figure 6. *Profiles of the mixing coefficients of water masses in the Eastern Mediterranean basin during the 2013 MedSeA cruise (right column) and the 2008 BOUM cruise (left column).*

3.2.1. Modified Atlantic Water, MAW

The proportion of the MAW increased from 2008 to 2013. During 2013, it is clear that its amount increases while propagating Eastward from the Sicily Strait (Fig.6). This water mass is initially made as a result of the mixing of comparatively fresh Atlantic water (S < 36.5) flowing via the Strait of Gibraltar into the Mediterranean Sea with the surface waters of the Alboran Sub-basin [57]. The incoming MAW is continuously modified by interactions with the atmosphere and mixing with older surface waters and with the waters underneath. All along its course, it is seasonally warmed or cooled, but overall its salt content increases and it becomes denser [58]. Thus, its increasing proportion could be attributed to the excessive evaporation trend in the context of global warming occurring in the Mediterranean Sea [59] [60]. Satellite observations from 1985–2006 indicate that in the last

two decades the temperature in the upper layer of the Mediterranean Sea has been increasing at an average (± SD) rate of $0.03 \pm 0.008°C$ yr^{-1} for the Western basin and $0.05 \pm 0.009°C$ yr^{-1} for the Eastern basin [60]. Nevertheless, these changes could also be due to the modifications in circulation patterns leading to blocking situations during the EMT (i.e. gyres, eddies, …), to the Adriatic-Ionian Bimodal Oscillating System (BiOS) [61] and/or to variations in the fresher water of Black Sea origin input through the Strait of Dardanelles [8].

3.2.2. Levantine Intermediate Water, LIW

The amount of this water mass has increased during the study period (2008-2013; Fig.6). Its proportion increased from 0.27 in 2008 to 0.37 in 2013 within the intermediate layers (~ 175 m) in the South of Cyprus. The increasing proportion could be related to the occurrence of high amount of MAW in this area which becomes denser. Thus, this latter sinks more quickly to participate in the formation of the LIW. This fact could be caused by the excessive evaporation and the decrease in precipitation and freshwater supplies to the Mediterranean Sea. This can be direct consequences of the global warming, well described in literature [62] [63] [64].

3.2.3. Cretan Intermediate Water, CIW

The proportion of this intermediate water mass has increased in the entire Eastern basin, particularly in the South of Crete where it has the highest proportions (~ 0.35 at ~ 170 m; Fig. 6). This fact is due to changes in circulation patterns leading to blocking situations concerning the MAW and the LIW, and to variations in the fresher water of Black Sea origin input through the Strait of Dardanelles [8] [42]. These results could also be attributed to specific atmospheric conditions which created large buoyancy fluxes from the Aegean Sub-basin, similarly to the ones mentioned by [65] during winters 1991/1992 and 1992/1993 (the "enhanced EMT winters"), intensifying intermediate and deep water production, although with modified characteristics [66].

3.2.4. Eastern Mediterranean Deep Water Originated in the Adriatic Sub-Basin, EMDW$_{Adr}$

During the study period, it is clear that the mixing coefficients of the EMDW$_{Adr}$ decreased in the Levantine Sub-basin from a range of 0.33-0.4 to 0.2-0.39 in deep layers (below 1000 m ; Fig. 6). However, its mixing proportion remains high in the South of the Ionian Sub-basin (~ 0.54). These results indicate that the EMDW$_{Adr}$ stays a minor deep water mass in the Levantine Sub-basin. Moreover, it shows that the EMDW originated from the Adriatic Sub-basin is reaching the South of the Ionian Sub-basin but maybe physical phenomena (like the more defined open ocean, free jet intensified structures mentioned by [55] : the Atlantic-Ionian Stream and the Mid-Mediterranean Jet in the Eastern basin) are disabling its dissemination all over the Eastern basin.

3.2.5. Eastern Mediterranean Deep Water Originated in the Aegean Sub-Basin, EMDW$_{Ag}$

The mixing coefficients of this water mass show a remarkable increase, particularly in the Levantine Sub-basin (from 0.29 to 0.54 at 1500 m ; Fig.6). The steady occurrence of this dominant deep water mass in the Levantine Sub-basin, instead of EMDW$_{Adr}$, could be due to many factors : 1- the waters formed in the Aegean Sub-basin still sufficiently denser than those originated in the Adriatic one, hence, it is able to enter the deep layers of the Eastern basin and not address the intermediate water range depth [67], 2- the deep water formation in the Adriatic Sub-basin seems to be significantly impacted by the EMT [68], thus it is not able to match its previous distribution in the Eastern basin, 3- the atmospheric and physical conditions favorable for the formation of this dense water still dominant in this area, 4- our results could also be attributed to the long residence time of deep water masses in the Levantine Sub-basin due the robust topography which could trap the deep waters for a long period (~ 100 years) [69] [70].

4. Conclusion

Overall, based upon data from the 2008 BOUM and 2013 MedSeA cruises, the present study shows a significant evolution of some water masses mixing coefficients, calculated via the MIX approach. Surface water mass proportions did not change significantly. However, intermediate and deep water mass mixing coefficients of both Eastern and Western basins were noticeably modified. Moreover, these results indicate that the mixing coefficient of the EMDW$_{Adr}$ is always high in the Ionian Sub-basin, while it remains low in the Levantine Sub-basin compared to the EMDW$_{Ag}$. Furthermore, the observed decrease of the mixing coefficient in the deep Western water masses (TDW, WMDW) is mainly attributed to changes in the deep Eastern water masses circulation. This study proves that the hypothesis of "steady state" situation for the Mediterranean Sea is far from being validated. This sea witnesses continuous and significant water masses changes [ex. The Eastern Mediterranean Transient, EMT [66] [71] [72] [73] [74], changes in circulation pattern in the Ionian Sub-basin which affect the contiguous sub-basins [61] [75]. Therefore, this work could be an incentive for further studies to innovate new oceanic models that take into considerations the unsteady state situation of the water masses circulation in the Mediterranean Sea (and probably in other oceanic areas). In the context of the global warming, further measurements of these water masses properties are necessary, to assess their evolutions and to evaluate the consequences of any modification on the global circulation.

Acknowledgements

This work was funded by the EC FP7 "Mediterranean Sea Acidification in a changing climate – MEDSEA" project (MedSeA ; grant agreement 265103 ; medsea-project.eu). The authors are pleased to thank the captains and the crew of the Spanish research vessel R/V Ángeles Alvariño. They would like furthermore to thank Mr. Michael GRELAUD for uploading the data of this cruise on Pangaea data repository [76, 77, 78, 79, 80]. Authors are grateful to the National Council for Scientific Research (CNRS) in Lebanon for the Ph.D. thesis scholarship granted to Mr. Abed El Rahman HASSOUN.

References

[1]   Mariotti A., Struglia M.V., Zeng N., Lau K.-M., 2002. The Hydrological Cycle in the Mediterranean Region and Implications for the Water Budget of the Mediterranean Sea. Journal of Climate, 15, 1674–1690.

[2]   Manca B., Burca M., Giorgetti A., Coatanoan C., Garcia M.-J., Iona A., 2004. Physical and biochemical averaged vertical profiles in the Mediterranean regions: an important tool to trace the climatology of water masses and to validate incoming data from operational oceanography. Journal of Marine Systems, 48, 83–116.

[3]   Castellari S., Pinardi N., Leaman K., 2000. Simulation of water mass formation processes in the Mediterranean Sea : influence of the time frequency of the atmospheric forcing. Journal of Geophysical research, 105 (C10), 24157-24181.

[4]   Malanotte-Rizzoli P., 1991. The Northern Adriatic Sea as a prototype of convection and water mass formation on the continental shelf. In: Chu, P.C., Gascard, J.C. (Eds.), Deep Convection and Deep Water Formation in the Oceans. Elsevier Oceanography Series. 57. Elsevier, Amsterdam, pp. 229– 239.

[5]   POEM Group, 1992. General circulation of the Eastern Mediterranean Sea. Earth-Science Reviews, 32, 285–308.

[6]   Moore W.S., 2006 a. The role of submarine groundwater discharge in coastal biogeochemistry. Journal of Geochemical Exploration, 88, 389–393, doi:10.1016/j.gexplo.2005.08.082.

[7]   Moore W. S., 2006 b. Radium isotopes as tracers of submarine groundwater discharge in Sicily. Continental Shelf Research, 26, 852–861, doi:10.1016/j.csr.2005.12.004.

[8]   Robinson A.R., Leslie W., Theocharis A., Lascaratos A., 2001. Mediterranean Sea circulation. Encyclopedia of Ocean Science, Vol. 3. Academic Press, San Diego, CA, pp. 1689–1705.

[9]   Lacombe H. and Richez C., 1982. The Regime of the Strait of Gibraltar. Hydrodynamics of Semi-Enclosed Seas. Proceedings of the 13th International Liege Colloquium on Ocean Hydrodynamics. Elsevier Oceanography Series, 34, 13–73.

[10] Bergamasco A. and Malanotte-Rizzoli P., 2010. The circulation of the Mediterranean Sea: a historical review of experimental investigations. Advances in Oceanography and Limnology, 1 (1), 11–28.

[11] Candela J., 2001. Mediterranean water and the global circulation. In: Siedler, G., Chuch, J., Gould, J., (Eds.), Ocean circulation and Climate. Observing and modeling the global ocean. Academic Press, New York, pp. 419–429.

[12] Malanotte-Rizzoli P. and the Pan-Med Group, 2012. Physical forcing and physical/biochemical variability of the Mediterranean Sea: A review of unresolved issues and directions of future research. Report of the Workshop "Variability of the Eastern and Western Mediterranean circulation and thermohaline properties: similarities and differences", Rome, 7-9 November, 2011, pp. 48.

[13] Cardin V., Bensi M., and Pacciaroni M., 2011. Variability of water mass properties in the last two decades in the South Adriatic Sea with emphasis on the period 2006–2009. Continental Shelf Research, 5 (31), 951–965.

[14] Kovačević V., Manca B.B., Ursella L., Schroeder K., Cozzi S., Burca M., Mauri E., Gerin R., Notarstefano G., and Deponte D., 2012. Water mass properties and dynamic conditions of the Eastern Mediterranean in June 2007. Progress in Oceanography, 104, 59–79.

[15] Schroeder K., Borghini M., Cerrati G., Difesca V., Delfanti R., Santinelli C., et Gasparini G.P., 2008. Multiparametric mixing analysis of the deep waters in the western Mediterranean Sea. Chemistry and Ecology, 24 (1), 47–56.

[16] Hainbucher D., Rubino A., Cardin V., Tanhua T., Schroeder K., and Bensi M., 2013. Hydrographic situation during cruise M84/3 and P414 (spring 2011) in the Mediterranean Sea. Ocean Science Discussion, 10, 2399–2432, doi:10.5194/osd-10-2399-2013

[17] Goyet C., Coatanoan C., Eischeid G., Amaoka T., Okuda K., Healy R., and Tsunogai S., 1999. Spatial variation of total CO_2 and total alkalinity in the northern Indian Ocean: A novel approach for the quantification of anthropogenic CO_2 in seawater. Journal Of Marine Research, 57, 135–163, doi: 10.1357/002224099765038599.

[18] BOUM (Biogeochemistry from the Oligotrophic to the Ultra-oligotrophic Mediterranean): http://www.com.univ-mrs.fr/BOUM/. Moutin T. and Torre M.P., Centre d'oceanologie de Marseille - Campus de Luminy.

[19] Moutin T., Van Wambeke F., and Prieur L., 2012. Introduction to the Biogeochemistry from the Oligotrophic to the Ultraoligotrophic Mediterranean (BOUM) experiment. Biogeosciences, 9, 3817–3825, doi:10.5194/bg-9-3817-2012

[20] MedSeA (Mediterranean Sea Acidification in a changing climate) project, 2015 : http://medsea-project.eu/

[21] 2013 MedSeA research cruise on ocean acidification and warming : http://medseaoceancruise.wordpress.com/

[22] Pujo-Pay M., Conan P., Oriol L., Cornet-Barthaux V., Falco C., Ghiglione J.-F., Goyet C., Moutin T., and Prieur L., 2011. Integrated survey of elemental stoichiometry (C, N, P) from the western to eastern Mediterranean Sea. Biogeosciences, 8, 883–899, doi: 10.5194/bg-8-883-2011.

[23] Crombet Y., Leblanc K., Quéguiner B., Moutin T., Rimmelin P., Ras J., Claustre H., Leblond N., Oriol L., and Pujo-Pay M., 2011. Deep silicon maxima in the stratified oligotrophic Mediterranean Sea. Biogeosciences, 8, 459–475, doi: 10.5194/bg-8-459-2011.

[24] Hansen H.P., 1999. Determination of oxygen. In: Grasshoff, K., Kremling, K., Ehrhardt, M., Methods of Seawater Analysis, 3rd Edition, Wiley-VCH, Weinheim, pp. 600.

[25] Outdot C., Gerard R., Morin P., 1988. Precise shipboard determination of dissolved oxygen (Winkler procedure) for productivity studies with commercial system. Limnology and Oceanography, 33, 146-150. Edition, Wiley-VCH, Weinheim, pp. 600.

[26] Grasshoff K., Kremling K., Ehrhardt M., 1999. Methods of Seawater Analysis, 3rd ed. Wiley-VCH, Weinheim, Germany.

[27] Tomczak M., 1981. A multiparameter extension of temperature/salinity diagram techniques for the analysis of non-isopycnal mixing. Progress in Oceanography, 10, 147–171.

[28] Tomczak M. and Large D.G.B., 1989. Optimum multiparameter analysis of mixing in the thermocline of the Eastern Indian Ocean. Journal of Geophysical Research, 94 (C11), 16141–16149.

[29] Budillon G., Pacciaroni M., Cozzi S., Rivaro P., Catalano G., Ianni C. and Cantoni C., 2003. An optimum multiparameter mixing analysis of the shelf waters in the Ross Sea. Antarctic Science, 15,105–118.

[30] Tomczak M. and Liefrink S., 2005. Interannual variations of water mass volumes in the Southern Ocean. Journal of Atmospheric and Ocean Science, 10, 31–42.

[31] Tomczak M. and Poole R., 1999. Optimum multiparameter analysis of the water mass structure in the Atlantic Ocean thermocline. Deep Sea Research Part I: Oceanographic Research Papers, 46, 1895–1921.

[32] Broecker W. S., 1974. "NO", a conservative water-mass tracer, Earth and Planetary Science Letters, 23, 100–107, doi: 10.1016/0012-821X(74)90036-3.

[33] Redfield A.C., Ketchum B.H., and Richards F.A., 1963. The influence of organisms on the composition of seawater, in: The Sea, Hill M. N., pp. 26–77.

[34] Gill P.E., Murray W., Wright M.H., 1991. Numerical Linear Algebra and Optimization, Addison Wesley.

[35] Touratier F., Guglielmi V., Goyet C., Prieur L., Pujo-Pay M., Conan P., Falco C., 2012. Distributions of the carbonate system properties, anthropogenic CO$_2$, and acidification during the 2008 BOUM cruise (Mediterranean Sea). Biogeosciences Discussions, 9, 2709-2753, doi:10.5194/bgd-9-2709-2012.

[36] Malanotte-Rizzoli P., Manca B.B., D' Alcala M.R., Theocharis A., Brenner S., Budillon G., and Ozsoy E., 1999. The Eastern Mediterranean in the 80s and in the 90s: the big transition in the intermediate and deep circulations. Dynamics of Atmospheres and Oceans, 29, 365–395, doi: 10.1016/S0377-0265(99)00011-1.

[37] Béthoux J.-P., 1980. Mean water fluxes across sections in the Mediterranean Sea, evaluated on the basis of water and salt budget and of observed salinities. Oceanologica Acta, 3, 79–88.

[38] Hopkins T.S., 1978. Physical Processes in the Mediterranean basins. In: Kjerfve, B. (Eds.), Estuarine Transport Processes. Univ. of South Carolina Press, Columbia, SC. pp. 269–310.

[39] Özsoy E., Hecht A., Unluata U., Brenner S., Sur H.I., Bishop J., Latif M.A., Rozentraub Z., Oguz T., 1993. A synthesis of the Levantine Basin circulation and hydrography, 1985– 1990. Deep-Sea Research Part II: Topical Studies in Oceanography, 40, 1075–1119.

[40] Marcos M. and Tsimplis M.N., 2008. Comparison of results of AOGCMs in the Mediterranean Sea during the 21st century. Journal of Geophysical Research: Oceans, 113 (C12), doi: 10.1029/2008JC004820.

[41] IPCC, 2014. Climate Change 2014: Impacts, Adaptation, and Vulnerability. The Fifth Assessment Report of the Intergovernmental Panel on Climate Change.

[42] Ludwig W., Dumont E., Meybeck M., Heussner S., 2009. River discharges of water and nutrients to the Mediterranean and Black Sea : Major drivers for ecosystem changes during past and future decades? Progress in Oceanography, 80, 199–217.

[43] Cozzi S. and Giani M., 2011. River water and nutrient discharges in the northern Adriatic Sea : current importance and long term changes. Continental Shelf Research, 31, 1881–1893.

[44] Sparnocchia S., Gasparini G.P., Astraldi M., Borghini M., Pistek P., 1999. Dynamics and mixing of the Eastern Mediterranean outflow in the Tyrrhenian Sea. Journal of Marine Systems, 20, 301–317.

[45] Fuda J.-L., Etiope G., Millot C., Favali P., Calcara M., Smriglio G. and Boschi E., 2002. Warming, salting and origin of the Tyrrhenian Deep Water. Geophysical Research Letters, 29 (19), 41–44, doi: 10.1029/2001GL014072.

[46] Bakun A. and Agostini V.N., 2001. Seasonal patterns of wind-induced upwelling/downwelling in the Mediterranean Sea. Scientia Marina, 65, 243–257.

[47] Estrada M., 1996. Primary production in the northwestern Mediterranean. Scientia Marina, 60 (2), 55–64.

[48] Lacombe H., Tchernia P., Gamberoni L., 1985. Variable bottom water in the Western Mediterranean basin. Progress in Oceanography, 14, 319–338.

[49] Béthoux J.P., Gentili B., Raunet J., and Tailliez D., 1990. Warming trend in the western Mediterranean deep water. Nature, 347, 660–662, doi: 10.1038/347660a0.

[50] Rohling E.J. and Bryden H.L., 1992. Man-induced salinity and temperature increases in Western Mediterranean deep water. Journal of Geophysical Research, 97, 11191-11198.

[51] Vargas-Yáñez M., Moya F., García-Martínez M.C., Tel E., Zunino P., Plaza F., Salat J., Pascual J., López-Jurado J. L., and Serra M., 2010. Climate change in the Western Mediterranean Sea 1900–2008. Journal of Marine Systems, 82, 171–176.

[52] Malanotte-Rizzoli P., Artale V., Borzelli-Eusebi G.L., Brenner S., Crise A., Gacic M., Kress N., Marullo S., Ribera d'Alcalà M., Sofianos S., Tanhua T., Theocharis A., Alvarez M., Ashkenazy Y., Bergamasco A., Cardin V., Carniel S., Civitarese G., D'Ortenzio F., Font J., Garcia-Ladona E., Garcia-Lafuente J.M., Gogou A., Gregoire M., Hainbucher D., Kontoyannis H., Kovacevic V., Kraskapoulou E., Kroskos G., Incarbona A., Mazzocchi M.G, Orlic M., Ozsoy E., Pascual A., Poulain P.-M., Roether W., Rubino A., Schroeder K., Siokou-Frangou J., Souvermezoglou E., Sprovieri M., Tintoré J., and Triantafyllou G., 2014. Physical forcing and physical/biochemical variability of the Mediterranean Sea: a review of unresolved issues and directions for future research. Ocean Science, 10, 281–322, doi: 10.5194/os-10-281-2014.

[53] PROOF program, 2015, http://www.obsvlfr.fr/jgofs/html/prosope/home.htm.

[54] Copin-Montégut C. and Bégovic M., 2002. Distributions of carbonate properties and oxygen along the water column (0-2000 m) in the central part of the NW Mediterranean Sea (Dyfamed site): influence of winter vertical mixing on air-sea CO2 and O2 exchanges. Deep-Sea Research Part II: Topical Studies in Oceanography, 49 (11), 2049–2066.

[55] Pinardi N., Zavatarelli M., Adani M., Coppini G., Fratianni C., Oddo P., Simoncelli S., Tonani M., Lyubartsev V., Dobricic S. and Bonaduce A., 2013. Mediterranean Sea large-scale low-frequency ocean variability and water mass formation rates from 1987 to 2007: A retrospective analysis. Progress in Oceanography, doi:10.1016/j.pocean.2013.11.003 (In Press).

[56] Millot C., 1979. Wind induced upwellings in the Gulf of Lions. Oceanologica Acta, 2, 261–274.

[57] Gascard J.C. and Richez C., 1985. Water masses and circulation in the western Alboran Sea and in the Strait of Gibraltar. Progress in Oceanography, 15, 157–216.

[58] Zavatarelli M., and Mellor G.L., 1995. A Numerical Study of the Mediterranean Sea Circulation. American Meteorological Society, 2013 GRID-Arendal.

[59] Millot C., Candela J., Jean-Luc F. and Youssef T., 2006. Large warming and salinification of the Mediterranean outflow due to changes in its composition. Deep Sea Research Part I: Oceanographic Research Papers, 53, 656–666.

[60] Nykjaer L., 2009. Mediterranean Sea surface warming 1985–2006. Climate Research, 39, 11–17, doi: 10.3354/cr00794

[61] Civitarese G., Gačić M., Lipizer M., and Eusebi Borzelli G. L., 2010. On the impact of the Bimodal Oscillating System (BiOS) on the biogeochemistry and biology of the Adriatic and Ionian Seas (Eastern Mediterranean). Biogeosciences, 7, 3987–3997, doi:10.5194/bg-7-3987-2010.

[62] Abboud-Abi Saab M., Romano J.-C., Bensoussan N., Fakhri M., 2004. Suivis temporels comparés de la structure thermique d'eaux côtières libanaises (Batroun) et françaises (Marseille) entre juin 1999 et octobre 2002. Comptes Rendus Geoscience, 336 (15), 1379–1390.

[63] IUCN and MedPanA, 2012. Changing mediterranean coastal marine environment under predicted climate-change scenarios. www.medpan.org and www.iucn.org/mediterranean

[64] Lelieveld J., Hadjinicolaou P., Kostopoulou E., El Maayar M., Hannides C., Lange M.A., Tanarhte M., Tyrlis E., Xoplaki E., 2012. Climate change and impacts in the Eastern Mediterranean and the Middle East. Climatic Change, 114 (3), 667–687, http://dx.doi.org/10.1007/s10584-012-0418-4.

[65] Romanski J., Romanou A., Bauer M., and Tselioudis G., 2012. Atmospheric forcing of the Eastern Mediterranean Transient by midlatitude cyclones. Geophysical Research Letters, 39 (L03703), doi: 10.1029/2011GL050298.

[66] Theocharis A., Klein B., Nittis K., Roether W., 2002. Evolution and status of the Eastern Mediterranean Transient (1997–1999). Journal of Marine Systems, 33–34, 91–116.

[67] Borzelli G. E., Gacic M., Lionello P., Malanotte-Rizzoli P., 2014. The Mediterranean Sea: Temporal Variability and Spatial Patterns. Geophysical Monograph Series. Editors: John Wiley & Sons, 2014, pp. 75-82.

[68] Vilibić I., Matijević S., Šepić J., and Kušpilić G., 2012. Changes in the Adriatic oceanographic properties induced by the Eastern Mediterranean Transient. Biogeosciences, 9, 2085–2097, doi: 10.5194/bg-9-2085-2012.

[69] Roether W., Manca B.B., Klein B., Bregant D., Georgopoulos D., Beitzel V., Kovačević V., and Luchetta A., 1996. Recent Changes in Eastern Mediterranean Deep Waters. Science, 271, 333–335, doi: 10.1126/science.271.5247.333.

[70] Stratford K., Williams R.G., and Drakopoulos P.G., 1998. Estimating climatological age from a model-derived oxygen-age relationship in the Mediterranean. Journal of Marine Systems, 18, 215–226.

[71] Gasparini G.P., Ortonab A., Budillon G., Astraldi M. and Sansone E., 2005. The effect of the Eastern Mediterranean Transient on the hydrographic characteristics in the Strait of Sicily and in the Tyrrhenian Sea. Deep Sea Research Part I: Oceanographic Research Papers, 52 (6), 915–935.

[72] Schröder K., Gasparini G.P., Tangherlini M., and Astraldi M., 2006. Deep and intermediate water in the western Mediterranean under the influence of the Eastern Mediterranean Transient. Geophysical Research Letters, 33 (L21607, doi: 10.1029/2006GL027121.

[73] Touratier, F. and Goyet, C., 2011. Impact of the Eastern Mediterranean Transient on the distribution of anthropogenic CO2 and first estimate of acidification for the Mediterranean Sea. Deep Sea Research Part I: Oceanographic Research Papers, 58, 1–15.

[74] Tanhua T., Hainbucher D., Schroeder K., Cardin V., Alvarez M., and Civitarese G., 2013. The Mediterranean Sea system: a review and an introduction to the special issue. Ocean Sciences, 9, 789–803, doi:10.5194/os-9-789-2013.

[75] Gačić M., Civitarese G., Eusebi Borzelli G.L., Kovačević V.,Poulain P.-M.,Theocharis A., Menna M., Catucci A., Zarokanellos N., 2011.On the relationship between the decadal oscillations of the northern Ionian Sea and the salinity distributions in the eastern Mediterranean. Journal of Geophysical Research: Oceans, 116 (C12002), doi: 10.1029/2011JC007280.

[76] Goyet C., Hassoun A.E.R., Gemayel E., 2015. Carbonate system during the May 2013 MedSeA cruise. Dataset #841933 (DOI registration in progress).

[77] Goyet C., Gemayel E., Hassoun A.E.R., 2015. Underway pCO2 in surface water during the 2013 MedSEA cruise. Dataset #841928 (DOI registration in progress).

[78] Ziveri P., and Grelaud M., 2013. Continuous thermosalinograph oceanography along Ángeles Alvariño cruise track MedSeA2013. Universitat Autònoma de Barcelona, doi:10.1594/PANGAEA.822153.

[79] Ziveri P., and Grelaud M., 2013. Physical oceanography during Ángeles Alvariño cruise MedSeA2013. Universitat Autònoma de Barcelona, doi:10.1594/PANGAEA.822162.

[80] Ziveri P., and Grelaud M., 2013. Physical oceanography measured on water bottle samples during Ángeles Alvariño cruise MedSeA2013. Universitat Autònoma de Barcelona, doi:10.1594/PANGAEA.822163.

4

A comparative analysis of Lajeunesse model with other used bed load models - effects on river morphological changes

Mohamed Gharbi[1], Amel Soualmia[1], Denis Dartus[2], Lucien Masbernat[2]

[1]Laboratory of Water Sciences and Technology, National Institute of Agronomy of Tunisia, University of Carthage, Tunis, Tunisia
[2]Institute of Fluid Mechanics of Toulouse, National Polytechnic Institute of Toulouse, University of Toulouse, Toulouse, France

Email address:
amel.inat@hotmail.fr (A. Soualmia), mohamed.gharbi@imft.fr (M. Gharbi), Denis.Dartus@imft.fr (D. Dartus),
lucien.masbernat@imft.fr (L. Masbernat)

Abstract: Among the phenomena that greatly influence the river morphology is the sediment transport, especially the bed load mode causing a significant changes in the river morphology. Indeed, the choice of a model or a methodology that can better quantify sediment transport, remains always poorly understood. In this context, a new approach to studying the morphological evolution of rivers is proposed by Charru in 2004, it is based on a model for the erosion and deposition of the particles under laminar flow. In 2006, Charru proposes an extension of this model to turbulent flow. In more advanced research, Lajeunesse in 2010, realizes an experimental study to support the erosion deposition model of Charru, and proposes a new formula to calculate the bed load transport rate. The current research focuses on the effects of bed load transport on the morphological changes in rivers. In the first part, a comparative analysis of empirical laws of bed load transport with experimental data was conducted, in order to test and validate the new bed load model proposed by Lajeunesse, then to check the grain size effect on the sediment transport capacity. In the second part, we are interested in the study of the morphological evolution in rivers. It was performed through numerical modeling using TELEMAC 2D coupled with SISYPHE. The aim is to understand and analyze the morphological changes in the channel bottom. The analysis of the results presented in this paper showed that through the calculated score, most formulas give satisfactory results. In particular at the grain scale, the new bed load transport relation of Lajeunesse, provides an excellent fit to the experimental data. Finally, we were interested in the study of the morphological changes in the channel bottom, it appers clear that the bed load transport has large impacts on river morphology.

Keywords: Bed Load Transport Models, River Morphology, Bed Load Transport, Erosion, Deposition

1. Introduction

The prediction of sediment transport in rivers is one of the most important tasks in fluvial hydraulics, because of its wide applications in nature. It is useful for solving the problems encountered in the river flows, such as turbulent flow in alluvial channels, deposition and erosion, etc... Indeed, the sediments entrained in a flow can be transported along the bed as bed load in the form of sliding and rolling grains, or in suspension as suspended load advected by the main flow [25]. The current research is about the sediment transport modeling and its effects on morphological changes in rivers. In particular, the bed load transport mode causing a significant change in the river morphology is considered here. In fact, the modeling of bed load has a great importance for the management and landscaping job streams and risk prevention. The purpose is to cover the basic phenomena related to sediments transport in order to fully understand the sediment transport processes [1].

In the first part of this paper, we are interested in a comparative analysis in order to test and validate the new bed load model proposed by Lajeunesse in 2010. Thereafter, several simulations were conducted in order to check the grain size effects on the sediment transport capacity [2]. The aim is to be able to understand and to interpret the exchange relations with the bottom (erosion and deposition) [23].

The second part is about the morphological changes in rivers. A model was set up to simulate the bed load transport rate for non-cohesive sediment (fine sand) using TELEMAC 2D coupled with SISYPHE. Several tests calculations were performed in this study to analyze and interpret morphological changes in the channel bottom.

2. Context and Issues

Morphological changes in river depend on climatic and geological conditions (soil type, flow rate, slope, particle size of the bottom, etc.). Various factors influence the morphology of a river system; among these factors that greatly influence the river morphology are water flow, flood, hydraulic structures, and largely the sediment transport, especially the bed load transport which may generate huge consequences on the morphology of the bed profile. In fact, the bed load consists of particules that roll, slide or saltate in a layer close to the bed [19]. The bed load transport has been studied exensively in open channel flows, rivers and flumes [13, 14, 16]. The accumulation of large datasets has led to the development of several laws for predicting the bed load transport [24]. Yet the choice of a such corresponding relation that can better quantify the bed load, remains always poorly understood.

In this study, we will focus on the fluid-sediment interaction using the new formula of Lajeunesse to estimate the bed load transport rate and its effects on river morphological changes [4].

In order to better predict the morphological changes in rivers, various methods to quantify the sediment transport rate have been reported from the literature [10]. Typically, there are several approaches to calculating bed load transport in river. Two main approaches can be distinguished. The first one is called the Bagnold Approach [19], defines the bed load transport as that in which the successive contacts of the particles with the bed are strictly limited by the effect of gravity. Whereas the second one, is based on a model for the erosion and deposition of the particles [1], where the bed load transport can also be defined as the product of the number of moving particles per unit area, the particle volume and the particle velocity.

Many formulas to predict the bed load transport rate are established in the Literature. One way to formulate the problem of sediment transport in river and to identify the relevant controlling parameters is to proceed to dimensional analysis [9]. Dimensional analysis of bed load involving three dimensions (length, time and mass), leads to the following relation:

$$q* = f\left(\tau*, R, S, \frac{H}{D}\right) \qquad (1)$$

One dimensionless form for dimensionless bed load transport q* where:

$$q^* = \frac{q_b}{\sqrt{RgD^3}} \qquad (2)$$

$$\tau = \rho g R_h S \qquad (3)$$

$$\tau^* = \frac{\tau}{(\rho_s - \rho)gD} \qquad (4)$$

$$S = \frac{U^2}{K_s^2 H^{4/3}} \qquad (5)$$

$$R = \frac{\rho_s - \rho}{\rho} \qquad (6)$$

Where q_b = bed load transport rate; τ = bed shear stress; τ^* = Shields number; H = water depth; U = flow velocity; g = the gravitational acceleration; R_h = the hydraulic radius; S = bed slope; K_s = Strickler coefficient; ρ = density of water usually 1000 kg m^{-3}; ρ_s = sediment density usually 2650 kgm^{-3}; and R = relative density [27].

In general, most of the bed load models involve a threshold valu ae of the Shields number τ^*_c below which no sediment is transported, [12]. They all predict the same dependence q* α $\tau*^{3/2}$. In the literature, there are two main groups; the first one predicts the bed load transport by "(7)":

$$q^* \propto \left(\tau^* - \tau_c^*\right)^{3/2} \qquad (7)$$

Such as Meyer Peter & Müller (M-P-M) [14]; Einstein [16]; Fernandez-Luque & Van Beek [18]; Wong [21]; and Recking [11].

Where τ_c^* is the critical Shields number:

$$\tau_c^* = \frac{\tau_c}{(\rho_s - \rho)gD} \qquad (8)$$

Whereas the second one proposes the following form "(9)":

$$q^* \propto \left(\tau^* - \tau_c^*\right)\left(\sqrt{\tau^*} - \sqrt{\tau_c^*}\right) \qquad (9)$$

Such as Ashida & Michiue [15]; Engelund & Fredsoe [17]; and Bridge & Dominic [20].

The next part is devoted to the description of the most commonly used formulas established in the literature for calculating bed load transport rate.

• *Meyer Peter & Müller (M-P-M) 1948*

$$q* = 8\left(\tau^* - \tau_c^*\right)^{3/2} \qquad (10)$$

Where $\tau_c* = 0.047$. This relation is empirical in nature [14]; it has been verified with data for uniform coarse sand and gravel. It was developed for Alpine streams in Switzerland [3].

• *Einstein 1950*

Einstein [16] defined a bed load formula as an equation linking the rate of bed load transportation with the properties of the grain and the flow causing the movement [26].

$$q^* = \begin{cases} \dfrac{K \exp(-0.391/\tau^*)}{0.465} & si \quad \tau^* < 0.182 \\ 40 K \tau^{*3} & si \quad \tau^* \geq 0.182 \end{cases} \qquad (11)$$

Where

$$K = \sqrt{\frac{2}{3} + \frac{36\,v^2}{g\,(R-1)D^3}} - \sqrt{\frac{36\,v^2}{D_*^3}} \qquad (12)$$

v = the kinematic viscosity, usually $v = 10^{-6}$ m^2/s.
- Van Rijn 1984

$$q^* = \begin{cases} 0.053(R-1)^{0.5}\,g^{0.5}\,D_{50}^{1.5}\,D_*^{-0.3}\,T^{1.5} & T < 3 \\ 0.1\,(R-1)^{0.5}\,g^{0.5}\,D_{50}^{1.5}\,D_*^{-0.3}\,T^{1.5} & T > 3 \end{cases} \qquad (13)$$

It's used to calculate bed load transport rates of particles with mean sizes in the range between 0.2 and 2.0 mm [3]. This equation is based on a dimensionless particle diameter and the transport stage parameter T, defined, respectively [13], as:

$$D_* = D_{50} \left(\frac{(s-1)g}{v^2} \right)^{1/3} \qquad (14)$$

$$T = \frac{(\tau - \tau_{cr})}{\tau_{cr}} \qquad (15)$$

- Lajeunesse 2010

In further research, Van Rijn submits an alternative way to treat the problem of bed load transport; it is to consider that q_b can be written:

$$q_b = \delta v.n.V \qquad (16)$$

Where n = the surface density of moving particles; V = the mean particle velocity; and δv = the volume of an individual particle [13]. This new alternative has motivated many researchers to study the trajectory of bed load particles [1, 4, 13].

Up to now, the majority of the bed load transport laws proposed in the literature have been focused on the establishment of a relation between the local volumetric flow rate of particles and the local shear stress exerted by the fluid flow on the bed. These relations implicitly consider that the particle flux is in equilibrium with the shear stress, and consequently ignore any relaxation effect [9]. In order to overcome this limitation, in 2004, Charru developed an erosion-deposition model from viscous flow experiments [1], which accounts for a relaxation effect related to the time needed for a particle to settle once it is entrained in the fluid flow. This model was applied for laminar flow and a small bed shear stress where the particles are transported as bed load. In 2006, Charru propose an extension of this model to turbulent flow [2].

Referring to "(16)", the bed load flux is proportional to n,

the surface density of moving particles. For steady and spatially uniform flow, this density corresponds to dynamical equilibrium between the particles eroded from the fixed bed and deposited onto it [2]. The variation of n(x, t) is governed by the conservation equation "(17)":

$$\frac{\partial n}{\partial t} = \dot{n}_e - \dot{n}_d - \frac{\partial Q}{\partial x} \quad where \quad Q = n.V \qquad (17)$$

"Equation (17) states that the variation rate of n is related to exchanges with the fixed bed, through the erosion rate ($\dot{n}_e$) and the deposition rate ($\dot{n}_d$); and to the divergence of the bed load flux".

In more advanced research, Lajeunesse in 2010 realize an experimental study to support the erosion deposition model of Charru [2], and allow the calibration of the involved coefficients [9].

In addition, based on the new approach of Charru "(17)" and experimental data, Lajeunesse propose a new bed load relation for a steady and uniform flow above a flat topography. The new law is given by the following equation:

$$\frac{q_{sat}}{\sqrt{RgD^3}} = 10.6 (\tau^* - \tau_c^*) \left(\sqrt{\tau^*} - \sqrt{\tau_c^*} + 0.025 \right) \qquad (18)$$

In the first part of this current research, we are interested in a comparative analysis of experimental data of Lajeunesse's experiment and those calculated through different empirical laws "(10)", "(11)", "(13)", and "(18)". The aim is to test the new bed load law proposed by Lajeunesse [9].

Thereafter, on the first hand in order to validate this new model another comparative study was conducted with other experimental data measured by Sequeiros 2010 [15].

On the other hand and in order to visualize the influence of the grain size on sediment transport; several performance simulation tests of the main sediment transport formulas on a set of data were investigated. It was considered the variation of the particle diameter ranging from 1 μm, to 2000 μm, (from silt-clay to coarse sand).

The second part is devoted to the numerical modeling using TELEMAC 2D coupled with SISYPHE; the objective is to study the morphological change of the channel bottom due to the bed load transport.

3. Materials and Methods

With the intention of testing and validating the new formula proposed by Lajeunesse in 2010, we conducted a comparative study [9], based on literature experimental data. In the first step we tested the model by using the results of the Lajeunesse experiment. In the second step, to validate this new model, we used the Sequeiros experiment [24]. But before that, we start here by giving a brief description of these experiments.

3.1. Lajeunesse Experiment 2010

An experimental study was conducted by Lajeunesse in 2010. The experiments were carried out in a rectangular inclinable flume of width W = 9.6 cm and length 240 cm "Fig. 1" partially filled with an erodible bed of quartz grains of density ρ_s = 2650 kg m^{-3} (R = 1.65).

Figure 1. Photo of the Lajeunesse experimental setup 2010 [8].

The bed slope S was measured with a digital inclinometer with accuracy 0.1°. Once the bed was ready, water was injected by a pump at the upstream flume inlet with constant discharge W.q$_w$, where q$_w$ is the water discharge per unit width. To prevent any disturbance of the bed, water was not injected as a point source but rather it overflowed smoothly onto the flume bed via a small reservoir. For all runs the discharge was high enough for the flow to fill the whole width of the flume.

The water depth H was measured from the deviation of a laser sheet. This depth was found to be constant along the section of the flume within the experimental precision. Assuming that the depth- averaged flow velocity was uniform in the spanwise direction, the velocity was given by U =q$_w$/H [8].

At the flume outlet, particles transported by the flow settled out in an overflow tank with constant water level. The tank rested on a high-precision scale connected to a computer that recorded the weight every ten seconds. The sediment discharge per unit river width q$_b$ was deduced from the sediment cumulative mass [8].

The influence of the grain size on sediment transport was investigated by performing three series of experiments. In the first series D is equal to 1 mm, in the second ones D=2.24 mm, and for the third ones D = 5.50 mm.

3.2. Sequeiros Experiment

An experimental study was conducted to studying bed load transport and to validate the modified formula of Mayer Peter by Sequeiros in 2010. The experience was carried out in a flume 15 m length, 0.45 m wide and 1.4 m deep, at the University of Illinois Ven Te Chow Hydrosystems Laboratory (V.T.C.H.L) "Fig. 2".

The tank was initially filled with fresh water. Dense mixtures of water were fed into the flume from a 4.5 m^3 mixing tank by means of a pump. The discharge rate was measured by a magnetic McCrometer flow meter. The sediment was injected uniformly along the flume width at its

upstream end [24]. A layer of plane bed sediment consisted of particles with a specific gravity of 2650 was placed on the bed and smoothed to a constant thickness before each run. This sediment served as the source for bed load. The water depth in the tank was 0.41 m at the upstream end of the sediment covered reach. At the downstream end, the water depth ranged from 0.71 to 0.43 m. A set of five sediment traps located immediately downstream of the movable bed ensured the collection of bed load [24]. The traps consisted of box like transversal slots, 4 cm high and 15 cm long, covering the whole width of the flume, located at the same level as the sediment bed, to avoid local erosion at the boundary between the traps and the bed. The velocity profiles were taken with an Acoustic Doppler Velocimeter (ADV). Several tests were conducted to measure each time the bed load transport rate and channel bed evolution

4. Results and Discussion

Let us summarize our results on the motion of bed load particles above a plane bed [9, 24]. The results may be divided into 4 parts: The first part is a comparative analysis of Lajeunesse data and simulations that we done with the different empirical laws "(10)", "(11)", "(13)", and "(18)". The second part is about the validation of the new formula proposed by Lajeunesse. For this, we used other experimental data, the Sequeiros experimental ones and simulations with the most commonly used models [9]. In the third part, we intend to visualize the particles size effects on the prediction of bed load transport rate. The last part is devoted to the river morphological changes under water flow.

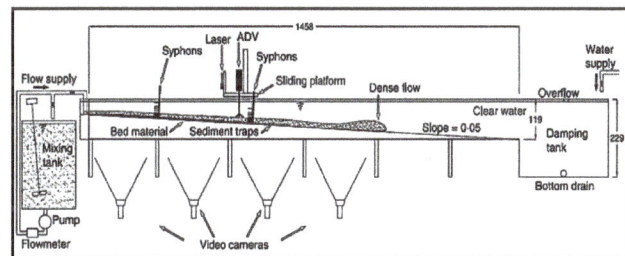

Figure 2. Sketch of the Sequeiros experimental set-up 2010 [24].

4.1. Tests of the Lajeunesse New Bed Load Model

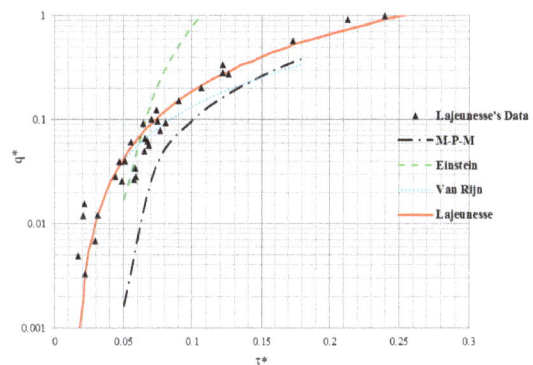

Figure 3. Simulations of the dimensionless transport rate q* versus the Shields number τ* for D = 1000 μm (Lajeunesse Data case).

We consider the same conditions of Lajeunesse experiment carried out in 2010. This mean the same hydraulic conditions (water flow, water depth, velocity) and also the same sediment properties (uniform non-cohesive sediments, the particles diameter D = 1000 μm). The critical Shields number τ^*_c is determined from the proposed formulas of Van Rijn "(13)" based on the Shields diagram [12].

Several simulations were carried out to test the new bed load transport model proposed by Lajeunesse in 2010. The results of our performed simulations are summarized in the "Fig. 3":

The analysis indicates that the proposed model of Lajeunesse gives satisfactory results. The simulated bed load rate is closely correlated with the experimental data. Furthermore, we note that the Van Rijn model [13] gives acceptable results but with a little lag. Each of the models used in calculating bed load transport rates has some failure to predict reliably the bed load rate. We also note that the Einstein [16] and Meyer Peter & Müller [14] models are unable to reproduce correctly the bed load transport rate.

To check the reliability of the results obtained by the different tested formulas used to quantify the bed load rate, we calculate the Percent Ratio given by the following equation :

$$r = \frac{Q_{sim}}{Q_{obs}} \qquad (19)$$

Where r= is the percent ratio within a given interval, Q_{obs}= observed solide discharge, and Q_{sim}= simulated solide discharge.

For example, a score of 50% shown for the interval [0.1 - 10] means that 50% of predictions are correct, which is a reasonable interval considering the natural fluctuations of bed load transport. For our case study, we have considered two intervals. The results are summarized in the Table 1 below.

Table 1. Scores (%) obtained with the tested formulas on the Lajeunesse data, in the considered intervals r

Formulas	0.5 < r < 4	0.1 < r < 10
Einstein	60	71
Meyer-Peter & Müller	64	75
Van Rijn	69	79
Lajeunesse	72	83

All the tested formulas have a score higher than 70% corresponding to the ratio r within the interval [0.1 - 10]. For the range 0.2 < r < 4, a good score (up 60%) is obtained by some formulas. The obtained scores confirmed again that the new law proposed by Lajeunesse give satisfactory results (up 83%).

4.2. Validation of the Lajeunesse Model

In order to check the validity of the Lajeunesse formula on other experimental cases, we applied it to the Sequeiros experiment carried out in 2010.

We desired to compare the experimental data measured during Sequeiros experiment [25] and the calculated bed load transport by various laws. The "Fig. 4" below summarizes the results of our performed simulations.

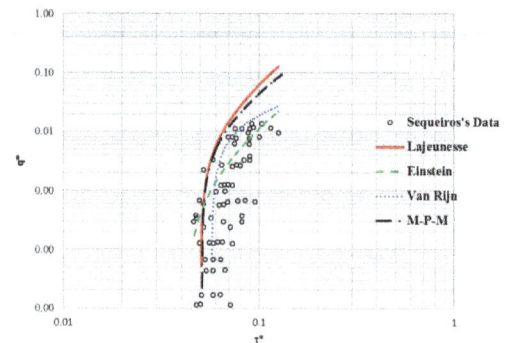

Figure 4. Simulations of the dimensionless transport rate q* versus the Shields number τ* (Sequeiros Data case).

It appears clear that the tested bed load transport relation fitted well the experimental data, it gives satisfactory results. Particular for a low bed shear stress the Lajeunesse model define a coherent relation with the experimental data.

We also note that the value of q* predicted by the Lajeunesse and M-P-M models are somewhat higher than the data trend especially for τ*>0.1, whereas the Van Rijn and Einstein relations provide a better fit with measured data.

In order to verify the reliability of the calculations performed with the different models on the Sequeiros experimental data, we calculate the Percent Ratio "(19)". The results are summarized in Table 2.

Table 2. Scores (%) obtained with the tested formulas on the Sequeiros data, in the considered intervals r

Formulas	0.5 < r < 4	0.1 < r < 10
Einstein	58	68
Meyer-Peter & Müller	69	79
Van Rijn	64	71
Lajeunesse	71	78

Most of the models have a score higher than 70% when the ratio r at the range of [0.1 - 10]. Concerning the Lajeunesse model, the calculated score is higher that 70%, it represents good correlations with the experimental data.

4.3. Effects of Particle Size on the Bed Load Transport Rate

The "Fig. 5" below show the particeles size effects on the prediction the bed load transport capacity.

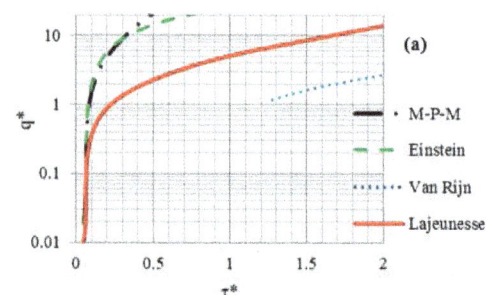

(a) Particles diameter 1 μm

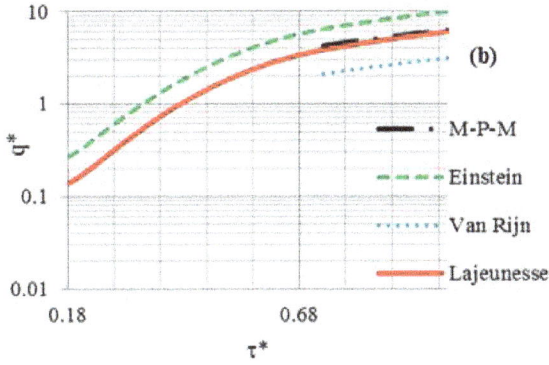

(b) Particles diameter 100 μm

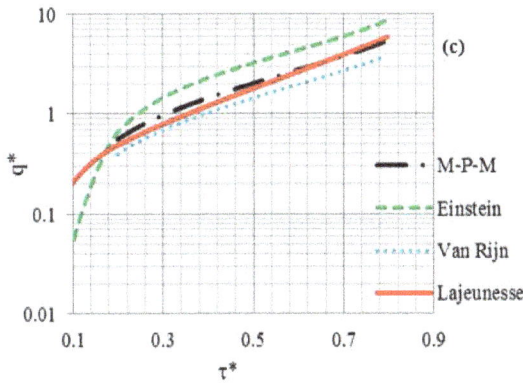

(c) Particles diameter 500 μm

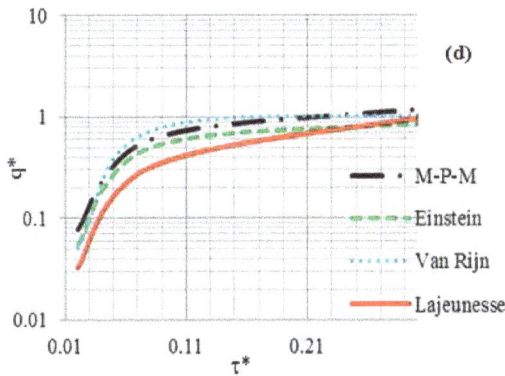

(d) Particles diameter 2000 μm

Figure 5. *Simulations of the dimensionless transport rate (q*) versus the Shields number (τ*): Influence of grain size (D).*

Here, it was considered the particle's diameter variation (from D = 1 μm to D = 2000 μm), to calculate the bed load transport rate by 4 of the most commonly used empirical laws given by the equations "(10)", "(11)", "(13)", and "(18)".

The "Fig. 5" illustrates the results of bed load transport rate calculated for several set of materials having different diameters ranging from 1 μm to 2000 μm (from silt-clay to coarse sand).

Four formulas were applied to predict the sediment transport rate by bed load. The aims is to visualize the effect of the particle diameter on the bed load transport rate. Thus we note that for a 25% variation of the particle diameter (D = 500 μm – D = 2000 μm) results in about 17% variation of the transport rate (2.95 10^{-4} – 0.425 10^{-4} m^2/s).

Our observations show that when the particle diameter is greater than 100 μm "Fig. 5c", "Fig. 5d", the results become acceptable. Whereas for the lower particle size such as D = 1 μm, we note that these models cannot replicate the reality because they are tested outside of their domain of validity "Fig. 5a", "Fig. 5b". We also note that when the grain size exceeds 100 μm, both formulas Van Rijn 1984 and Lajeunesse 2010 give remarkable agreement. Finally, it appears clear that the bed load transport rate depends strongly on the material diameter.

4.4. Bed Load Transport Effects on Morphological Changes in River

Several phenomena may greatly influence the river morphology, in particular the sediment transport. Generally, rivers are constantly seeking equilibrium between the shape of their bed and their flow rates. Sediment deposits tend to offset their wrenched off. In fact, rivers are always looking for a dynamic equilibrium between the two operating processes of erosion and deposition [3].

This current research focuses on the study of morphological changes in channel bottom, by using the Lajeunesse's experimental data. In order to achieve this, a modeling of sediment transport was performed using TELEMAC 2D coupled with SISYPHE to calculate the bed load transport rate, and subsequently to deduce the channel bottom evolution.

TELEMAC 2D is the hydrodynamic module, it solve the two dimensional Saint Venant equations to simulate free surface flows in two dimensions of horizontal space [6, 7]. At each point of the mesh, the program calculates the water depth and the two velocity components. Whereas SISYPHE is the sediment transport module, it can be used to model complex morphodynamics processes in diverse environments. To calculate the bed evolution, SISYPHE solves the Exner equation:

$$(1-p)\frac{\partial Z_f}{\partial t} + Div\left(\overrightarrow{q_b}\right) = 0 \qquad (20)$$

Where p = the bed porosity (p~0.4 for non-cohesive sediment); Z_f = the bottom elevation; q_b = the solid bed load transport per unit width [22].

The Exner "equation (20)" states that the variation of sediment bed thickness can be derived from a simple mass balance. "Equation 20 is strictly valid for bed load only". In this contribution, the effect of the bed slope will be considered by SISYPHE in computing the bed load transport rate. In truth, SISYPHE offers several empirical formulas to calculate the bed load transport rate, but not the Lajeunesse formula.

Therefore, we programmed this new law "(18)" in SISYPHE in order to compare the results given by the different models.

The study area is a rectangular flume as shown in "Fig. 1". We used a triangular mesh of 3528 regular elements, leading to mesh size 1 cm. The experimental set-up has a flat bed of sediment consisting of sandy uniform diameter D = 1000 μm. At the initial step, the bottom is attached to the elevation z = 0 m, and the friction coefficient is constant, corresponding to Strickler coefficient of $K_s = 60$ m$^{1/3}$/s

In TELEMAC 2D, we imposed on the upstream of the channel a flow rate $q_w = 0.001$ m^3/s for a water depth imposed on the downstream of the channel H= 0.01 m (Lajeunesse's experiment conditions). In SISYPHE, the sediment transport at the entrance of the channel is calculated assuming equilibrium conditions. The shear stress is calculated by TELEMAC 2D at each time step. During the Lajeunesse experience, the bed load was the sole mode of sediment transport.

Two dimensional river modeling and especially its applications to channel bed morphology can be considered as a relatively new technique [5]. The purpose is a comparative study of 4 bed load laws, Meyer Peter & Müller [14]; Einstein [16]; Van Rijn [13]; and Lajeunesse [9], was conducted. The aims are to compute the sediment transport rate, and to determine the morphological evolution of the considered channel.

In a first step, we determine the bed load transport rate by the most commonly used bed load transport models. The "Fig. 6" below shows the results.

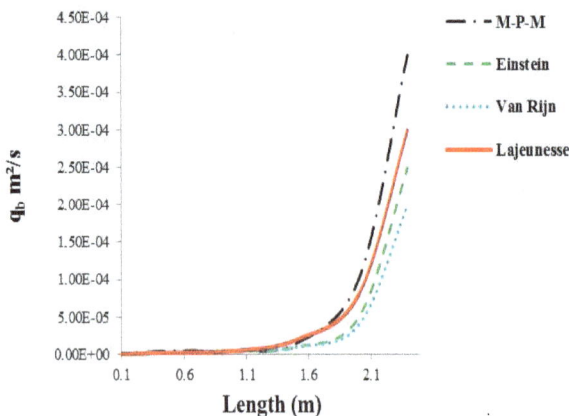

Figure 6. *Evolution of the bed load transport along the channel for sediments diameter D = 1000μm.*

We note that the different tested models lead to acceptable results compared with the sediment discharge measurements, with the exception of Meyer Peter & Müller model (M-P-M) [14] that presents an overestimation of the bed load rate compared with the measured data. These results confirm again the first part of this work, showing that both of Van Rijn and the Lajeunesse model give acceptable results [13, 9].

In a second step, we are interested in the study of the channel bed elevation changes referring to the numerical modeling by using TELEMAC 2D coupled with SISYPHE. In fact, once the bed load has been determined, the resulting change of the bed level Z_f is calculated from "(19)". SISYPHE is coupled (internally) with the hydrodynamic model TELEMAC 2D, where it's called inside this hydrodynamic model, the data will be exchanged directly between the two programs.

The results of the numerical modeling using TELEMAC 2D coupled with SISYPHE show that there is a longitudinal

evolution of the channel bed. We note that the Meyer Peter & Müller model show an overestimation of sediment transport rate, hence there is excessive digging in the channel bottom that exceeds 4 cm into the sand layer.

In the "Fig. 7" is showed the channel bottom evolution under the flow effect; using different models to calculate the bed load transport (not including suspended sediment).

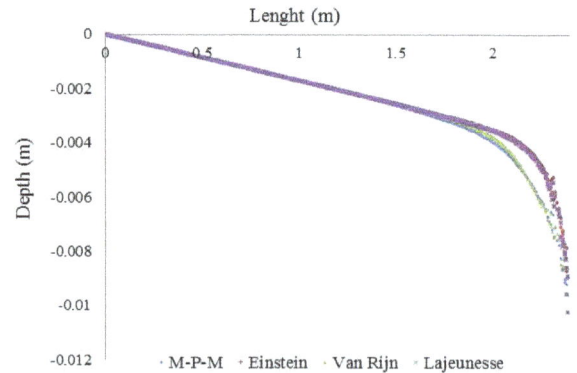

Figure 7. *Morphological evolution of the channel bottom (Lajeunesse Experiment).*

Our observation show that the Einstein and Lajeunesse models gives practically the same values of the bed load rate, hence we observed that there was a digging of the channel bottom under the flow effect. For the different cases, we note that there is a digging in the sand layer of about 3 cm in average. These changes in the morphological channel bed are induced by the bed load transport.

5. Conclusion

Several formulas have been established in literature to calculate bed load transport, particularly the Lajeunesse model, based on the Charru approach. Model intercomparaisons, and comparaisons between model predictions and field data, have been made in order to test and validate this new bed load transport model. The analysis shows that most of the formulas have a score greater than 70% when the ratio r is tested in the range [0.1 - 10]. In particular the proposed new formula of Lajeunesse has a score that exceeds 80%. Furthermore the simulations analysis showed that the sediment transport capacity depends strongly on the particles diameter. Yet, the use of a such model still requires accurate data on the considered site, especially for higher accuracy predictions.

Regarding the river morphological evolution, it appears clear from the numerical modeling realized by TELEMAC 2D coupled with SISYPHE that the changes in the channel bottom depend greatly on the bed load transport rate.

In perspective, we will focus on the inclusion of the Charru's full model in SISYPHE, and its application, to a laboratory set (Both suspended and bed load transport) then to a real case, which is the Medjerda River situated in the north west of Tunisia.

Acknowledgments

The authors acknowledge the cooperation of the National Polytechnic Institute of Toulouse (INPT), and particulary the members of HydroEco groups worked under the Institute of Fluid Mechanics of Toulouse (IMFT), France.

Financial support for this study was provided by a grant from the Ministry of Higher Education and Scientific Research in Tunisia.

References

[1] F. Charru, H. Mouilleron, and O. Eiff, "Erosion and deposition of particles on a bed sheared by a viscous flow," Journal of Fluid Mechanic, Vol. 519, pp. 55–80, 2004.

[2] F. Charru, "Selection of the ripple length on a granular bed sheared by a liquid flow," Journal Physics of Fluids, Vol. 18, pp. 121508-1, 2006.

[3] E. M. Franklin, "Dynamique de dunes isolées dans un écoulement cisaillé, " Thesis, University of Toulouse, 2008.

[4] M. García, "Sedimentation Engineering: Processes, Measurements, Modeling, and Practice, in Examining the Confluence of Environmental and Water Concerns," World Environmental and Water Resource Congress, Graham, pp. 91-94, 2006.

[5] P. Belleudy, "Restoring flow capacity in the Loire River bed," Hydrological Processes. Hydrol, Process 03. 1220-1233, 2000.

[6] M. Gharbi, and A. Soulamia, "Modélisation de la prévision des crues éclair au niveau du bassin versant de la Medjerda," Revue de Mécanique Appliquée et Théorique, Vol. 2, 6, pp.585-595, 2013.

[7] M. Gharbi, and A. Soulamia, "Simulations des crues éclair au niveau de la haute vallée de la Medjerda," Deuxième Congrès Tunisien de Mécanique COTUME, Sousse, Tunisie, pp.81-86, 2012.

[8] M. Houssais, and E. Lajeunesse, "Bed load transport of a bimodal sediment bed," Journal of Geophysical Research, pp 115-1, 2012.

[9] E. Lajeunesse, L. Malverti, and F. Charru, "Bed load transport in turbulent flow at the grain scale: Experiments and modeling," Journal of Geophysical Research, Vol. 115, 2010.

[10] A. Recking, "An analysis of nonlinearity effects on bed load transport prediction," Journal of Geophysical Research: Earth Surface, Vol. 118, pp.1-18, 2013.

[11] A. Recking, P. Frey, A. Paquier, and P. Belleudy, "An experimental investigation of mechanisms involved in bed load sheet production and migration," Journal of Geophysics Research Vol.114, 2009.

[12] A.F. Shields, Application of similarity principles and turbulence research to bed-load movement, Mitteilungen der Preussischen Versuchsanstalt fur Wasserbau und Schiffbau, Germany," pp 5–24, 1936.

[13] L. C. Van Rijn, "Principles of Sediment Transport in rivers, Estuaries and Coastal Seas," Aqua Publications in Netherlands: pp.1-612, 1993.

[14] E. Meyer-Peter, and R. Müller, "Formulas for bed-load transport, " The 2nd Meeting of International Association for Hydraulic Research," Int. Assoc. for Hydraul, Stockholm, Germany, 1948.

[15] K. Ashida, and M. Michiue, "Studies on bed-load transport rate in open channel flows," International Association for Hydraulic Research International, Symposium on River Mechanics, Bangkok, Thailand, pp.407–417, 1973.

[16] H. Einstein, "The bed-load function for sediment transportation in open channel flows," Tech. Bull, U.S, Departement of Agriculture, Washington, D. C, 1950.

[17] F. Engelund, and J. Fredsoe, "A sediment transpot model for straight alluvial channels," Nord. Hydrol, Vol 7(5), pp.293–306, 1976.

[18] R. Fernandez-Luque, and R. Van Beek, "Erosion and transport of bed load sediment," Journal of Hydraulic Research, Vol.14, pp.127–144, 1976.

[19] R. Bagnold, "The flow of cohesionless grains in fluids," Philos. Trans. R. Soc. London A, Vol.249, pp.235–297, 1956.

[20] J. S. Bridge, and D. F. Dominic, "Bed load grain velocity and sediment transport rates," Water Resourc. Res, Vol.20, pp. 476–490, 1984.

[21] M. Wong, "Does the bedload equation of Meyer Peter and Müller fit its own data," 30th Congress of the International Association for Hydraulic Research, Int. Assoc. of Hydraul. Res., Thessaloniki, Greece, 2003.

[22] C Villaret, and J. M. Hervouet, "Comparaison croisée de différentes approches pour le transport sédimentaire par charriage et suspension, " Laboratoire National d'Hydraulique et Environnement. IXèmes Journées Nationales Génie Civil – Génie Côtier, 2006.

[23] Wu Weiming, W Rodi, and T. Wenka, "3D Numerical Modeling of Flow and sediment transport in open Channels," Journal of Hydraulic Enginee.

[24] O. E. Sequeiros, B. Spinewine, R. T. Beaubouef. T. Sun, M. H. Garcia, and G. Parker, "Bed load transport and bed resistance associated with density and turbidity currents," Journal of International Association of Sedimentologists, Sedimentology, Vol. 57, pp.1463–1490, 2010.

[25] H. Chanson, "The Hydraulics of Open Channel Flow: An Introduction," Butterworth-Heinemann, 2nd edition, Oxford, UK, pp. 3-8, 2004.

[26] S. Talukdar, Bimlesh Kumar, and S. Dutta, "Predictive capability of bedload equations using flume data," Journal of Hydrol. Hydromech, Vol. 60, pp. 45–56, 2012.

[27] M. S. Yalin, "Mechanics of Sediment Transport," Pergamon, New York, 1977.

5

Bacterial Abundance – Chlorophyll *a* Concentration Relationships in Cross River Basin, Southeastern Nigeria: An Evaluation of Empirical Bacterial Abundance – Chlorophyll *a* Models Using a Multivariate Analysis

Inyang, Aniefiok I*, **Antai, Ekpo E, Dan, Monica U**

Department of Marine Biology, Akwa Ibom State University, Ikot Akpaden Mkpat Enin, Nigeria

Email address:

aniefiokinyang@yahoo.com (A. Inyang), ekpo.eyoa@yahoo.com (E. Antai), kyrianmonica@yahoo.com (M. Dan)

Abstract: The relationship between bacteria and chlorophyll *a* was investigated for three months from July to September, 2008 during flood and ebb tide periods at two sites in the Cross River basin, southeastern Nigeria, to ascertain the reliance of bacteria on autochthonous organic matter. There was no significant difference between flood water and ebb water samples at both stations for the entire variable measured. Bacterial cell abundance showed a weak positive correlation with chlorophyll *a* concentration at Station 1 ($r^* = 0.132$, $P > 0.05$) and a strong positive correlation at Station 2 ($r^* = 0.599$, $P > 0.5$). The micronutrients at both stations showed a weak negative correlation with chlorophyll *a* concentration. Linear regression analysis together with standardized coefficient value revealed a high bacteria/chlorophyll *a* relationship at Station 2 and a weak bacteria/chlorophyll *a* relationship at Station 1. Principal Component Analysis was applied to the measured variable, and a high correlation was observed between bacterial cell abundance and chlorophyll *a* concentration at Station 2. A strong reliance of bacteria on chlorophyll *a* was established at Station 2 whereas Station 1 showed a weak correlation indicating the strong dependence of bacteria on allochthonous organic matter. This bacterial abundance could serve as a good source food for protozoans in transferring energy to classic food chain.

Keywords: Bacteria, Chlorophyll *a*, PCA, Protozoan

1. Introduction

Bacteria are decomposers, decomposing dead organic matter either on land or in aquatic environment for their food. In the aquatic ecosystem, particulate organic matter or carbon is available to this organism in two ways; land driven (allochthonous materials) and water generated (autochthonous material), mainly derived from phytoplankton, excretory products of other organisms and dead organisms. The assumption that planktonic bacteria depend on phytoplankton in aquatic environments devoid of allochthonous input of organic matter is now widely recognized. This statement is supported by the correlation which exists between bacterial biomass and chlorophyll *a* concentration (1, 2, 3, 4, 5) or between bacterial production and both chlorophyll *a* concentration and planktonic primary production (6), over a large range of temperature and tropical aquatic environments. Even though there is limited

direct evidence, the relationship is often thought to reflect the reliance of planktonic bacteria on algae for their organic carbon requirements. (7), assumed that, up to 50% of algal primary production is released as Dissolved Organic Carbon (DOC) (8). Algal-released carbon may support up to 95% of bacterial production (9, 8). This co-variation could also hold from similar responses of both groups of organisms to common regulating factors (10), such as the supply of inorganic nutrients (11, 12). Some studies have reported strong algal-bacteria coupling, whereas some studies show weak relationship or undetectable (13, 14). This is because most of these studies were conducted in systems that received a large amount of allochthonous DOC, and the reduced relative importance of algal-released DOC. For instance, in the Hudson River estuary, (15), found no statistical significant correlation between bacteria density and chlorophyll concentrations, which ranged from about 4 - 15µgl⁻¹. In contrast, several cross-system studies detected significant correlations between algal-released DOC and bacterial density. Its chlorophyll concentration ranged from 0.2 - 189µgl^{-1} (1); 0.05 - 120µgl^{-1} (2); and less than 1µgl^{-1} to more than 100µgl^{-1} (6). Bacterial carbon can often exceed that of the phytoplankton carbon in low chlorophyll *a* waters, and is commonly 2 – 3 times greater (16, 17). This is due to reliance of bacteria on allochthonous DOC. (18), found a positive relationship between algae and bacteria in streams. In a comparison of 69 streams, (12) found that epilithic bacterial numbers were related to chlorophyll concentration over a broad range (< 1 to > 100µgcm^{-2}) but such a co-variation was not apparent when only streams with low chlorophyll concentration were included (< 5µgcm^{-2}). (19), found a positive co-variation between chlorophyll concentrations and epilithic bacteria number in one stream and negative relationship in another stream in the same watershed. It is in this disparity that this study was carry out in this ecosystem to ascertain if there will be any relationship between bacterial density and chlorophyll *a* concentrations. The significance of this study was to reveal the important of carbon to bacteria population in this area.

2. Materials and Methods

2.1. Study Area

The study areas located in Figure1 are; (040^0 57′ N, 080^0 18′ E), Station1 and (040^0 45′ N, 080^0 35′ E), Station2. Cross River takes its origin from Oban hills of the south eastern Nigeria and meanders south-north. It is one of the tidal tributaries of the Cross River Estuary draining through heavily forested land scape of south-eastern Nigeria with catchment area of approximately 16662km^2. Cross River Estuary is the largest estuary along the Gulf of Guinea (20,21) covering an estimated area of 54,000km^2 and 39,000km^2 in Nigeria while the remaining 14,000km^2 lies in Cameroon (20). It also has a long coastline with fringing mangrove and a characteristic muddy bottom. It is the largest in the West African sub-region and is approximately 25km wide at the mouth and more than 440km long with a tidal flushing of 1.83 billion cm^3 per day (20). Its climate and hydrology have been reported by (22, 23). The average temperature of the estuarine surface water is 26.7°C. The river discharge upstream (Itu) is 879m^{3}s^{-1} (dry season) and 2533m^{3}s^{-1} in wet season (24). Nypa palms with some dotted mangrove plants fringing the coastal waters and other inundation. The local climate of this area is characterized by typical humid regions of southern Nigeria, with a wet season from April – October, and dry season from November – March each year. As in other riverine area, settlement of people and location of industries close to the coast is not excepted. Coastal dwellers engaged in both commercial and artisanal fishery activities and other activities such as logging, sand mining, and petty trading. Station 1 is noted for a commercial boat activity that normally conveys people to Oron town.

2.2. Hydroclimatic Analysis

At each location, surface water samples were collected by hand-fill method on each occasion with respect to flooding and ebbing tides, once a week for three months, with a sterile 1L plastic bottle. Surface water temperatures were measured insitu using hand-held seaward immersion thermometer with a range value of -10 to 110^{0}C and with a percentage error of ±1%. The collected water sample was transported to the laboratory in an ice chest within two hours. Nitrate, phosphate and chlorophyll *a* were analyzed as described by (25).

2.3. Estimation of Bacterial Cell Abundance

From the 1L surface water sample, 250mL were subsample each. The subsample was placed in larger beaker, thereafter were sized fraction using negative filtration method. 1ml replicate of each sample was diluted to 10^{-4} for subsequent analysis. 1ml of each replicate was incubated for three hours thereafter subsample for analysis. Bacterial cells present in the subsample were determined directly by epiflorescence microscope using 4, 6, diaminido-2 –phenyl indole (DAPI) as described by (26).

2.4. Statistical Analysis

Statistical analysis was carried out with the aid of XLSTAT (2014 version), SPSS (version 17.0) and Microsoft Excel (2010 version). T – test analysis was used to ascertain the significant difference between flood water sample and ebb water sample for all the variables. Pearson correlation coefficient was used to evaluate relationship between bacterial cells abundance and some other environmental variables (chlorophyll *a* concentration, nitrate concentration, phosphate concentration and surface water temperature). Principal Component Analysis (PCA) was used to authenticate the correlating behavior between variables and especially with bacteria. Linear regression analysis was used to model a quantitative dependent variable (bacterial cell abundance) with chlorophyll *a* concentration and with other variables. Since there was no significant difference (P < 0.05, 0.01) between flood water sample and ebb water sample at both stations for the entire variables measured, therefore N = 22 in all the statistical analysis used.

3. Results

The summary statistics for station 1 and station 2 are given in table 1, while the variation in the measured variables is given in figure 2 to figure 5 in respect to flood tide and ebb tide water sample. T – test analysis showed no significant difference between flood water sample and ebb water sample at both stations for the entire variables measured (P < 0.05, 0.01).

However, the Pearson correlation matrix for the entire measured variables is given in table 2. Thus, bacterial cell abundance showed a strong negative correlation with phosphate concentration at station 1 (r* = -0.587, P < 0.05) and a weak positive correlation with chlorophyll *a* concentration (r* = 0.132, P < 0.05). At station 2, bacterial cell abundance showed a strong positive correlation with chlorophyll *a* concentration (r* = 0.599, P < 0.05) and at station 1, it showed a weak positive correlation with nitrate concentration (r* =

0.389, P < 0.05). The micronutrients variables at both stations showed a weak negative correlation with chlorophyll *a* concentration.

Moreover, the linear regression plot for bacteria and chlorophyll *a* concentration together with standardized coefficient value for both stations are given in Figure 6 to Figure 13. At station 1, the standardized coefficient value for bacteria/chlorophyll *a* relationship showed a lower value ($\beta = 0.132$) whereas at station 2, a higher value was recorded ($\beta = 0.599$).

Furthermore, the Principal Component Analysis (PCA) was applied to the five variables collected during the period of sampling (bacteria abundance, chlorophyll *a* concentration, nitrate concentration, phosphate concentration and surface water temperature). The highest correlation exists between bacterial cell abundance and chlorophyll *a* concentration at station 2 whereas at station 1 bacterial cell abundance showed a negative correlation with phosphate concentration Figure 15 and Figure 17. A scree plot presented in Figure 14 and Figure 16 shows the sorted eigenvalues from large to small as the function of principal components number for station 1 and station 2 respectively. Results of factor analysis including factor-loading matrix, eigenvalues, variability and cumulative variance values are given in table 3. The factor analysis derived from PCA generated three significant factors which explained 84.43% of the variance in data sets. Factor 1 (F1) explained 43.91% and 36.58% of the variance at station 1 and station 2 respectively. The F1 revealed a high positive loading with respect to bacterial cell abundance and surface water temperature, and a high negative loading with phosphate concentration which were 0.819, 0.674 and -0.800 respectively for station 1. At station 2, F1 revealed a high positive loading with respect to bacterial cell abundance, chlorophyll *a* concentration and surface water temperature which were 0.846, 0.536 and 0.676 respectively.

4. Discussion

The stipulated period of study coincided with the wet season normally encountered in Nigeria that is from April to October. (27) reported a major influence of this season on the aquatic physiochemical characteristics and periphyton community. Surface water temperature showed a little fluctuation at both stations. (28, 29) reported a positive effect of temperature on aquatic bacterial production and growth rate. Therefore, the observed highest value of surface water temperature in the month of September 2008, during flood tide at both stations could be as a result of time of collection and specific heat capacity of the water that could speed up bacterial multiplication and growth. The increased value of chlorophyll *a* concentration in the month of July and August at both station coincided with the increased level of the micronutrients in this month. The weak Pearson correlation matrix of chlorophyll *a* with the micronutrients at both stations could explain the importance of flood water discharge and storm water in the introduction of phyto-planktonic cells, and the time lag in the remineralization of the introduced nutrients into the system. A strong relationship between

phosphorus, chlorophyll and water clarity have been reported for fresh water systems around the world (30). Thus the weak chlorophyll a – phosphate relationship observed in station 1 could explain the important of this nutrient to phytoplankton growth and been a limiting one in this area.

However, increase in bacteria population in water after rainfall has been recorded frequently. At station 1 and station 2, bacterial population may hold a positive correlation with precipitation if noted. It is likely that nutrients brought into the river basin were rapidly utilized, so a direct effect of rainfall was confined to areas near the inflows. After every heavy rainfall, an immediate change in the bacterial population could be detected. During this sampling period, there was unusual high rainfall. The observed peaks at station 1 and station 2 could be attributed to a combined effect of large amount of organic matter available for bacterial decomposition together with the effect of inflow of nutrients and bacteria on bacterial cells populations near to the shore where samples were taken. The bacterial cell abundance correlates significantly with micronutrients level mostly at station 1 than station 2. This corresponds to increase in bacterial cell abundance with nutrient level in the surface water at both study sites. The high abundance of bacterial cell was due to availability of carbon source. Bacteria could sever as a good food material for the protozoans thereby transferring energy back into the classic food chain.

Heterotrophic bacteria in coastal water have two sources of organic matter at their disposal – local ones comprising excreta and the remains of phytoplankton and zooplankton; and the external ones in the form of terrigenous, allochthonous organic matter. Therefore bacteria could depend directly on phytoplankton itself, or excreted products from animals (31) or indirectly from allochthonous input. It is likely, however, that both mechanisms occur simultaneously.

The release of dissolved organic carbon (DOC) by healthy phytoplankton is a small percentage of the photosynthetic rate (5 – 10%), (32). Considering this low release rate of DOC makes it necessary for bacteria to depend on allochthonous input of DOM for their production and multiplication. Therefore, the high abundance of bacterial cells reported at station 1 which correlates weakly with chlorophyll *a* concentration show a reliance of bacteria on allochthonous input of organic cardon. Its lower standardized coefficient value with chlorophyll *a* could also confirm their weak relationship. This observation might be attributed to the unusual precipitation that brought enormous allochthonous carbon into the system during the study period. The discharge of large amount of organic matter together with the direct effect of inflow of nutrients and bacterial cells on bacteria population at station 1 may also be a reason for the weak relationship. At station2 the correlation is high indicating the dependency of bacterial cell on chlorophyll *a* concentration.

In summary, a weak statistical correlation between bacterial cells abundance and chlorophyll *a* concentration at station 1 confirmed the dependency of bacteria on allochthonous influx of particulate and dissolved organic carbon especially during wet seasons. It is obvious that bacteria population utilized organic substrate for its growth which could be phytoplankton exudates or other organic materials. This high bacterial density observed, reflects its percentage biomass that may serve as food source for protists which would in turn serve as food to the next trophic level and thus link the DOC/POM to classic food chain in the environment.

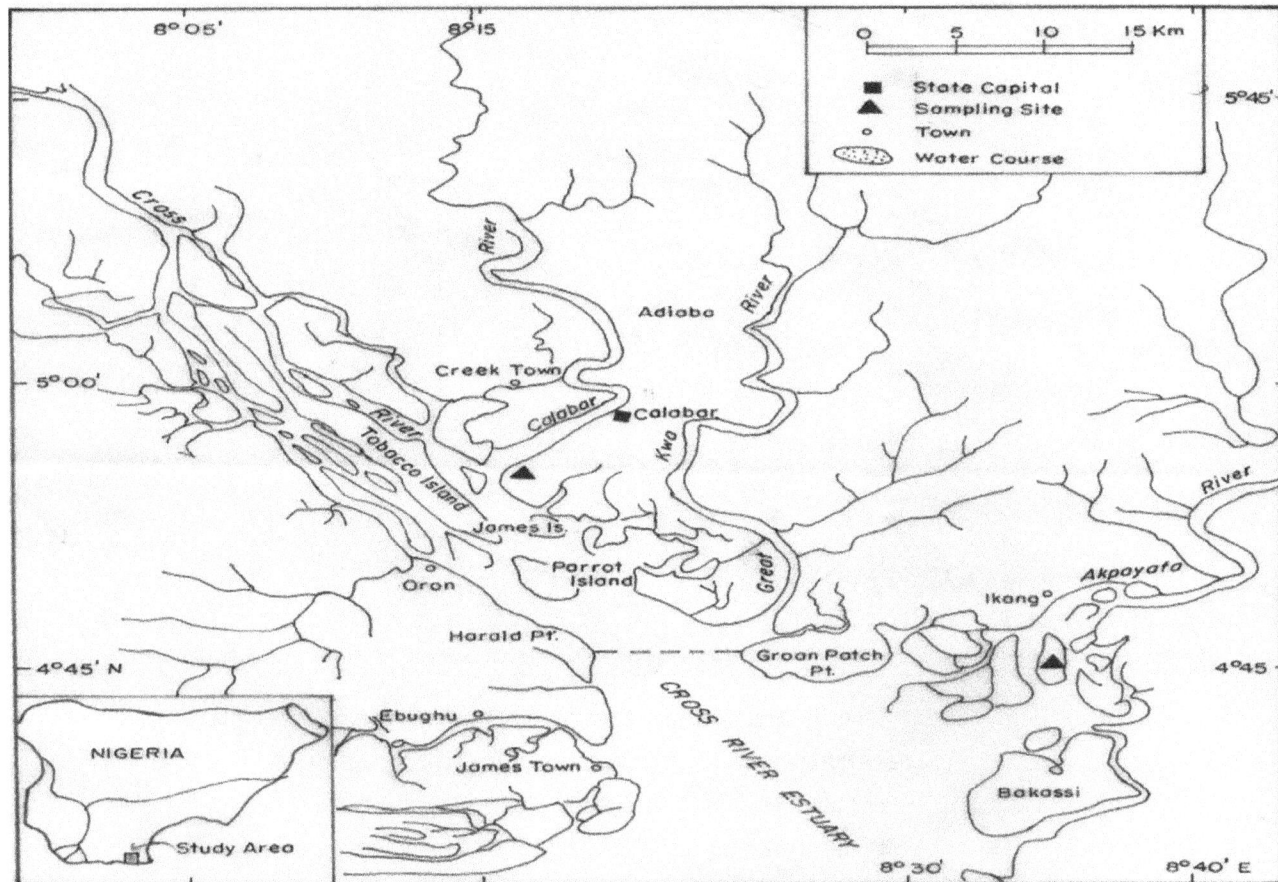

Figure 1. *Map of the cross river basin southeastern Nigeria showing the sampling site.*

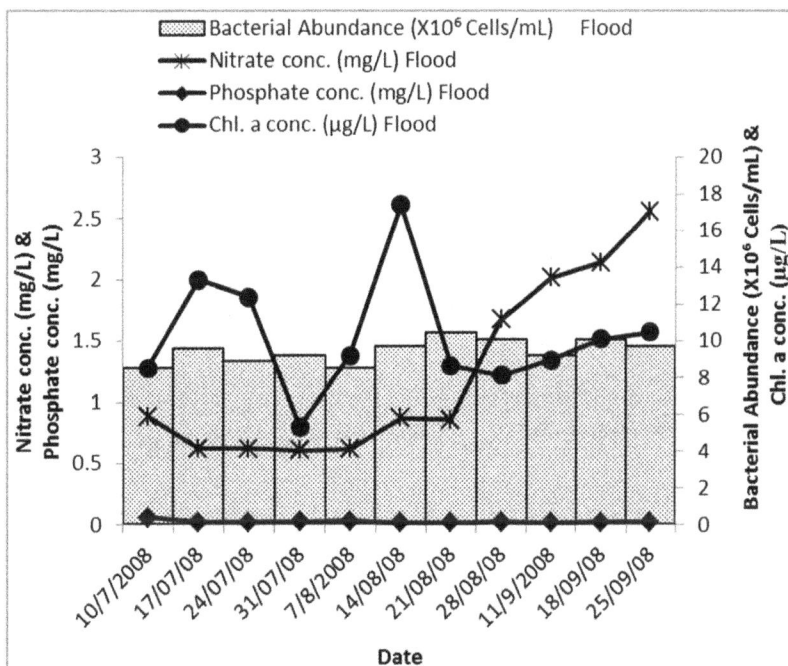

Figure 2. *Variation in micro nutrient, chlorophyll a concentration and bacterial cells abundance for station 1 at flood tide from July, 2008 to September, 2008.*

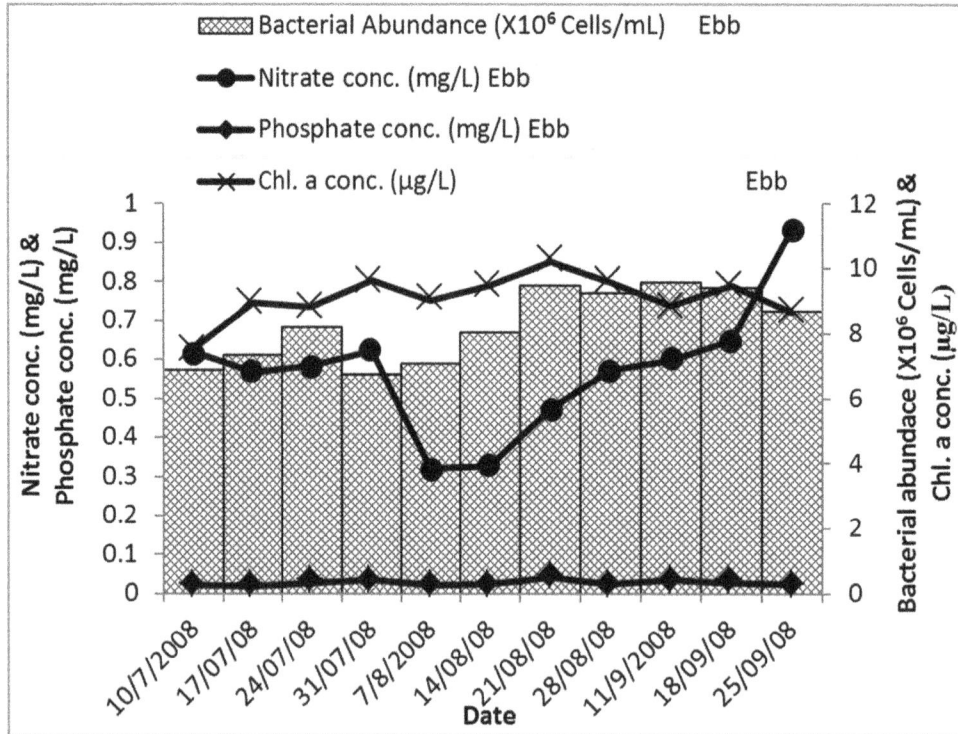

Figure 3. *Variation in micro nutrient, chlorophyll a concentration and bacterial cells abundance for station 1 at ebb tide from July, 2008 to September, 2008.*

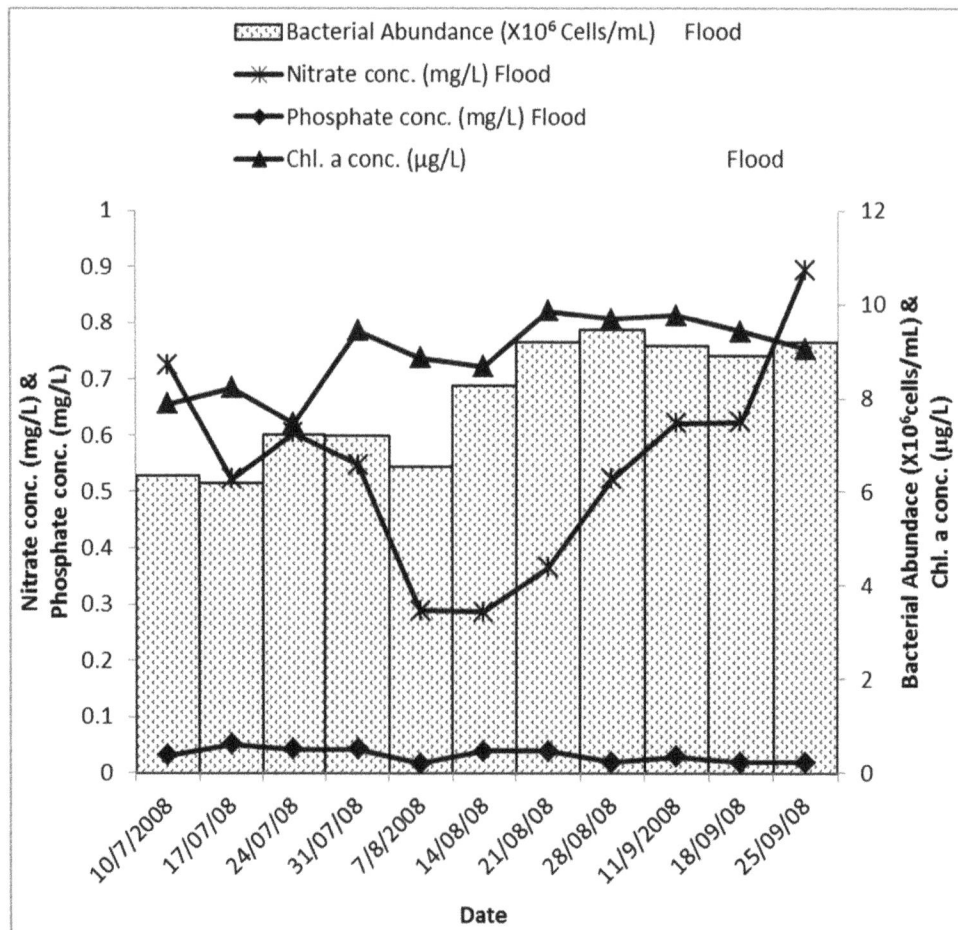

Figure 4. *Variation in micro nutrient, chlorophyll a concentration and bacterial cells abundance for station 2 at flood tide from July, 2008 to September, 2008.*

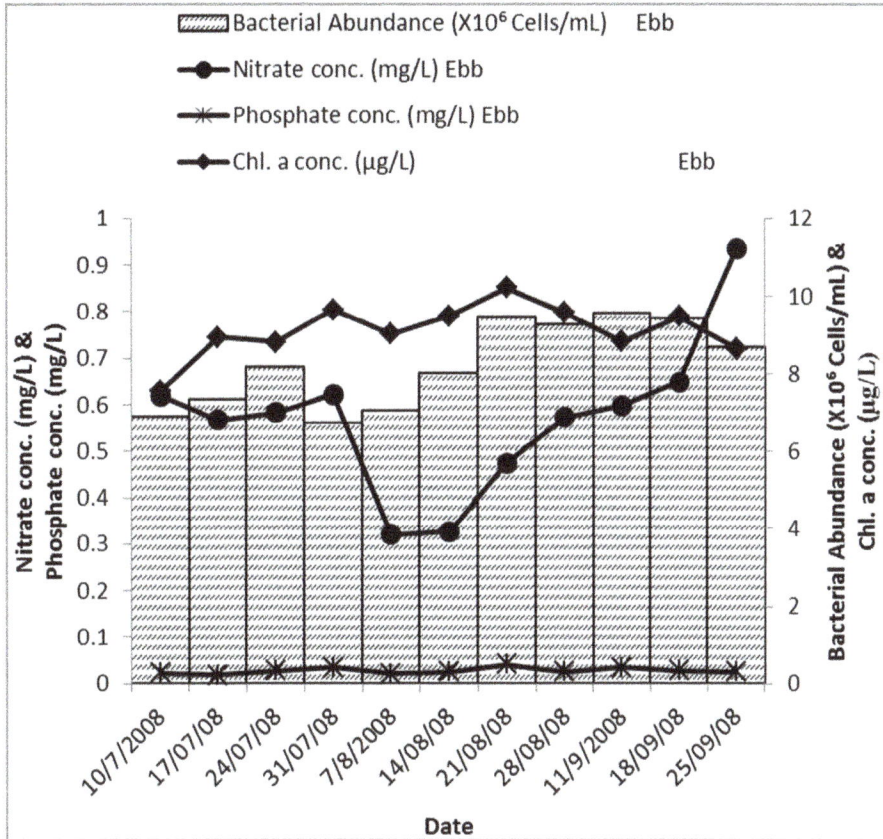

Figure 5. Variation in micro nutrient, chlorophyll a concentration and bacterial cells abundance for station 2 at ebb tide from July, 2008 to September, 2008.

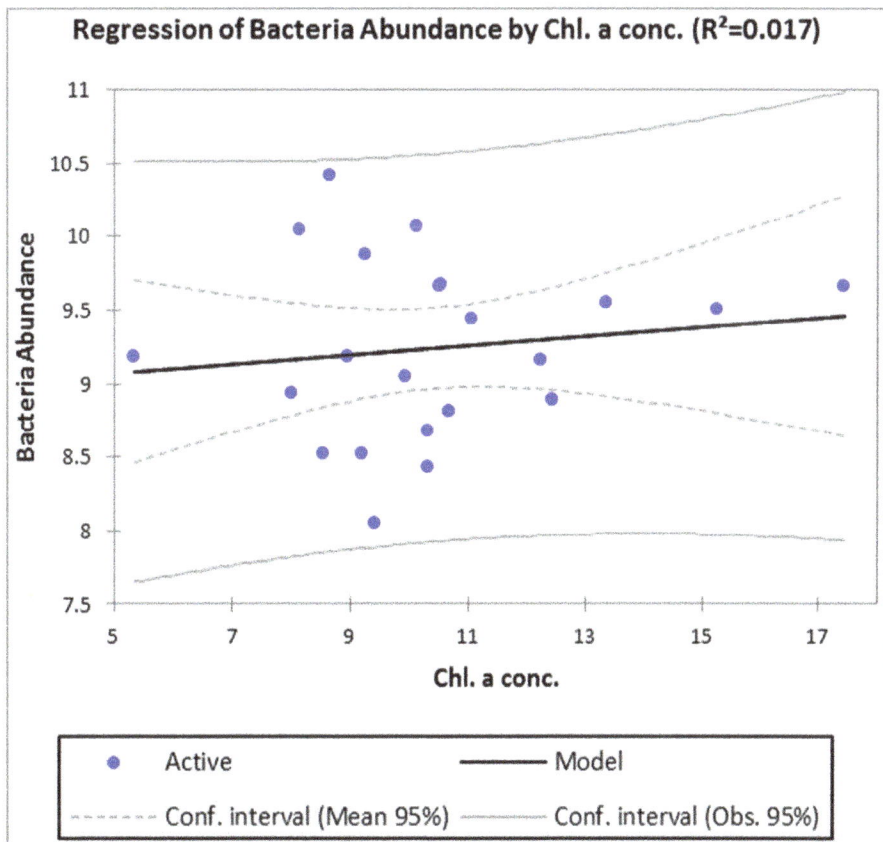

Figure 6. Regressional plot for bacterial cells abundance and chlorophyll a conc. at station 1 from July, 2008 to September, 2008. Bacteria = 8.917 + 0.0312Chl a. {$R^2 = 0.017$, $r^* = 0.132$, $N = 22$, $P < 0.05$}.

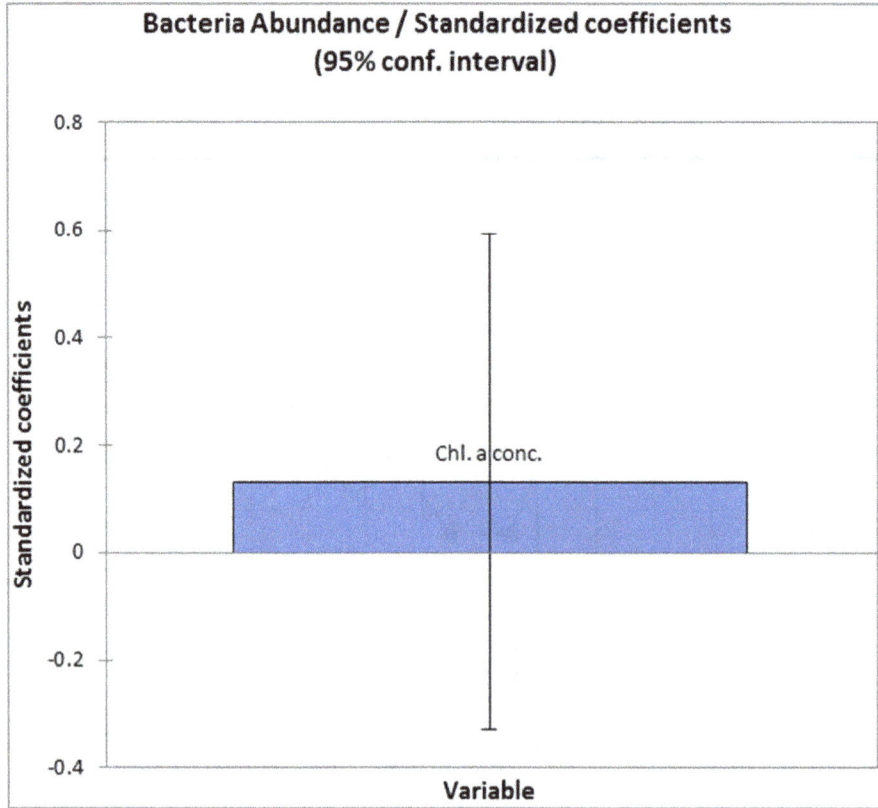

Figure 7. *Bacteria/Chl. a standardized plot at station 1 from July, 2008 to September, 2008. {Standardized coefficient = 0.132, Standard error = 0.222, t = 0.594, Pr > |t| = 0.559, lower bound = -0.331, upper bound = 0.594, N = 22.*

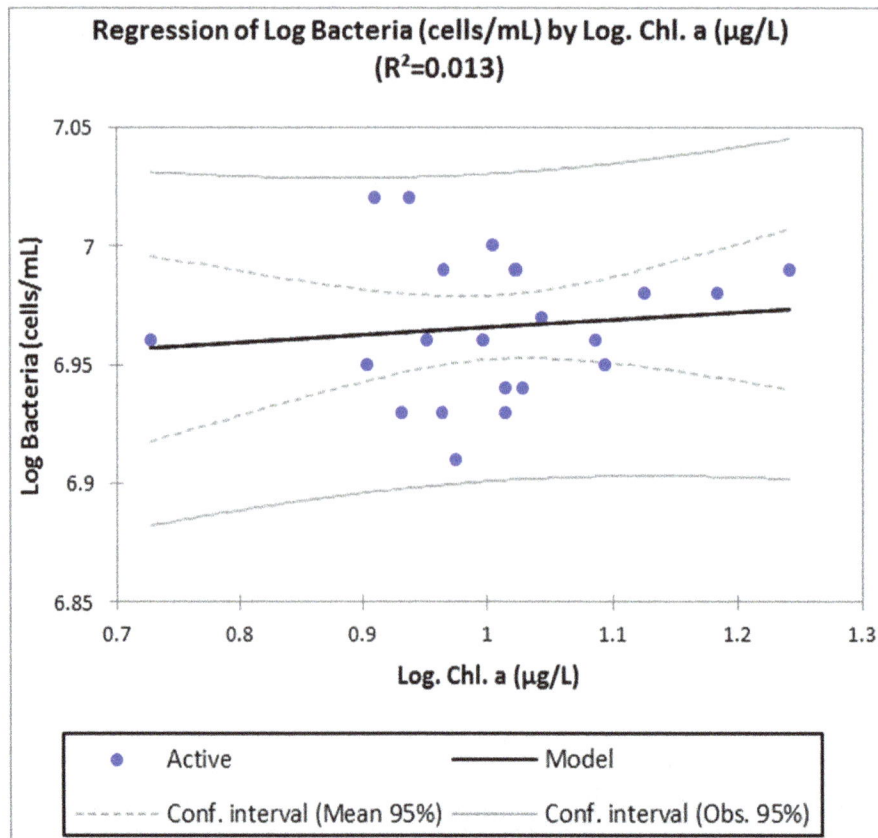

Figure 8. *Regressional Log plot for bacterial cells abundance and chlorophyll a conc. at station 1 from July, 2008 to September, 2008. Log. Bacteria = 6.933 + 0.032Log Chl a {N = 22, P < 0.05, r* = 0.114, R² = 0.013}.*

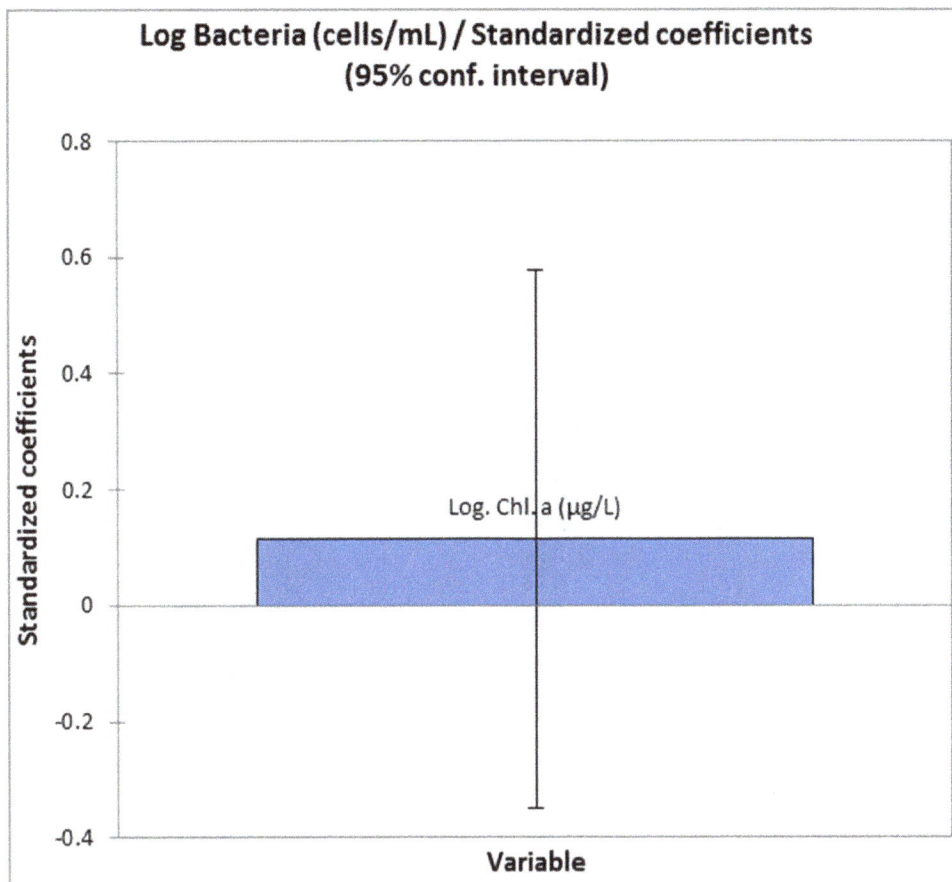

Figure 9. *Log Bacteria/Log Chl. a standardized plot at station 1 from July, 2008 to September, 2008. {Standardized coefficient = 0.114, Standard error = 0.222, t = 0.513, Pr > |t| = 0.614, lower bound = -0.349, upper bound = 0.577, N = 22}.*

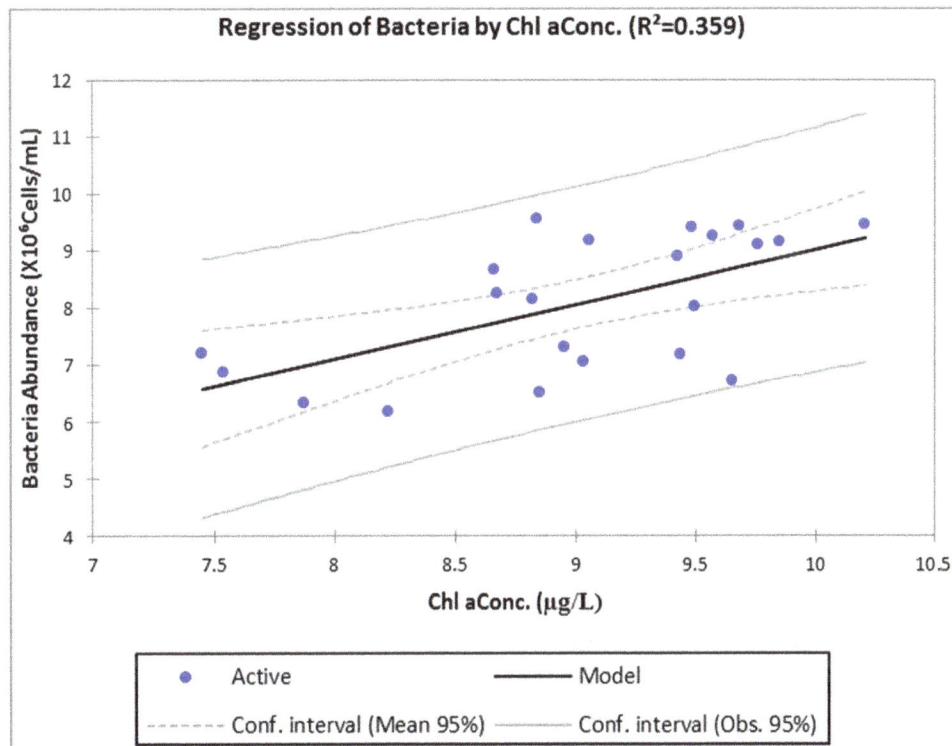

Figure 10. *Regressional plot for bacterial cells abundance and chlorophyll a conc. at station 2 from July, 2008 to September, 2008. Bacteria = 0.953Chl. a – 0.506. {N = 22; R^2 = 0.359; P < 0.05, r* = 0.599}.*

Bacteria / Standardized coefficients
(95% conf. interval)

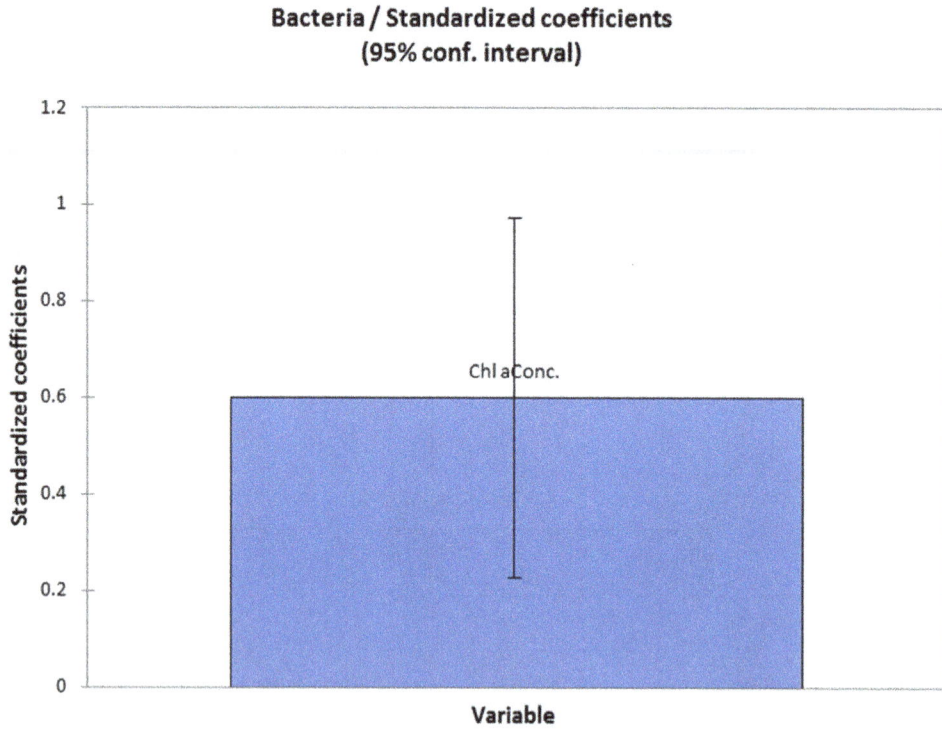

Figure 11. Bacteria/Chl. a standardized plot at station 2 from July, 2008 to September, 2008. {Standardized coefficient = 0.599, t = 3.348, Pr > |t| = 0.003, lower bound (95%) = 0.226, upper bound (95%) = 0.973, P < 0.05}.

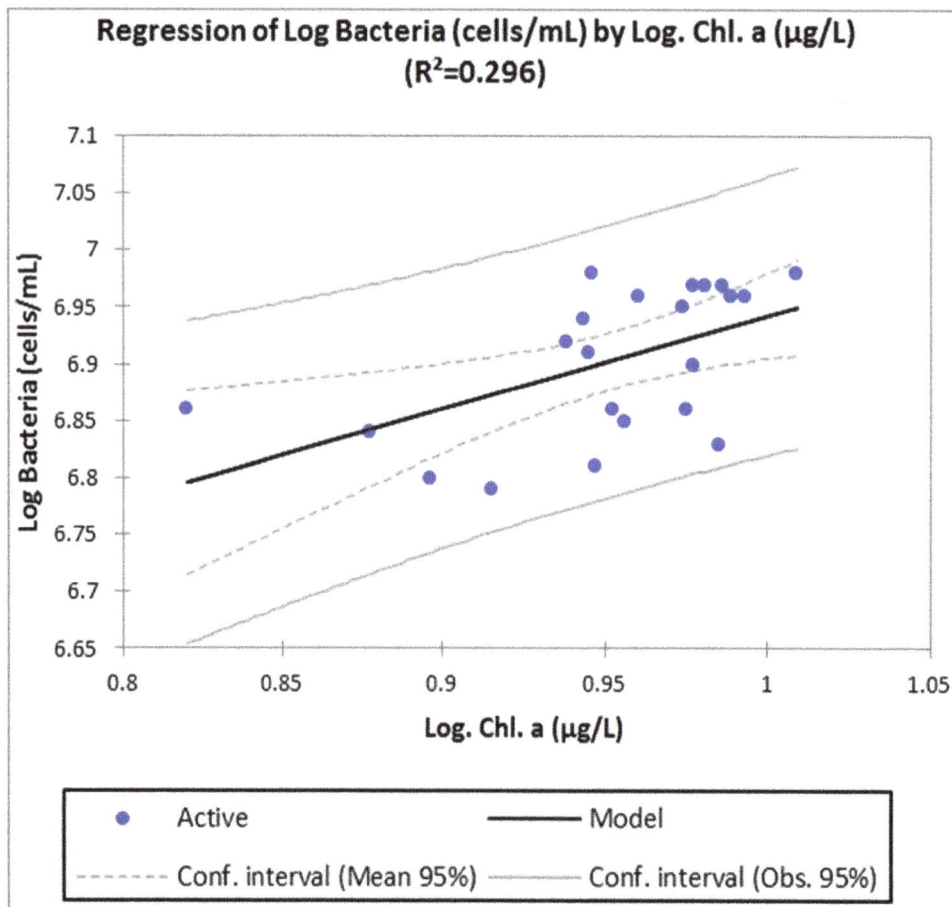

Figure 12. Regressional Log plot for bacterial cells abundance and chlorophyll a conc. at station 2 from July, 2008 to September, 2008. Log Bacteria = 6.129 + 0.813Log Chl. a {R^2 = 0.296, N = 22, r^* = 0.544, P < 0.05}.

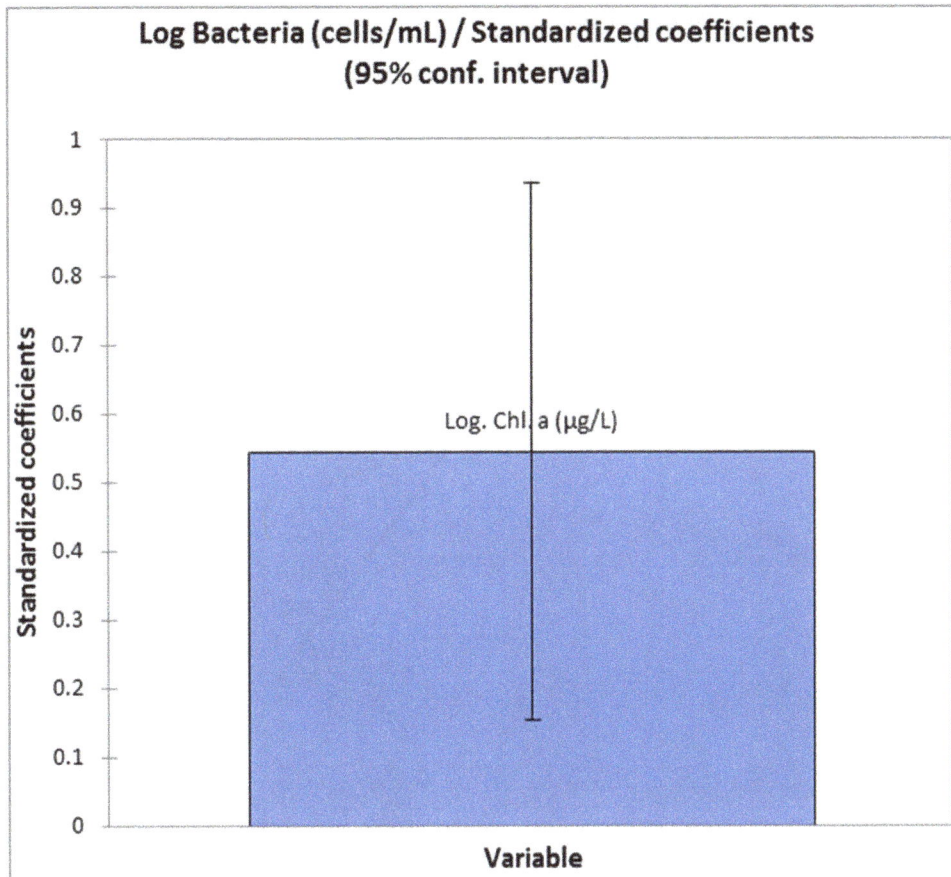

Figure 13. *Log Bacteria/Log Chl. a standardized plot at station 2 from July, 2008 to September, 2008. {Standardized coefficient = 0.544, t = 0.188, Pr > |t| = 0.009, lower bound = 0.153, upper bound = 0.935, P < 0.05}.*

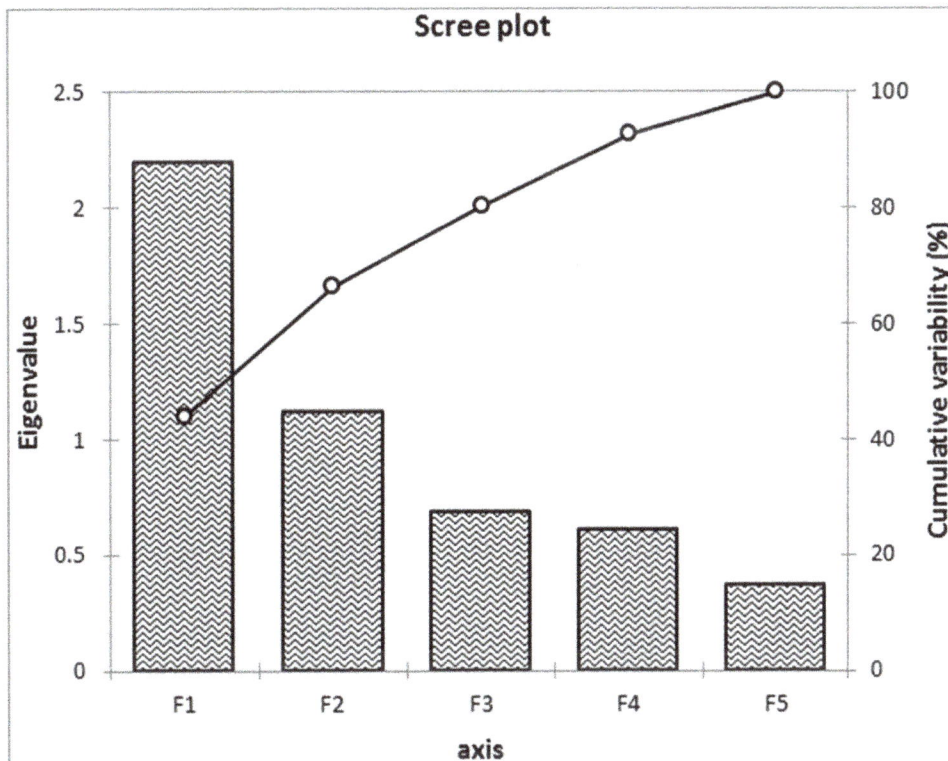

Figure 14. *Scree plot of the eigenvalue for station 1 from July, 2008 to September, 2008.*

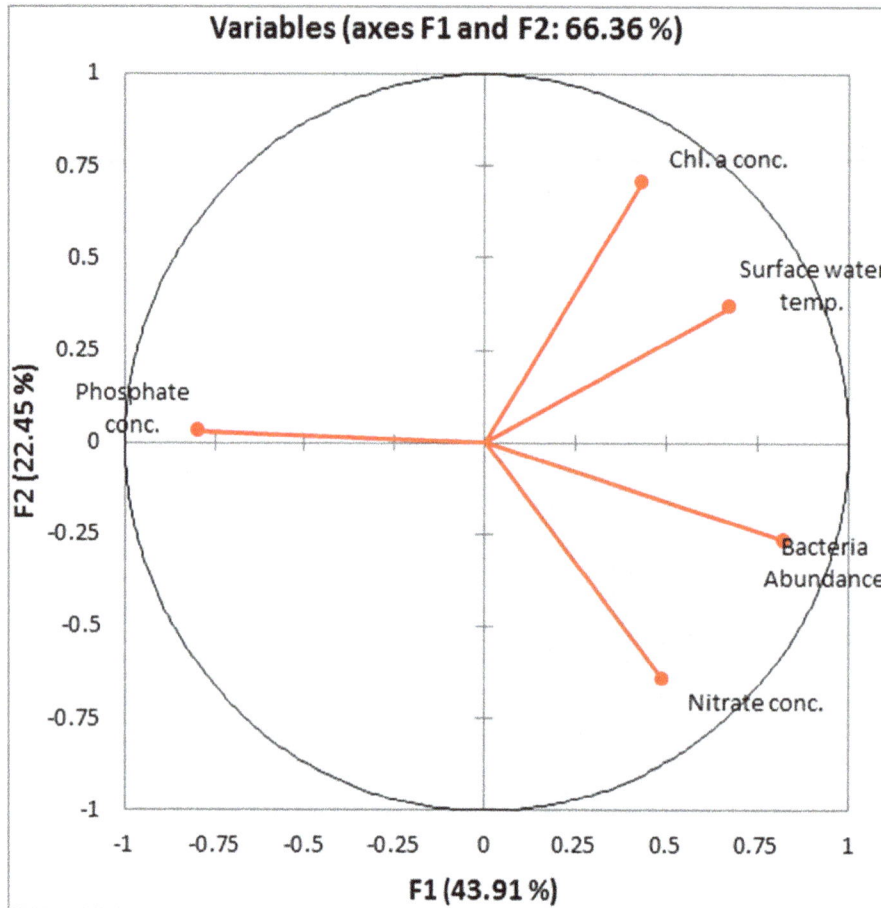

Figure 15. PCA plot of variables for station 1 from July, 2008 to September, 2008.

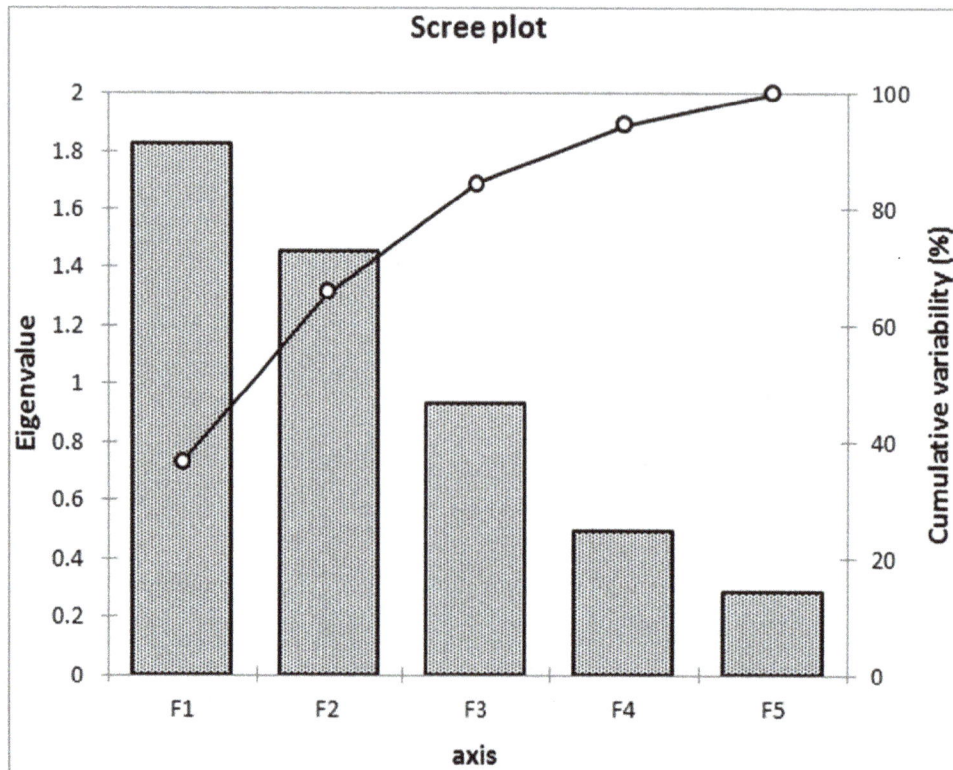

Figure 16. Scree plot of the eigenvalue for station 2 from July, 2008 to September, 2008.

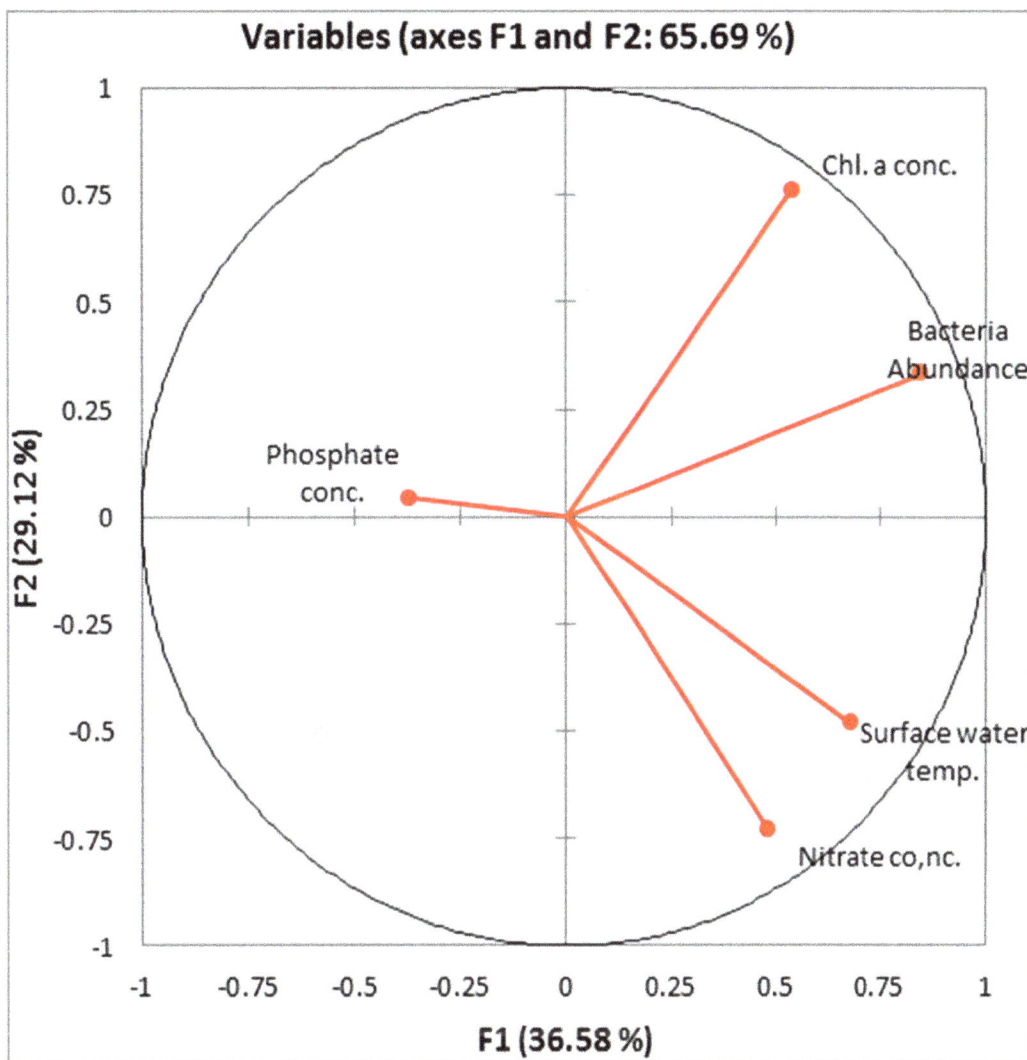

Figure 17. PCA plot of variables for station 2 from July, 2008 to September, 2008.

Table 1. Summary statistics for station 1 and station 2 from July, 2008 to September, 2008.

Variable	Station1				Station 2			
	Min.	Max.	Mean	Std. Dev.	Min.	Max.	Mean	Std. Dev.
Bacterial Abundance	8.050	10.42	9.243	0.609	6.180	9.560	8.091	1.178
Chl. a conc.	5.340	17.430	10.441	2.568	7.450	10.210	9.022	0.741
Nitrate conc.	0.253	2.862	1.256	0.773	0.286	0.934	0.558	0.171
Phosphate conc.	0.018	0.071	0.027	0.013	0.017	0.051	0.030	0.010
Surface water temp.	25.0	30.0	27.545	1.224	25	29	26	0.913

Table 2. Pearson correlation matrix for station 1 and station 2 $\{N = 22, P < 0.05\}$.

	Variables	Bacterial Abundance	Chl. a conc.	Nitrate conc.	Phosphate conc.	Surface water temp.
	Bacterial Abundance	1				
	Chl. a conc.	0.132	1			
Station 1	Nitrate conc.	0.389	-0.014	1		
	Phosphate conc.	-0.587	-0.247	-0.249	1	
	Surface water temp.	0.387	0.326	0.113	-0.365	1
	Bacterial Abundance	1				
	Chl. a conc.	0.599	1			
Station 2	Nitrate conc.	0.174	-0.21	1		
	Phosphate conc.	-0.153	-0.0093	-0.132	1	
	Surface water temp.	0.354	0.005	0.464	-0.098	1

Values in bold are different from 0 with a significance level alpha=0.05

Table 3. Factor-loadings, eigenvalues, variability and cumulative value for station 1 and station 2.

	F1	F2	F3	F1	F2	F3
Bacterial Abundance	0.819	-0.263	-0.204	0.846	0.335	0.151
Chl. a conc.	0.434	0.710	0.500	0.536	0.763	0.07
Nitrate conc.	0.492	-0.642	0.552	0.484	-0.727	0.047
Phosphate conc.	-0.800	0.033	0.259	-0.373	0.045	0.924
Surface water temp.	0.674	0.370	-0.170	0.676	-0.482	0.231
Eigenvalue	2.195	1.123	0.692	1.829	1.456	0.937
Variability (%)	43.908	22.450	13.844	36.576	29.117	18.73
Cumulative (%)	43.908	66.358	80.202	36.576	65.693	84.43

{N = 22}

Acknowledgement

I would like to appreciate Dr. E.E. Antai, Institute of Oceanography, University of Calabar for his encouragement and support throughout the period I carried out this work, and to all the laboratory staff of the Biological Oceanography Department, University of Calabar, Nigeria for their support during the analysis.

References

[1]   Aizaki M, Otsuki A, Fukushima T, Hosomi M and Muraoka L, 1981. Application of Carlson's trophic state index to Japanese lakes and relations between the index and other parameters. *Verhandlungen des Internationalen Verein* Limnologie 12: pp. 675 – 681.

[2]   Bird DF and Kalff J, 1984. Empirical relationships between bacterial abundance and chlorophyll concentration in fresh and marine waters. *Canadian Journal of Fisheries and Aquatic Sciences* 41: pp. 1051 – 1023.

[3]   Rob Smith, and Julie Hall, 1997. Bacterial abundance and production in different water masses around South Island, New Zealand. *Marine & Freshwater Research* 31: pp. 515 – 524.

[4]   Kapiris K, Kormas, KA, Maria T, and Nicolaidou A, 1998. Quantitative relationships between phytoplankton, bacteria & Protists in an Acgean semi-enclosed embayment of Maliakos Gulf, Greece. Aquatic Microbial Ecology 15: pp. 255 – 264.

[5]   Tada K, Monaka K, Morishita M, and Hashimoto T, 1998. Standing stocks and Production Rate of Phytoplankton and Bacteria in the Seto Inland Sea, Japan. *Oceanography* 54: pp. 285 – 295.

[6]   Cole JJ, Findlay S, and Pace ML, 1988. Bacterial production in fresh and salt water ecosystems: a cross-system over-view. *Marine Ecology Progress Series* 43: pp. 1 – 10.

[7]   Gasol JM, and Duarte CM, 2002. Comparative analyses in aquatic microbial ecology: how far do they go? *FEMS Microbial Ecol*ogy 31: pp. 99 – 106.

[8]   Lyche A, Adersen T, Christoffersen K, Hessen DO, Hansen PHB, and Klysner A, 1996. Mesocosm tracer studies. 2. The fate of primary production and the role of consumers in the pelagic carbon cycle of a mesotrophic lake. *Limnology and Oceanography* 41: pp. 475 – 487.

[9]   Coveney MF, 1982. Bacterial uptake of photosynthetic carbon from freshwater phytoplankton. *Oikos* 38: pp. 8 – 20.

[10]   Coveney MF, and Wetzel RG, 1995. Biomass, Production, and Specific growth rate of bacterioplankton and coupling to phytoplankton in an oligotrophic lake. *Limnology and Oceanography* 40: pp. 1187 – 1200.

[11]   Le J, Wehr JD, and Campbell L, 1994. Uncoupling of bacterioplankton and phytoplankton production in fresh waters is affected by inorganic nutrient limitation. *Applied Environmental Microbiology*. 60: pp. 2089 – 2093.

[12]   Rier ST, and Stevenson RJ, 2001, Relation of environmental factors to density of epilithic lotic bacteria in 2 ecoregions. *North American Benthological Society* 20: pp. 520 – 532.

[13]   Coffin RB, and Sharp JH, 1987. Bacterial production in fresh and salt water ecosystems: a cross system overview. *Marine Ecology Progress Series* 43: pp. 1 – 10.

[14]   Findlay S, Pace ML, David L, and Cole JJ, 1991. Weak coupling of bacterial and algal production in a heterotrophic ecosystem: The Hudson River estuary. *Limnology and Oceanography* 36: pp. 268 – 278.

[15]   Findlay S, Howe K, and Fontvielle D, 1993. Bacterial-algal relationship in streams of the Hubbard Brook Experimental Forest. *Ecology* 74: pp. 2326 – 2336.

[16]   Fuhrman JA, Sleeter TD, Carlson CA, and Proctor LM, 1989. Dominance of bacterial biomass in the Sargasso Sea and its ecological implications. *Marine Ecology Progress Series* 57: pp. 207 – 217.

[17]   Cho, BC, and Azam F, 1990. Biogeochemical significance of bacterial biomass in the ocean's euphotic zone. *Marine Ecology Progress Series* 63: pp. 253 – 259.

[18]   Rier ST, and Stevenson RJ, 2002. Effects of light dissolved organic carbon, and inorganic nutrients on the relationship between algae and heterotrophic bacteria stream periphyton. *Hydrobiologia* 489: pp. 179 – 184.

[19]   Geesey G, Mutch R, and Costerton JW, Green RB, 1978. Sessile bacteria: An important component of the microbial population in a small mountain stream. *Limnology and Oceanography* 23: pp.1214 – 1223.

[20]   Nawa IG, 1982. *An ecological study of the Cross River Estuary*. PhD Thesis, University of Kiel. F.R. Germany. 162pp.

[21]   Enyenihi UK, 1991. The Cross River Basin: Soil Characteristics, Geology, Climate, Hydrology & Pollution, International Workshop on Methodology & Quantitative Assessment of Pollution Load of Coastal Environment (FAO/UNAP/IOC/WHO), 17pp.

[22]   Akpan ER, 1994. Seasonal variation in phytoplankton biomass in relation to physicochemical factors in Cross River Estuary of South Eastern Nigeria, PhD thesis, University of Calabar.

[23] Asuquo FE, 1998. Physiochemical characteristics and Anthropogenic Pollution of the surface water of Calabar River, Nigeria. *Global Journal of Pure & Applied Science* 5: pp.595 – 600.

[24] Lowenberg U, and Kunzel T, 1991. Investigation on the Trawl Fishery of the Cross River Estuary, Nigeria. *Journal of Applied Ichthyology* 7: pp.44 – 53.

[25] APHA, *Standard Methods for the Examination of Water Waste Water*, American Public and Health Association, American Water, Works Association and Water Environment Federation (WEF), 20th edition, 1998.

[26] Feig YS, and Porter KG, 1980. The use of DAPI for identification and counting aquatic microflora. *Limnology and Oceanography* 25: pp.943 – 948

[27] Inyang AI, Sunday KE, and Nwankwo, DI, 2015. Composition of Periphyton Community on Water Hyacinth (*Eichhornia crassipes*): In Analysis of Environmental Characteristics at Ejirin Part of Epe Lagoon in Southwestern Nigeria. *Journal of Marine Biology* 2015: pp.1 – 9.

[28] Ratkowsky DA, Olley J, McMeekin TA, and Ball A, 1982. Relationship between temperature and growth rate of bacterial cultures. *Journal of Bacteriology* 149: pp.1-5

[29] Adams HE, Crump BC, and Kling GW, 2010. Temperature controls on aquatic bacterial production and community dynamics in arctic lakes and streams. *Environmental Microbiology* 12: pp.1319 – 1333.

[30] Brown CD, Hoyer MV, Bachmann RW, and Canfield DE Jr. 2000. Nutrient-chlorophyll a relationships: an evaluation of empirical nutrient-chlorophyll a models using Florida and northern temperate lake data. *Canadian Journal of Fisheries and Aquatic Sciences* 57: pp.1574 – 1583.

[31] Larson U, and Hagstron A, 1979. Phytoplankton exudates release an energy source for the growth of pelagic bacteria. *Marine Biology* 52: pp.199 – 206.

[32] Mague T, Friberg E, Hughes DJ, and Morris I, 1980. Extracellular release of carbon by marine phytoplankton: a physiological approach. *Limnology and Oceanography* 25: pp.262 – 279.

6

Seasonal Variations of Wind Driven Mixing off Port Said, Egypt

Sayed Sharaf El-Din[1], Mohamed Elsharkawy[2]

[1]University of Alexandria, Faculty of Science, Department of Oceanography, Alexandria, Egypt
[2]National Institute of Oceanography and Fisheries, Physical Laboratory, Alexandria, Egypt

Email address:
ecosalex@hotmail.com (S. S. El-Din), sharkawy70@gmail.com (M. Elsharkawy)

Abstract: The wind contribution to the process of vertical mixing in the near surface oceanic layers is empirically, estimated at a deep station (500 m) off Port Said, Egypt. The dimensional less parameter Ru which scales wind contribution is greatest during summer with values as high as 0.6. Calculations show that wind stress over the sea surface generally has a significant role in the dynamical process of vertical turbulent mixing except during winter.

Keywords: Vertical Mixing, Wind Stress, Wind Driven Mixing

1. Introduction

Tidal currents and atmospheric forcing over the upper sea surface are major producers of vertical turbulent mixing. The lunar semi-diurnal tide (M2) is estimated to produce about 2.5 TW which is mostly dissipated by bottom friction in the boundaries of the water basins. The dynamical process of vertical turbulent mixing in coastal regions and continental slopes is strengthened by dissipation of tidal energy [1]. The wind driven mixing energy is estimated to be only 1 TW. The equator to pole heat flux, about 2000 TW, together with continuous processes of upwelling and down welling maintain the equilibrium in the water basin [2].

Numerous coupled atmosphere – ocean models have been implemented to study the impact of barotropic tides in the North Atlantic Ocean. The surface density and salinity are predicted to increase by the effect of barotropic tides [3]. Surface wave breaking is an important source of shear production of turbulent kinetic energy. Momentum and energy transport across the air – sea interface also affect the distribution of viscous dissipation. These physical parameters have been used successfully to model wave enhanced turbulence in the upper sea layer [4]. This model has also been applied to study mixing rate and its dependence on buoyancy, turbulence, shear and dissipation off the Eastern Egyptian Mediterranean Coast [5].

Mixing parameters, as diffusive flux of salt, and momentum, off Port Sudan, Red Sea have been computed. The study showed that their magnitudes decrease with depth and that the water column tends to be more stable at deeper layers [6]. The analysis of the turbulent structure off the Eastern Egyptian Mediterranean Coast showed that instability occurs intermittently along the water column. The dynamical instability associated with negative values of squared buoyancy frequency occurred in the upper layers. Also, the study showed that Reynolds stress was greatest during winter [7].

In this study, the contribution of wind stress over the sea surface in the dynamical process of vertical mixing production was investigated. Seasonal variations of the non-dimensional parameter Ru were analyzed to compare between tidal driven and wind driven mixing in the water column.

2. Data Availabilty

A series of daily records of wind and current field have been taken at an offshore fixed station off Port Said, Egypt with a geographic location (31.99 N, 32.07 E), and depth 500 m. Wind data was collected by a floating Anderra weather station, while the surface current was measured by a moored acoustic Doppler current profiler. The atmospheric-oceanic parameters were collected at time intervals corresponding to the four annual seasons. Information about the available data set is presented in table 1.

Table 1. *Information about available data set.*

	Spring	Summer	Autumn	Winter
Start date	21/3/99	21/6/99	8/10/99	6/1/99
End date	17/5/99	19/7/99	28/11/99	21/1/99
No. of records	58	29	52	16

3. Estimates of Wind Stress

The zonal component of wind stress τ_x, and the meridional component of wind stress τ_y, are related to their corresponding components of wind speed V_x and V_y respectively by the formula:

$$\tau_x = \rho_a\, C_D\, V_x;\ \tau_y = \rho_a\, C_D\, V_y \qquad (1)$$

where ρ_a is the air density (1.024 kg m^{-3}), C_D is the drag coefficient defined in terms of the magnitude of the wind speed W as in [8,9]:

$$C_D = (0.75 + 0.067\, W)\ \mathrm{X}\ 10^{-3} \qquad (2)$$

Temporal distribution of wind stress over the sea surface is shown in figures (1-4)

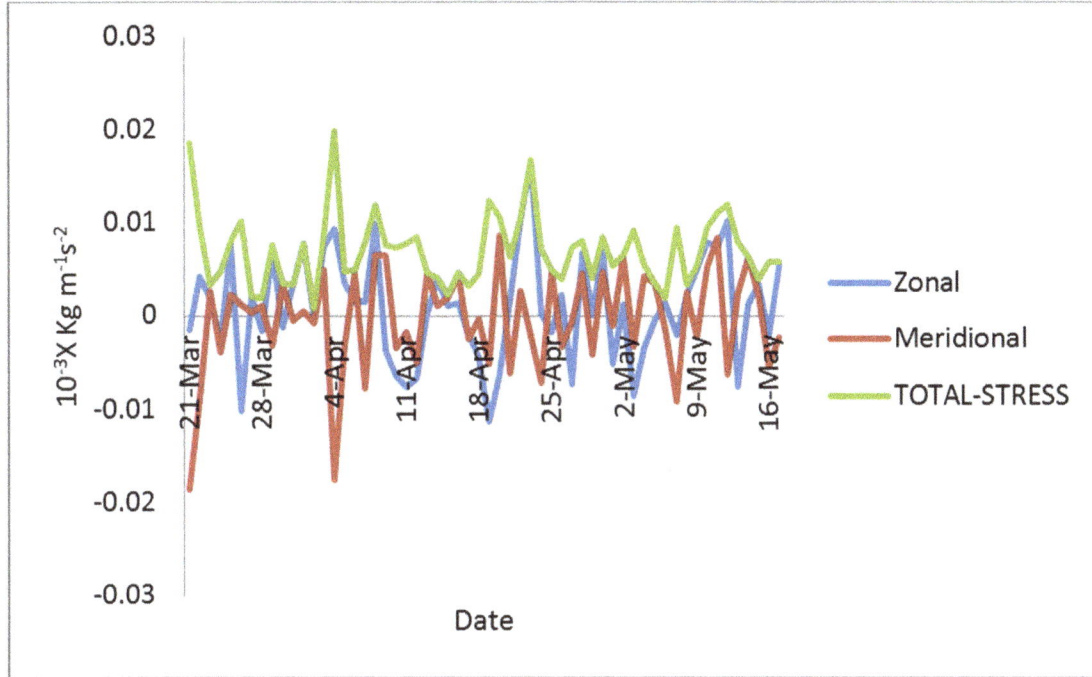

Fig. 1. *Time series of wind stress components during spring survey.*

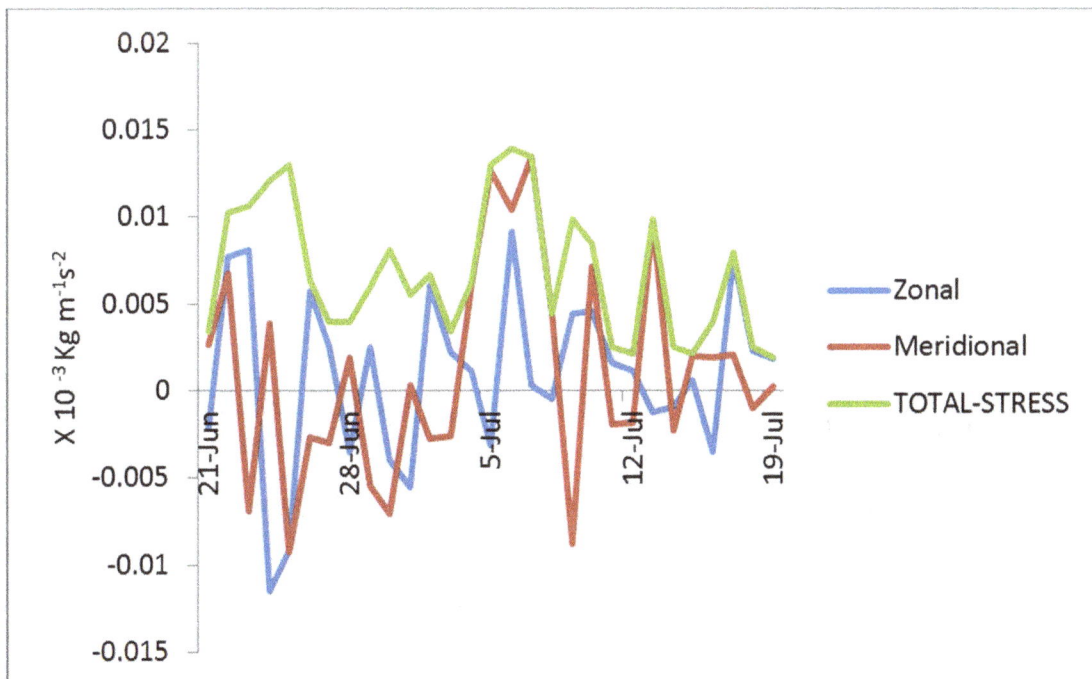

Fig. 2. *Time series of wind stress components during summer survey.*

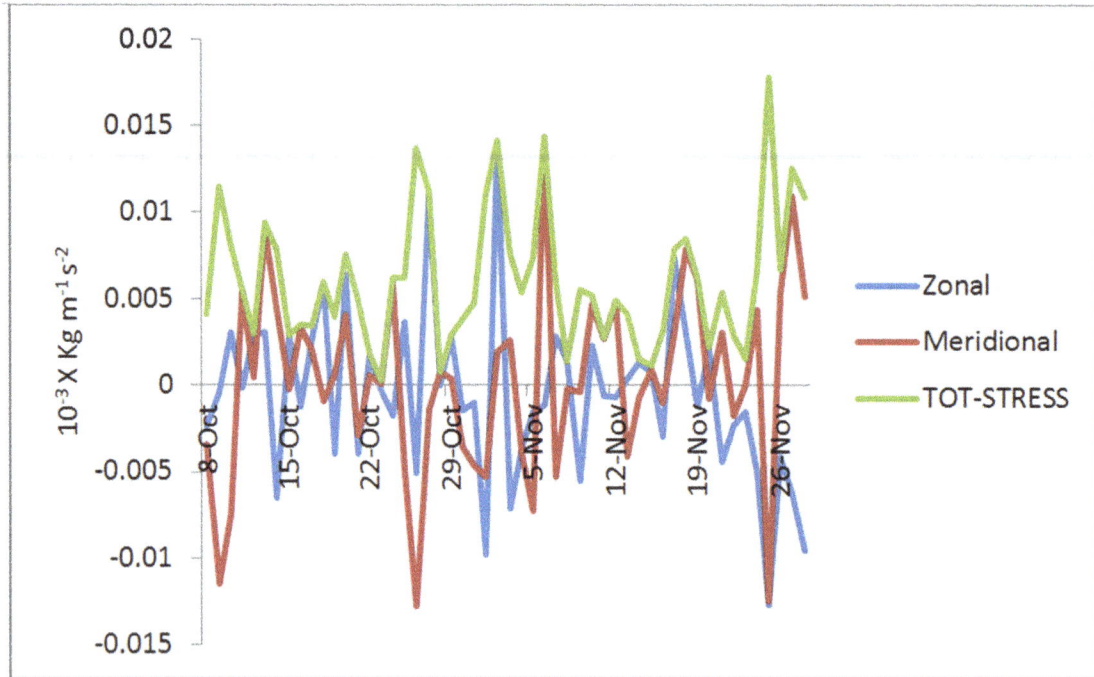

Fig. 3. *Time series of wind stress components during autumn survey.*

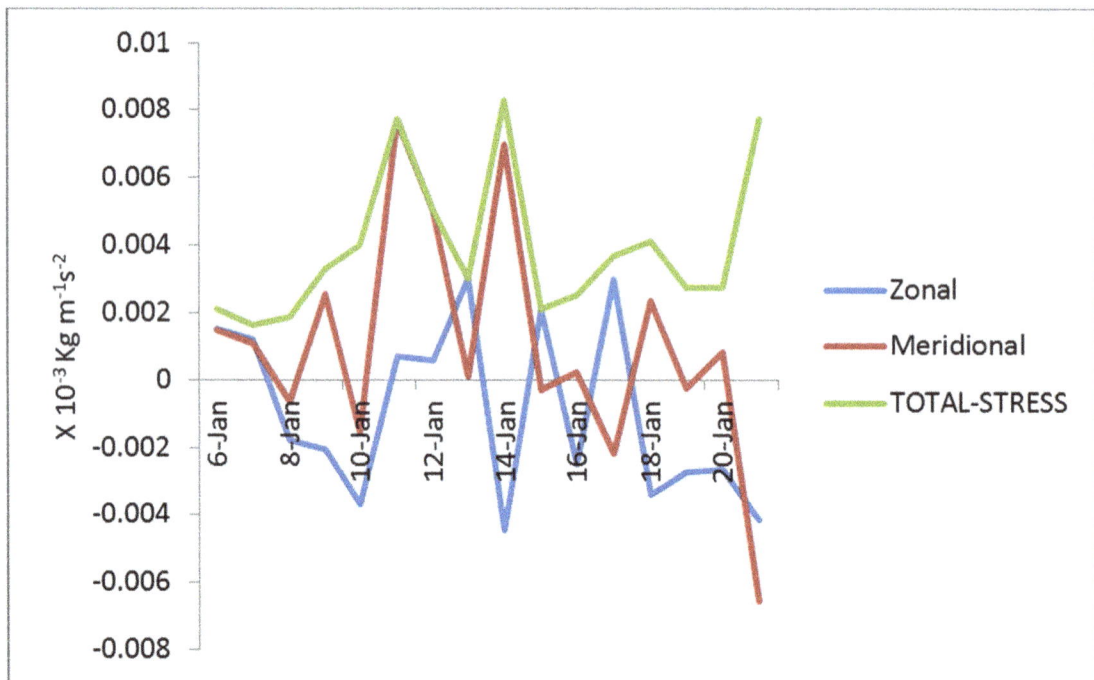

Fig. 4. *Time series of wind stress components during winter survey.*

4. Estimates of Wind Contribution

The contribution of atmospheric forcing to the dynamical process of vertical turbulent mixing is estimated by the empirical formula [10]:

$$Ru = \frac{\frac{\rho_a}{\rho_s}W^2}{U_{max}^2 + \frac{\rho_a}{\rho_s}W^2} \qquad (3)$$

where U_{max} is the greatest magnitude of the surface current,

ρ_s is the mean sea water density (1028 kg m^{-3}). The non-dimensional parameter Ru has values ranging from 0 to 1, where values close to 1 indicate strong wind contribution to vertical mixing in the upper layers of the sea and vice versa.

Generally, wind contribution to vertical mixing exhibits an oscillatory behavior as shown in figures (5-8). The contribution of wind stress over the sea surface in vertical mixing production is greatest during summer. This is most probably a result of the meteorological fact that mean wind stress is greatest during summer due to the north westerly prevailing wind. Values of *Ru* attain the maximum at 0.6, with a mean value of 0.3, while wind contribution is weakest during winter with a maximum value of *Ru* = 0.2, and mean value of *Ru* = 0.1. Statistically, autumn and summer have similar characteristics as wind contribution varies from zero to 0.5, with a mean value of 0.2.

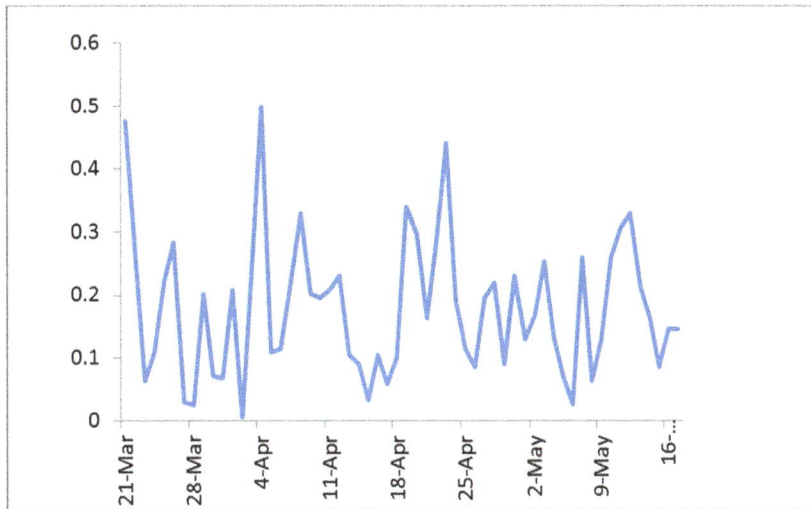

Fig. 5. *Time series of wind contribution during spring survey.*

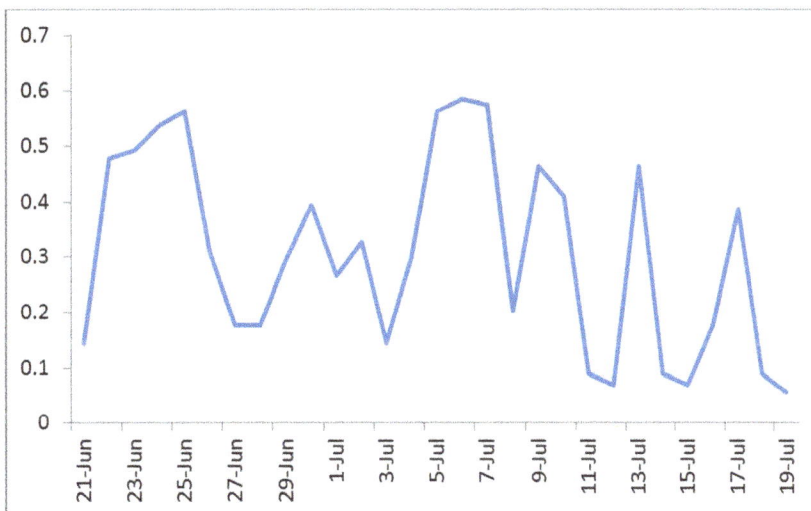

Fig. 6. *Time series of wind contribution during summer survey.*

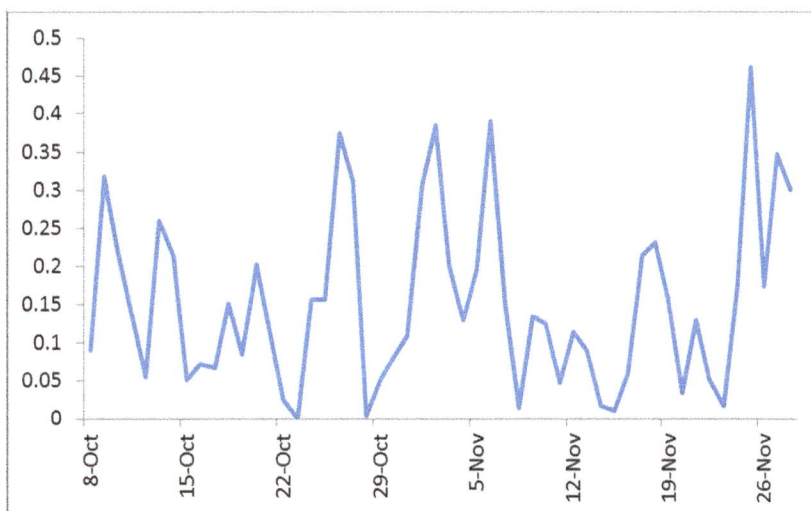

Fig. 7. *Time series of wind contribution during autumn survey.*

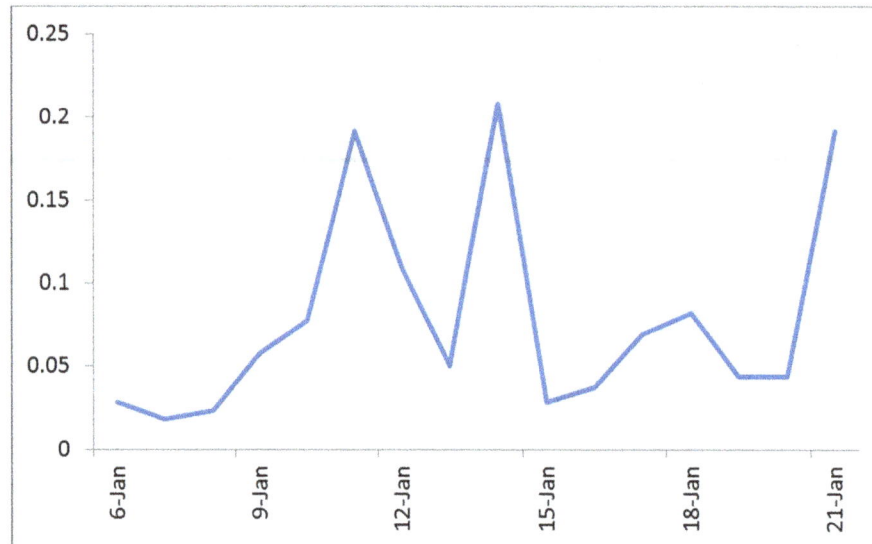

Fig. 8. *Time series of wind contribution during winter survey.*

Table 2. *Basic statistics for the index of wind contribution to vertical mixing (Ru).*

	Mean	Range
Spring	0.2	0 - 0.5
Summer	0.3	0.1 - 0.6
Autumn	0.2	0 - 0.5
Winter	0.1	0 – 0.2

5. Conclusions

Generally, wind stress played a significant role in the turbulent process of vertical mixing over most of the year. Winter was an exception, and atmospheric forcing had a small role in vertical turbulent mixing as *Ru* never exceeded 0.2, and had a mean value of 0.1. The small tidal range at Port Said (less than 1 m) produced an insignificant role for tidal forcing, particularly, at a deep offshore station. In conclusion, wind was a significant factor, besides surface wave breaking and tidal forcing, in the production of vertical mixing in the near surface layers off Port Said. Further studies with longer durations, and more sampling stations are needed for a comprehensive understanding of the role of atmospheric forcing over the sea surface in production of vertical mixing.

Acknowledgements

I express my appreciation to the National Institute of Oceanography and Fisheries, Egypt for providing research facilities. Also, I thank Dr. Ahmed Radwan for his valuable help.

References

[1]   Lien, R. C., Greg, M. C., Observations of turbulence in a tidal beam and across a coastal ridge. Jour. Geophys. Res., vol. 106, pp. 4575-4591. 2001.

[2]   Munka, W, Wunschb, C., Abysal Recipes II: Energetics of tidal and wind mixing, Jour. Deep sea Res., part 1, 45(12), pp. 1977-2010, 1998.

[3]   Lee, H-C, Rosati, A., Spelman, M. J., Barotropic tidal mixing effects in a coupled climate model: Oceanic conditions in the Northern Atlantic, Ocean modeling, vol. 11, pp. 464-477.2006.

[4]   Ly, L. N., Garwood, R. W., Numerical modeling of wave enhanced turbulence in the oceanic upper layer, Jour. Phys. Ocean., Vol. 56, pp. 473-483.2000.

[5]   Sharaf El-Din, S. H., Eid, F. M., Saad, N. N., Alam El-Din, K. A., Elsharkawy, M. S., A numerical study of turbulent mixing parameters at the nile delta, north of the egyptian coast. Inter. Jour. Bio. Cons. Vol. 2 (12), pp. 422-428. 2010.

[6]   Saad, N. N., Elsharkawy, M. S., Evaluation of some dynamical parameters at the central red sea during early summer. JKAU: Mar. Sci., vol. 21(1), pp. 115-122, 2010. DOI: 10. 4197/Mar. 21-1.6.

[7]   Sharaf El-Din, S. H., Eid, F. M., Saad, N. N., Alam El-Din, K. A., Elsharkawy, M. S., Variation of turbulent mixing parameters at the Egyptian Mediterranean Coast. Inter. Jour. Bio. Cons. Vol. 2 (5), pp. 114-129. 2010.

[8]   Hellerman, S., Rosentein, M., Normal wind stress over the world ocean with error estimates. Jour. Phys. Ocean. Vol. 13, pp. 1093-1104. 1983.

[9]   Said, M. A., Variations of current, wind and water fluxes in the south –eastern Mediterranean off the Egyptian Coast during winter and spring seasons. Mahasagar, Vol. 27 (1), pp. 1-16.1994.

[10] Verspecht, F., Rippeth, T., Howrath, M., Souza, A., Simpson, J., Burchard, H., Processes impacting on stratification in a region of fresh water influence: application to Liverpool Bay, Jour. Geophys. Res. vol. 114, pp. 1-12. 2009.

7

Modeling of the Total Alkalinity and the Total Inorganic Carbon in the Mediterranean Sea

Abed El Rahman Hassoun[1,2,3,*], Elissar Gemayel[1,2,3], Evangelia Krasakopoulou[4,5], Catherine Goyet[2,3], Marie Abboud-Abi Saab[1], Patrizia Ziveri[6], Franck Touratier[2,3], Véronique Guglielmi[2,3], Cédric Falco[2,3]

[1]National Council for Scientific Research, National Center for Marine Sciences, Batroun, Lebanon
[2]IMAGES_ESPACE-DEV, Université de Perpignan Via Domitia, Perpignan, France
[3]ESPACE-DEV, UG UA UR UM IRD, Maison de la télédétection, 500 rue Jean-François Breton, Montpellier, France
[4]Institute of Oceanography, Hellenic Centre for Marine Research, Anavyssos, Greece
[5]University of the Aegean, Department of Marine Sciences, University Hill, Mytilene, Greece
[6]ICREA - Institute of Environmental Science and Technology (ICTA), Universitat Autònoma de Barcelona, Ed. Z, ICTA-ICP, Carrer de les columnes, E- 08193 Bellaterra, Barcelona, Spain

Email address:

abedhassoun@gmail.com (A. E. R. Hassoun), elissargemayel@hotmail.com (E. Gemayel), ekras@marine.aegean.gr (E. Krasakopoulou), cgoyet@univ-perp.fr (C. Goyet), mabisaab@cnrs.edu.lb (M. Abboud-Abi Saab), patrizia.ziveri@uab.cat (P. Ziveri), touratier@univ-perp.fr (F. Touratier), veronique.guglielmi@univ-perp.fr (V. Guglielmi), cedric.falco@univ-perp.fr (C. Falco)

Abstract: Measurements of the CO_2 system parameters in the Mediterranean Sea are relatively scarce and not representative for all its sub-basins. High quality data collected on May 2013 during the 2013 MedSeA cruise covering the whole basin were used to provide for the first time linear relationships estimating the total alkalinity (A_T) and the total dissolved inorganic carbon (C_T) from salinity in each Mediterranean basin and sub-basin at different depth layers. These correlations show that a substantial quantity of alkalinity is added to the seawater during its residence time in the Mediterranean Sea, whereas the biological processes, the air-sea exchange and the high remineralization rate are responsible of the high C_T concentrations in this sea. Moreover, these fits could be used to estimate the A_T and C_T from salinity where there are not available measurements of the carbonate system parameters.

Keywords: Carbonate System, Total Alkalinity, Total Dissolved Inorganic Carbon, Fits, Mediterranean Sea

1. Introduction

Aiming to understand and quantify the carbonate system in the Mediterranean Sea, several studies have been realized (e.g. [1], [2], [3], [4], [5], [6], [7]). However, the amount of high quality measurements of the carbonate system properties, particularly the total alkalinity (A_T) and the total dissolved inorganic carbon (C_T), through the whole Mediterranean Sea remained scarce. Recently, during several oceanographic cruises carried out from 2001 until now, the measurement of carbonate system parameters was included [5] [8] [9] [10] [11]. However, many Mediterranean sub-basins remain out of coverage and need to be studied in order to have a better understanding of the carbonate system in this enclosed sea.

The measurement of the A_T and C_T parameters could be furthermore used to calculate the carbon budget in the Mediterranean and to estimate the acidification variation and the concentrations of the anthropogenic carbon dioxide sequestered in this sea.

The present paper is based on A_T and C_T data measured during the 2013 MedSeA cruise which covered during the same month almost the entire Mediterranean Sea from the West to the East and from the South to the North. In this study, we present and discuss, for the first time, the A_T-Salinity and C_T-Salinity relationships in each Mediterranean basin and sub-basin at different depth layers. These correlations could be used to estimate the A_T and C_T from salinity data where there is a lack in the measurements of the carbonate system parameters.

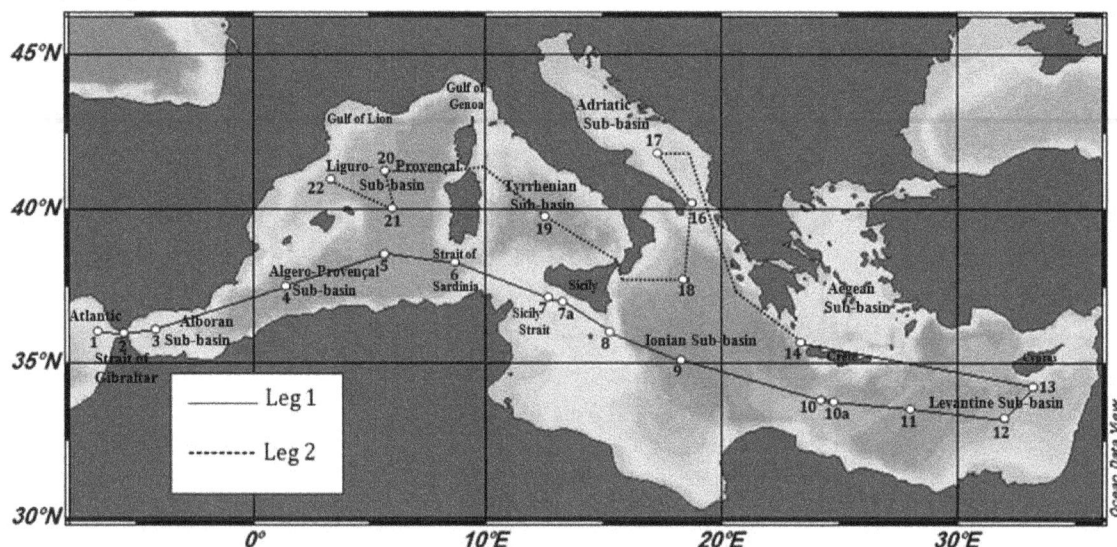

Fig. 1. Tracks of the 2013 MedSeA cruise in the Mediterranean Sea. The numbers from 1 to 22 correspond to the sampled stations.

2. Study Area and Methodology

2.1. MedSeA Cruise

During the MedSeA (Mediterranean Sea Acidification In A Changing Climate) cruise realized on board of the Spanish R/V Angeles Alvariño, from May 2nd to June 2nd 2013, 23 stations along the Mediterranean Sea were sampled throughout the water column. The overall goal and scientific objectives of the MedSeA project are well described in the following links [12 and 13]. The full cruise track (more than 8000 km long) consisted of two almost longitudinal legs. During the first leg, samples were collected from Atlantic waters off Cadiz harbor, Spain to the Levantine Sub-basin in the Eastern Mediterranean Sea [3879 km long, 15 stations, 279 sampled points, maximum sampled depth = 3720 m]. The second leg was conducted in the Northern part of the Mediterranean from the Western Cretan Straits in the Eastern Mediterranean basin to Barcelona, Spain in the North Western Mediterranean basin, passing through the South of the Adriatic Sub-basin [3232.5 km long, 8 stations, 183 sampled points, maximum sampled depth = 3000 m] (Figure 1).

2.2. Sampling and Measurement

The salinity (S) was measured *in situ* using a Sea-Bird Electronics CTD system (SBE 911 plus) associated with a General Oceanic rosette sampler, equipped with twenty four 12 L Niskin bottles. The precision of salinity measurements is ± 0.0003.

For the determination of A_T and C_T, seawater samples were collected, at all stations and depths, into washed 500 ml borosilicate glass bottles, according to standard operational protocol. A small headspace (< 1 %) was adjusted to prevent pressure build-up and loss of CO_2 during storage. Few drops of a saturated solution of $HgCl_2$ were added to the samples in order to avoid any biological activity. Then, the samples were

stored in the dark at constant temperature (~ 4 °C) until their analysis on shore (IMAGES laboratory, Perpignan, France). The measurement of these two parameters was performed simultaneously by potentiometric acid titration using a closed cell. The principle and procedure of measurements, as well as a complete description of the system used to perform accurate analysis can be found in [14]. The precision of A_T and C_T analysis was determined to be ± 2 µmol kg^{-1} for A_T and ± 4 µmol kg^{-1} for C_T, by titration of 261 samples, collected at the same conditions of temperature and S, from Banyuls Sur Mer, South France. The accuracy of A_T and C_T measurements was determined to be ± 1 µmol kg^{-1} for A_T and ± 4 µmol kg^{-1} for C_T by analyzing a total of 26 bottles of three different batches of Certified Reference Material (Andrew Dickson, CA, USA, batches 85, 86 and 128). More information about the carbonate system measurements and precisions are detailed by [15].

The carbonate system and hydrographic data of the 2013 MedSeA cruise are almost available in Pangaea data repository [16, 17, 18, 19 and 20].

3. Results and Discussion

3.1. A_T-S Relationships in the Mediterranean Sea, its Basins and Sub-Basins

Termed as "evaporation basin", the Mediterranean Sea is characterized by elevated salinities in relation to the adjacent Atlantic Ocean. As a consequence of the surface heat loss and the excessive evaporation [21] [22], the general pattern of salinity in the Mediterranean is an Eastward global increase (Fig.2). The highest salinity values were measured in the Eastern Mediterranean basin (Max. 39.18 in front of Nile Delta, at ~ 5 m), while the lowest salinities were detected at the surface of both the Western Mediterranean basin and the Strait of Gibraltar (Min. 36.29 at the surface of the Strait of Gibraltar, ~ 20 m).

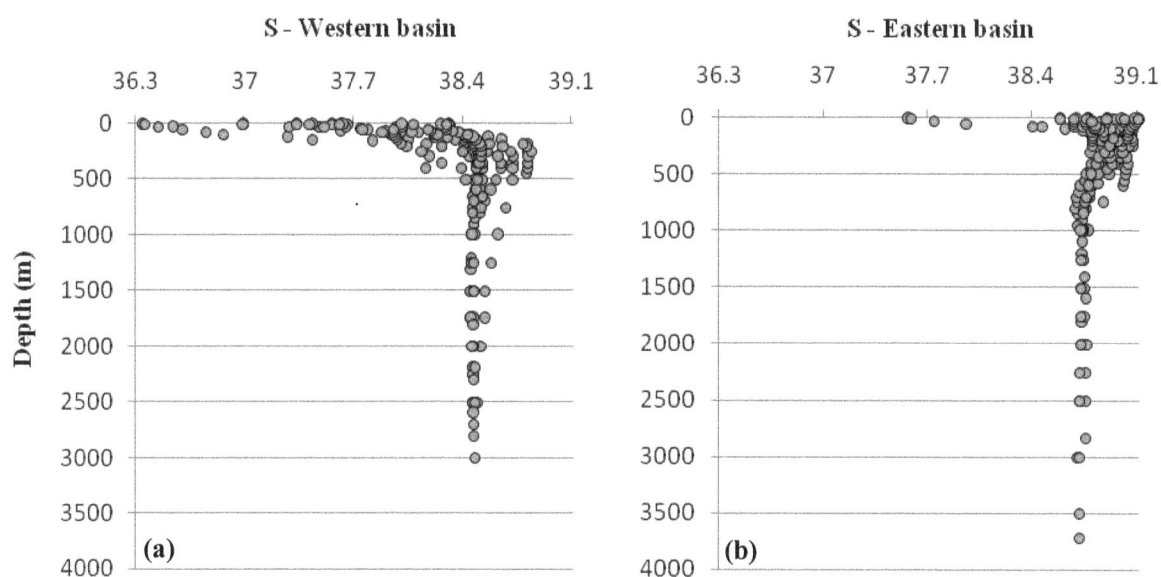

Fig. 2. *Salinity (S) as a function of depth in the Western (a) and in the Eastern (b) Mediterranean basins.*

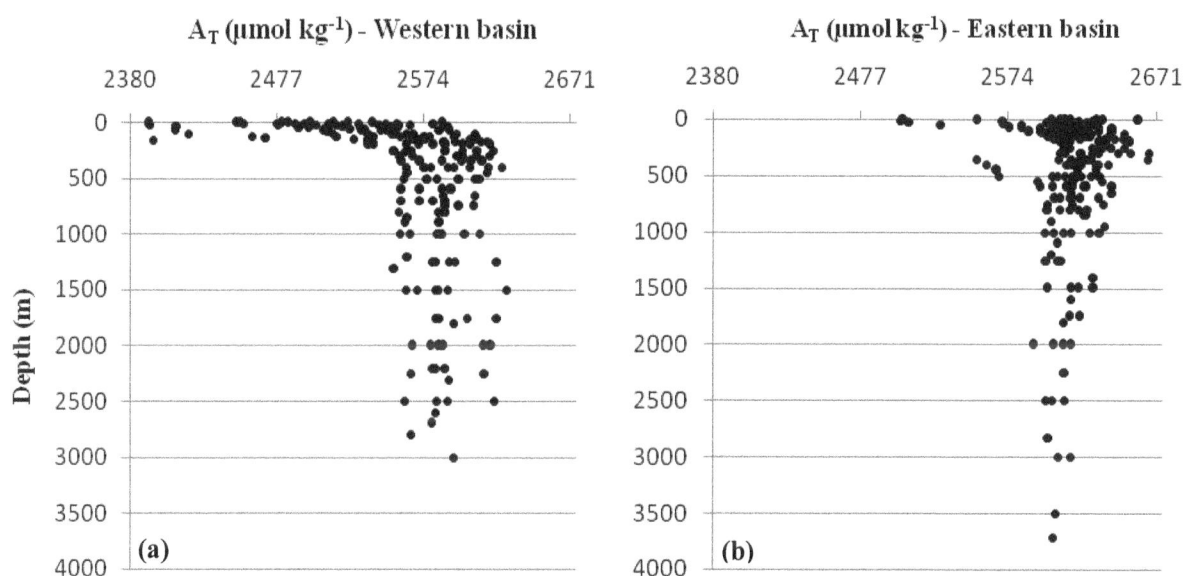

Fig. 3. *Total alkalinity ($\mu mol\ kg^{-1}$) as a function of depth in the Western (a) and in the Eastern (b) Mediterranean basins.*

High total alkalinity concentrations were recorded in the Mediterranean Sea (Mean = $2588 \pm 46\ \mu mol\ kg^{-1}$). The highest A_T concentrations were measured in the Eastern Mediterranean basin (Max. $A_T = 2666.0\ \mu mol\ kg^{-1}$, at 300 m in the area of Western Cretan Straits, whereas the lowest A_T concentrations were measured in both the Strait of Gibraltar and the surface waters of the Western Mediterranean basin (Min. $A_T = 2377.0\ \mu mol\ kg^{-1}$ at 25 m in the Strait of Gibraltar). An Eastward increasing tendency for the total alkalinity was also well noticeable, likewise the salinity trend (Fig.3).

Based on the carbonate system parameters data collected during the 2013 MedSeA cruise, an Eastward increasing trend of A_T was remarked in parallel with the salinity one [15]. Our results indicate the presence of significant correlations between A_T and S in the Mediterranean Sea globally, its main basins and many of its sub-basins as well. Table 1 presents the Model II linear regression between these two parameters in each basin and sub-basin and at different depth layers (surface, intermediate and deep layers). These equations were obtained using the Excel Microsoft Office program (2007). For all the equations derived in this study, the root mean square deviation (RMSD), which is a good measure of the accuracy of the fit, the coefficient of correlation (Pearson coefficient ; r), and the number of data pairs (n) used to derive each correlation, are also mentioned in the Table 1 for each A_T-S fit.

Evaporation would drive a steady increase in salinity and total alkalinity during the propagation of surface waters toward the Eastern part of the Mediterranean Sea. Therefore, the negative intercepts obtained in most A_T-S fits (Table 1) may be explained by the high evaporation in the sea along with the influence of high A_T inputs by freshwater contributions from the local rivers and the Black Sea. These two systems carry very high total alkalinities (between 2000 µmol kg^{-1} and 6500 µmol kg^{-1}), at low or zero salinity, to shelf areas, especially in the Eastern basin, where surface S and A_T tend to be higher due to the cumulative effects of evaporation [5]. Moreover, our results indicate the absence of a significant A_T-S correlation in the intermediate and deep layers of the Eastern basin, especially in the Levantine Sub-basin in which we detected no significant A_T-S relationship at all depth layers (Table1).

Table 1. The A_T-S relationships in the different layers of the main basins and sub-basins in the Mediterranean Sea during May 2013(R.M.S.D. = root mean square deviation, r = coefficient of correlation (Pearson's coefficient), n = number of data pairs used to derive each relationship, [-] means that no significant relationship was found).

Basin/Sub-basin	Depth	A_T-S relationships	R.M.S.D	r	n	Number of equation
Mediterranean	All	$A_T = 98.48*S - 1208$	± 19	0.9	428	Eq. 1
	Surface (0-25 m)	$A_T = 89*S - 845$	± 18	0.96	58	Eq. 2
	Intermediate (>25-400 m)	$A_T = 98*S - 1191$	± 18.5	0.91	208	Eq. 3
	Deep (> 400 m)	$A_T = 134*S - 2578$	± 17	0.57	160	Eq. 4
Western basin	All	$A_T = 99*S - 1225.5$	± 15.6	0.94	215	Eq. 5
	Surface (0-25 m)	$A_T = 92*S - 956$	± 12.5	0.97	30	Eq. 6
	Intermediate (>25-400 m)	$A_T = 101.5*S - 1322$	± 15.6	0.94	105	Eq. 7
	Deep (> 400 m)	-	-	-	-	-
Eastern basin	All	$A_T = 111.42*S - 1713$	± 19	0.54	212	Eq.8
	Surface (0-25 m)	$A_T = 89*S - 846$	± 19	0.86	27	Eq. 9
	Intermediate (>25-400 m)	-	-	-	-	-
	Deep (> 400 m)	-	-	-	-	-
Alboran Sub-basin	All	$A_T = 81.15*S - 567.7$	± 4.4	0.99	23	Eq. 10
	Surface (0-25 m)	$A_T = 10.65*S + 2005$	± 0.07	0.98	3	Eq. 11
	Intermediate (>25-400 m)	$A_T = 82*S - 605.6$	± 5	0.99	11	Eq. 12
	Deep (> 400 m)	$A_T = 171*S - 4020$	± 2	0.7	8	Eq. 13
Algero-Provencal Sub-basin	All	$A_T = 98*S - 1190$	± 11.7	0.95	73	Eq. 14
	Surface (0-25 m)	$A_T = 100.72*S - 1282.6$	± 14	0.89	9	Eq. 15
	Intermediate (>25-400 m)	$A_T = 119.38*S - 2009.7$	± 11.8	0.92	33	Eq. 16
	Deep (> 400 m)	-	-	-	-	-
Liguro-Provencal Sub-basin	All	$A_T = 126*S - 2252$	± 15	0.84	71	Eq. 17
	Surface (0-25 m)	$A_T = 93.33*S - 1008$	± 12.5	0.9	9	Eq. 18
	Intermediate (>25-400 m)	$A_T = 128*S - 2327$	± 11.6	0.89	31	Eq. 19
	Deep (> 400 m)	-	-	-	-	-
Tyrrhenian Sub-basin	All	$A_T = 95.4*S - 1084.5$	± 5.4	0.97	21	Eq. 20
	Surface (0-25 m)	$A_T = 339.73*S - 10384$	± 3	-0.96	3	Eq. 21
	Intermediate (>25-400 m)	$A_T = 96.72*S - 1139$	± 4	0.98	11	Eq. 22
	Deep (> 400 m)	-	-	-	-	-
Ionian Sub-basin	All	$A_T = 112*S - 1730$	± 15	0.88	57	Eq. 23
	Surface (0-25 m)	$A_T = 97.56*S - 1173.5$	± 17	0.95	8	Eq. 24
	Intermediate (>25-400 m)	$A_T = 140*S - 2818$	± 16	0.82	27	Eq. 25
	Deep (> 400 m)	-	-	-	-	-
Adriatic Sub-basin	All	-	-	-	-	-
	Surface (0-25 m)	$A_T = 239.5*S - 6653.7$	± 8	-0.84	6	Eq. 26
	Intermediate (>25-400 m)	-	-	-	-	-
	Deep (> 400 m)	-	-	-	-	-
Western Cretan Straits	All	-	-	-	-	-
	Surface (0-25 m)	$A_T = 81.28*S - 569$	± 0.13	0.99	3	Eq. 27
	Intermediate (>25-400 m)	-	-	-	-	-
	Deep (> 400 m)	$A_T = 1236.95*S - 45690$	± 6.3	-0.88	4	Eq. 28
Levantine Sub-basin	All	-	-	-	-	-
	Surface (0-25 m)	-	-	-	-	-
	Intermediate (>25-400 m)	-	-	-	-	-
	Deep (> 400 m)	-	-	-	-	-

In addition, our results reveal the presence of a very strong positive A_T-S correlation (Eq.27) in the surface layer of the Western Cretan Straits, whereas in the deep layer a significant negative A_T-S correlation is observed (Eq.28). It is also shown the presence of a negative and significant A_T-S fit exclusively in the surface layer of the Adriatic Sub-basin (Eq.26; Table 1). These remarks demonstrate that the relationship between S and A_T does not follow the Eastward strict evaporation trend. However, it reflects the mixture of waters characterized by high A_T concentrations with low salinity waters in the Eastern part of the Eastern Sub-basins and in the Adriatic Sub-basin. This fact is explained by the mixing, between high salinity surface water in the Eastern Sub-basins with waters from rivers flowing through carbonate-dominated terrains and/or the Black Sea, that change the characteristics of surface waters which become less saline with higher A_T concentrations.

Our A_T-S equation (Eq.1), that takes into account all available data pairs of both parameters in the entire

Mediterranean basin, is very similar to the one published by [23]: $A_T = 99.6*S - 1238.4 \pm 4.5 \mu mol \; kg^{-1}$ (Eq.I), obtained using all the available data in the DYFAMED site in the North-Western basin of the Mediterranean Sea from 1998 to 2005. The reason for the slightly less steep slope and the less negative intercept in our correlation (Eq.1) compared to the other one (Eq.I) could be attributed to the difference of sampling locations (different specific total alkalinities between the Western and the Eastern basins). Located in the North-Western Mediterranean basin (43° 25'N, 07°52'E), the DYFAMED site is more influenced by the contributions of waters coming from the continents than most of the stations of the 2013 MedSeA cruise (Fig.1). In the coastal zone, total alkalinity inputs by rivers (ex. the nearby Rhone has an $A_T =$ 2885 $\mu mol \; kg^{-1}$ [5]) and potentially by sediments, may induce the steep slope of the regression line.

In the Alboran Sub-basin, a positive and highly significant A_T-S correlation was detected (Table 1). This is attributed firstly to the parallel increase of salinity and A_T of the surface Atlantic waters incoming towards the Mediterranean Sea, and secondly, to the high S and A_T in the outgoing Mediterranean waters toward the Atlantic at the deepest layers. In addition, it was noted that the A_T-S fit (Eq.10) obtained based on all the data collected in the Alboran Sub-basin, is different from the one mentioned by [4] in the same sub-basin: $A_T = 94.85*S - 1072.6 \; \mu mol \; kg^{-1}$ (Eq.II). This may be due to the large difference in the number of sampled points (23 in our study vs. 440 in [4]) and the time difference of ~25 years between the sampling dates of the two studies. Therefore, the observed differences between Eq.10 and the equation of [4] could also be attributed to the temporal variability of the temperature and S, related to the climate change, in the Mediterranean Sea [24] [25] [26] [27].

Similarly, a positive and highly significant A_T-S correlation was noted in the deepest and most isolated sub-basin of the Western basin, the Tyrrhenian Sub-basin. The circulation patterns of the Mediterranean Sea show generally that most of the saline and high-A_T water, coming from the East to the Western basin, enters the Tyrrhenian Sub-basin [28]. Below the surface modified Atlantic waters, the Tyrrhenian Sub-basin is the first Western Sub-basin reached by the Levantine Intermediate Waters (LIW), coming from the Eastern basin through the Sicily Channel. The LIW enters the Tyrrhenian Sub-basin along the slope of Sicily and leaves it along the Sardinian coast [29]. The fact that the intermediate layer of this sub-basin is occupied by the LIW explains the positive and very significant correlation between A_T and S in the intermediate depths of the Tyrrhenian. Our A_T-S fit (Eq.20), obtained by all the available data of the Tyrrhenian Sub-basin, is similar to the one recorded by [8]: $A_T = 96.62*S - 1139.1 \; \mu mol \; kg^{-1}$ (Eq.III). The slight differences in the slopes and intercepts between the two equations are due firstly to the different number of stations and sampled points (one station and n = 21 in our study vs. six stations and n = 320 in the other one), and secondly to the date of sampling (May 2013 in our study, November 2006, February, April and July 2007 and February 2008 in the other one).

The A_T-S correlation (Eq.2) for the surface layer of the entire Mediterranean Sea differs from the regression reported by [5]: $A_T = 73.7*S - 285.7 \pm 8.20 \; \mu mol \; kg^{-1}$ (Eq. IV). The reason for the steeper slope and the more negative intercept in Eq.2 compared to Eq.IV, could be attributed to the different sampling locations : our track passed through the Strait of Gibraltar toward the Eastern part of the Mediterranean Sea and the stations were equally distributed between the two main Mediterranean basins, while the track covered in the other study was conducted from the South-East of Sardinia toward the Levantine Sub-basin and the sampled stations were concentrated mainly in the Eastern basin. In addition, the sampling periods were different: the data of the present study were collected in May 2013, whereas the data of [5] were sampled in October/November 2001. The differentiation in Eq.2 and Eq.IV could thus be attributed to the temporal variability of temperature and S in the Mediterranean Sea, which is in part associated to the climate change (as mentioned above for the Alboran Sub-basin). Equation 2 also differs from the recently reported A_T-S fit (Eq.V) for Mediterranean surface waters reported by [30]: $A_T = 79.84*S - 510$ (Eq.V) that was based exclusively on measurements made at 5m below the sea surface.

Furthermore, the A_T-S relationships noted for the surface waters of the entire Mediterranean Sea (Eq.2) and specifically for the surface waters of the Western basin (Eq.6), are very close to the relationship: $A_T = 93.996*S - 1038.1 \pm 2.5 \; \mu mol \; kg^{-1}$ (Eq.VI), calculated for samples located between the surface and the maximum of salinity (LIW horizon) at the DYFAMED site [31].

The abovementioned similarities of A_T-S correlations in the different Mediterranean areas, imply that the A_T could considered to be conservative in the Mediterranean Sea, as previously was stated by [32]. The conservative behavior of A_T with respect to salinity occurs because HCO_3^- (which is, together with CO_3^{2-} and $B(OH)_4^-$, one of the most important bases for seawater) is a major constituent of seawater and the ratio of HCO_3^- to the salinity or the chlorinity is nearly constant [8]. Equations 5 and 8 describing the A_T-S relationship in the Western and Eastern Mediterranean basins, respectively, are considerably different although both have negative intercepts. The slope and the absolute intercept values are higher in the Eastern basin, indicating that the salinity-specific alkalinity (i.e. A_T/S ratio) increases with salinity. The specific alkalinity of the Eastern Mediterranean waters is therefore higher than that of less saline Western ones. As the regression lines do not pass through the origin, the different values of the intercept between the different A_T-S equations (Table 1) reflect the different specific total alkalinities of the Eastern and Western Mediterranean water masses due to the contributions to A_T besides the salinity. However, excessive inputs of A_T from rivers in the Adriatic Sub-basin [33] result in the absence of a significant correlation between A_T and S at all the depths of this sub-basin, except on the surface layer where a significant but negative A_T-S fit was detected. The mixing of saline surface water, coming from the Eastern Sub-basins with local waters discharged by the nearby rivers and characterized by very high A_T concentrations, shaping the specific fit.

3.2. C_T-S Relationships in the Mediterranean Sea, its Basins and Sub-Basins

Total dissolved inorganic carbon (C_T) varied between a minimum of 2095 μmol kg^{-1} at the surface layers (~ 5 m) of the Alboran Sub-basin in the Western Mediterranean basin and a maximum of 2359.0 ± 0.4 μmol kg^{-1} in the intermediate waters (~ 350 m) of the Eastern Mediterranean basin (Mean = 2308.5 ± 22 μmol kg^{-1}, Fig.4). However, it is obvious that the intermediate, deep and bottom layers of the Western basin are characterized by the highest C_T concentrations (2321 ± 12 μmol kg^{-1} in the intermediate layers of the Algero-Provencal Sub-basin and 2322.0 ± 0.0 μmol kg^{-1} in the deep layers of the Liguro-Provencal Sub-basin) compared to the Eastern basin.

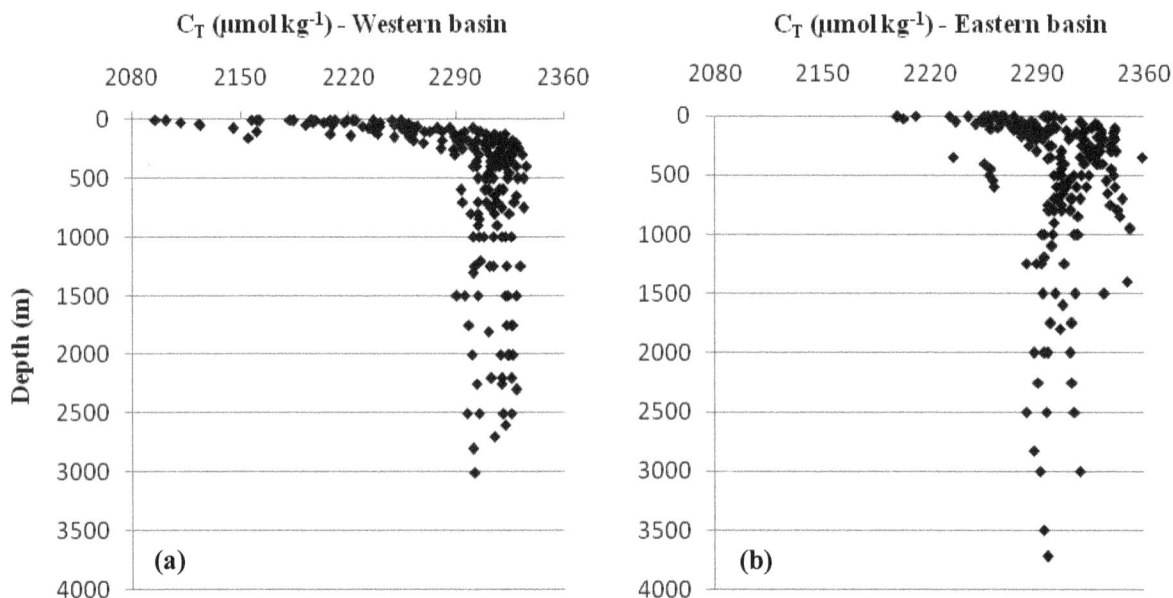

Fig. 4. Total inorganic carbon (μmol kg^{-1}) as a function of depth in the Western (a) and in the Eastern (b) Mediterranean Basins.

The presence of high C_T concentrations in conjunction with high salinities (Fig.2 and 4) in both the Eastern and the Western Mediterranean basins explains the general significant correlation obtained between these two parameters in the entire Mediterranean Sea. Up to our knowledge, the C_T-S equations are not assessed for all the Mediterranean basins and sub-basins and at different depth layers (surface, intermediate and deep). The Table 3 presents the C_T-S fits in the different layers of the main basins and sub-basins in the Mediterranean Sea obtained by linear regression of Model II, using the data of the 2013 MedSeA cruise.

The variability of C_T, apart of salinity, is also controlled by biological processes (i.e. precipitation and dissolution of calcium carbonate, photosynthesis, oxidation of organic matters), and the air-sea exchange of carbon dioxide. Due to the additional impact of these non-conservative processes on C_T, significant C_T-S correlations are not obtained at all depth layers. In the Eastern basin, there is no significant relationship between these two parameters in the intermediate layer, but we have noted a significant and negative C_T-S fit (Eq.8) in the deep layers. The deep waters of this basin are characterized by high C_T concentrations and relatively low salinities in relation to the overlying layers, which explain the negative C_T-S correlation. In the Mediterranean Sea, the convective processes and the consequent advection of dense waters assume a relevant role in sustaining the amount of remineralization in deep layers and appear to be more important than the sinking of particulate matter from the upper layers [34]. It is evident that the active overturning circulation of the Mediterranean Sea fuels the deep layers with labile carbon entrained in the newly formed deep waters inducing enhanced production of respiratory CO_2. The involvement of remineralization explains the high C_T concentrations measured in the deep layers of both the Eastern and the Western basins and the absence of a significant C_T-S correlation in the deep layers of many Mediterranean sub-basins (Table 2).

Our C_T-S correlations at the surface (Eq.17) and intermediate (Eq.18) layers of the Liguro-Provencal Sub-basin are different from the ones reported by [32]. However, it has to be mentioned that the C_T-S relationships obtained by the two studies are not totally comparable. The differences could be attributed to the different sampling strategy ; the equations of [32] are based on data collected in one fixed station in a coastally-influenced area (DYFAMED site) that was sampled monthly for a 2-years period, whereas the equations of the present study derived from data collected in May 2013 in three sampling stations. In addition, the data used in the study of [32] are originating from different depth intervals than ours: the relationship [C_T = 74.53*S-555.2 μmol kg^{-1}] corresponds to depths below the salinity maximum (LIW horizon) down to the bottom and the equation [C_T =155.17*S − 3662.6 μmol kg^{-1}] was calculated for samples located between the surface and the maximum of salinity in wintertime.

Table 2. The C_T-S relationships in the different layers of the main basins and sub-basins in the Mediterranean Sea during May 2013 (R.M.S.D. = root mean square deviation, r = coefficient of correlation (Pearson's coefficient), n = number of data pairs used to derive each relationship, [-] means that no significant relationship was found).

Basin/Sub-basin	Depth	C_T-S correlation	R.M.S	r	n	Number of equation
Mediterranean	All	$C_T = 90.91*S - 1213$	± 29	0.72	428	Eq. 1
	Surface (0-25 m)	$C_T = 63.65*S - 198$	± 18	0.93	58	Eq. 2
	Intermediate (>25-400 m)	$C_T = 80.75*S - 822$	± 26.5	0.69	207	Eq. 3
	Deep (> 400 m)	-	-	-	-	-
Western basin	All	$C_T = 107.68*S - 1836.6$	± 19	0.93	215	Eq. 4
	Surface (0-25 m)	$C_T = 74*S - 583.5$	± 14	0.94	30	Eq. 5
	Intermediate (>25-400 m)	$C_T = 99.3*S - 1515$	± 18	0.91	105	Eq. 6
	Deep (> 400 m)	-	-	-	-	-
Eastern basin	All	-	-	-	-	-
	Surface (0-25 m)	$C_T = 66*S - 292.6$	± 20	0.68	28	Eq. 7
	Intermediate (>25-400 m)	-	-	-	-	-
	Deep (> 400 m)	$C_T = 269.6*S - 8145.8$	± 16	-0.53	81	Eq. 8
Alboran Sub-basin	All	$C_T = 97.84*S - 1456.5$	± 7.6	0.99	23	Eq. 9
	Surface (0-25 m)	$C_T = 148*S - 3282$	± 4	0.83	3	Eq. 10
	Intermediate (>25-400 m)	$C_T = 102*S - 1611$	± 5	0.99	11	Eq. 11
	Deep (> 400 m)	$C_T = 244*S - 7083$	± 2	0.87	8	Eq. 12
Algero-Provencal Sub-basin	All	$C_T = 123.78*S - 2445.5$	± 13.7	0.95	73	Eq. 13
	Surface (0-25 m)	$C_T = 85.61*S - 1012$	± 12	0.88	9	Eq. 14
	Intermediate (>25-400 m)	$C_T = 136.62*S - 2934.5$	± 12	0.93	32	Eq. 15
	Deep (> 400 m)	-	-	-	-	-
Liguro-Provencal Sub-basin	All	$C_T = 143*S - 2904$	± 18	0.83	70	Eq. 16
	Surface (0-25 m)	$C_T = 56.89*S + 66$	± 5.6	0.95	8	Eq. 17
	Intermediate (>25-400 m)	$C_T = 113.5*S - 2059$	± 13.6	0.8	33	Eq. 18
	Deep (> 400 m)	-	-	-	-	-
Tyrrhenian Sub-basin	All	$C_T = 126.6*S - 2285$	± 8	0.96	21	Eq. 19
	Surface (0-25 m)	$C_T = 335.17*S - 10537$	± 3	-0.95	3	Eq. 20
	Intermediate (>25-400 m)	$C_T = 103.24*S - 1683.5$	± 5.26	0.97	11	Eq. 21
	Deep (> 400 m)	-	-	-	-	-
Ionian Sub-basin	All	$C_T = 111*S - 2003$	± 21	0.77	57	Eq. 22
	Surface (0-25 m)	$C_T = 72.71*S - 540$	± 16	0.91	8	Eq. 23
	Intermediate (>25-400 m)	$C_T = 149*S - 3481$	± 23.6	0.61	27	Eq. 24
	Deep (> 400 m)	-	-	-	-	-
Adriatic Sub-basin	All	-	-	-	-	-
	Surface (0-25 m)	-	-	-	-	-
	Intermediate (>25-400 m)	-	-	-	-	-
	Deep (> 400 m)	-	-	-	-	-
Western Cretan Straits	All	-	-	-	-	-
	Surface (0-25 m)	-	-	-	-	-
	Intermediate (>25-400 m)	-	-	-	-	-
	Deep (> 400 m)	$C_T = 104.18*S - 1805$	± 0.8	-0.68	4	Eq. 25
Levantine Sub-basin	All	-	-	-	-	-
	Surface (0-25 m)	$C_T = 39.22*S + 1082$	± 4	-0.54	4	Eq. 26
	Intermediate (>25-400 m)	-	-	-	-	-
	Deep (> 400 m)	-	-	-	-	-

Our results showed that the total alkalinity and total inorganic carbon were higher in the Mediterranean outflow than in the Atlantic inflow, in agreement with [35]. It seems that there is an important flux of C_T from the Mediterranean to the Atlantic (at the Strait of Gibraltar, C_T is equal to 2327 ± 2μmol kg^{-1} in the outflowing Mediterranean waters vs. 2123 ± 7 μmol kg^{-1} in the inflowing Atlantic waters). This is probably due to the supply of carbon by the rivers and the Black Sea, to the transformation of 40% of the organic carbon entering the Mediterranean to inorganic carbon [4] and to the high remineralization rates in the Mediterranean deep layers [34]. Moreover, our results confirm that the Mediterranean Sea exports C_T to the Atlantic Ocean. These findings are in agreement with those of [36] whom estimated that there is a net export of inorganic carbon from the Mediterranean Sea to the Atlantic Ocean varying from 0.02 to 0.07 pg C yr^{-1}, whereas [37] estimated that this export amounted to 0.025 PgCyr^{-1}.

4. Conclusion

Based on high quality and recent carbonate system data collected on May 2013 during the MedSeA cruise, this paper provides for the first time A_T-S and C_T-S fits in each Mediterranean basin and sub-basin and at different depth layers. These equations could be used to estimate, based on salinity data, the carbonate system parameters in cases where there is a lack in this kind of measurements. This study show that a substantial quantity of alkalinity is added to the seawater during its residence time in the Mediterranean Sea, whereas the biological processes, the air-sea exchange and the high remineralization rate are responsible of the high C_T concentrations in this sea.

A continuous monitoring of the CO_2 system parameters in the main sub-basins of the Mediterranean Sea is recommended to evaluate the spatial and temporal evolution of this system in the context of climate change and ocean acidification.

Acknowledgements

This work is an essential part of the European project "Mediterranean Sea Acidification in a changing climate – MedSeA" (http://medsea-project.eu/), funded by the EC FP7 Cooperation program (grant agreement 265103). The authors are pleased to thank the captain and the crew of the Spanish research vessel R/V Ángeles Alvariño. Authors are grateful to the National Council for Scientific Research (CNRS) in Lebanon for the PhD thesis scholarship granted to A.E.R. HASSOUN and E. GEMAYEL.

References

[1]   Alekin O.A., 1972. Saturation of Mediterranean Sea water with calcium carbonate. *Geochemistry*, 206, 239–242.

[2]   Chernyakova A.M., 1976. Elements of the carbonate system in the Straits of Sicily (Tunis Strait) area. *Oceanology*, 16, 36–39.

[3]   Millero F.J., Morse J., Chen C.-T.,1979. The carbonate system in the western Mediterranean Sea. *Deep Sea Research Part A : Oceanographic Research Papers*, 26 (12), 1395–1404.

[4]   Copin-Montégut C., 1993. Alkalinity and carbon budgets in the Mediterranean Sea. *Global Biogeochemical Cycles,* 7 (4), 915–925, doi:10.1029/93GB01826.

[5]   Schneider A., Wallace D.W.R., and Körtzinger A., 2007. Alkalinity of the Mediterranean Sea. *Geophysical Research Letters,* 34, L15608, doi:10.1029/2006GL028842.

[6]   Schneider A., Tanhua T., Körtzinger A., Wallace D.W.R., 2010. High anthropogenic carbon content in the eastern Mediterranean. *Journal of Geophysical Research*, 115, C12050, doi:10.1029/2010JC006171.

[7]   De Carlo E.H., Mousseau L., Passafiume O., Drupp P.S., Gattuso J., 2013. Carbonate chemistry and air-sea CO2 flux at a fixed point in a NW Mediterranean bay over a four-year period: 2007-2011. *Aquatic Geochemistry*, 19, 399–442.

[8]   Rivaro P., Messa R., Massolo S., Frache R., 2010. Distributions of carbonate properties along the water column in the Mediterranean Sea: Spatial and temporal variations. *Marine Chemistry*, 121, 236–245.

[9]   Touratier F. and Goyet C., 2011. Impact of the Eastern Mediterranean Transient on the distribution of anthropogenic CO2 and first estimate of acidification for the Mediterranean Sea. *Deep Sea Research Part I: Oceanographic Research Papers*, 58,1–15.

[10]   Pujo-Pay M., Conan P., Oriol L., Cornet-Barthaux V., Falco C., Ghiglione J.F., Goyet C., Moutin T., and Prieur L., 2011. Integrated survey of elemental stoichiometry (C, N, P) from the western to eastern Mediterranean Sea. *Biogeosciences*, 8, 883–899.

[11]   Álvarez M., Sanleón-Bartolomé H., Tanhua T., Mintrop L., Luchetta A., Cantoni C., Schroeder K., and Civitarese G., 2014. The CO2 system in the Mediterranean Sea : a basin wide perspective. *Ocean Science*, 10, 69–92.

[12]   MedSeA (Mediterranean Sea Acidification in a changing climate) project, 2015: http://medsea-project.eu/

[13]   2013 MedSeA research cruise on ocean acidification and warming: http://medseaoceancruise.wordpress.com/

[14]   DOE, 1994. Handbook of methods for the analysis of the various parameters of the carbon dioxide system in sea water ; version 2, A. G. Dickson & C. Goyet, eds., ORNL/CDIAC-74.

[15]   Hassoun A.E.R., 2014. Analyse et Modélisation de l'Acidification en Mer Méditerranée. Thèse de doctorat soutenue le 29 septembre 2014 à l'Université de Perpignan Via Domitia, laboratoire IMAGES : https://tel.archives-ouvertes.fr/tel-01083406.

[16]   Goyet C., Hassoun A.E.R., Gemayel E., 2015. Carbonate system during the May 2013 MedSeA cruise. Dataset #841933 (DOI registration in progress).

[17]   Goyet C., Gemayel E., Hassoun A.E.R., 2015. Underway pCO2 in surface water during the 2013 MedSEA cruise. Dataset #841928 (DOI registration in progress).

[18]   Ziveri P., and Grelaud M., 2013. Continuous thermosalinograph oceanography along Ángeles Alvariño cruise track MedSeA2013. Universitat Autònoma de Barcelona, doi:10.1594/PANGAEA.822153.

[19]   Ziveri P., and Grelaud M., 2013. Physical oceanography during Ángeles Alvariño cruise MedSeA2013. Universitat Autònoma de Barcelona, doi:10.1594/PANGAEA.822162.

[20]   Ziveri P., and Grelaud M., 2013. Physical oceanography measured on water bottle samples during Ángeles Alvariño cruise MedSeA2013. Universitat Autònoma de Barcelona, doi:10.1594/PANGAEA.822163.

[21]   Mariotti A., Struglia M.V., Zeng N., Lau K.-M., 2002. The Hydrological Cycle in the Mediterranean Region and Implications for the Water Budget of the Mediterranean Sea. *Journal of Climate,* 15, 1674–1690.

[22]   Bergamasco A. and Malanotte-Rizzoli P., 2010. The circulation of the Mediterranean Sea: a historical review of experimental investigations. *Advances in Oceanography and Limnology*, 1 (1), 11–28.

[23]   Touratier F. and Goyet C., 2009. Decadal evolution of anthropogenic CO2 in the northwestern Mediterranean Sea from the mid-1990s to the mid-2000s. *Deep Sea Research Part I: Oceanographic Research Papers*, 56 (10), 1708–1716.

[24] Béthoux J. P., Gentili B., Raunet J., and Tailliez D., 1990. Warming trend in the western Mediterranean deep water. *Nature*, 347, 660–662, doi: 10.1038/347660a0.

[25] Cacho I., Grimalt J.O., Canals M., Sbaffi L., Shackleton N.J., Schönfeld J., and Zahn R., 2001. Variability of the western Mediterranean Sea surface temperature during the last 25,000 years and its connection with the Northern Hemisphere climatic changes. *Paleoceanography*, 16 (1), 40–52, doi:10.1029/2000PA000502.

[26] Tsimplis M.N., and Rixen M., 2002. Sea level in the Mediterranean Sea -The contribution of temperature and salinity changes. *Geophysical Research Letters*, 0094-8276.

[27] Skliris N., Sofianos S., Gkanasos A., Mantziafou A., Vervatis V., Axaopoulos P. and Lascaratos A., 2011. Decadal scale variability of sea surface temperature in the Mediterranean Sea in relation to atmospheric variability. *Ocean Dynamics*, 62 (1), 13–30, doi: 10.1007/s10236-011-0493-5.

[28] Sparnocchia S., Gasparini G.P., Astraldi M., Borghini M., Pistek P., 1999. Dynamics and mixing of the Eastern Mediterranean outflow in the Tyrrhenian Sea. *Journal of Marine Systems*, 20, 301–317.

[29] Budillon G., Gasparini G.P. and Schröder K., 2009. Persistence of an eddy signature in the central Tyrrhenian basin. *Deep Sea Research Part II: Topical Studies in Oceanography*, 56, 713–724.

[30] Jiang Z.-P., Tyrrell T., Hydes D.J., Dai M., and Hartman S.E., 2014. Variability of alkalinity and the alkalinity-salinity relationship in the tropical and subtropical surface ocean, *Global Biogeochemical Cycles*, 28, doi:10.1002/2013GB004678.

[31] Copin-Montégut C. and Bégovic M., 2002. Distributions of carbonate properties and oxygen along the water column (0 – 2000 m) in the central part of the NW Mediterranean Sea (Dyfamed site). Influence of winter vertical mixing on air – sea CO2 and O2 exchanges. *Deep-Sea Research, II. Topical studies in oceanography*, 49 (11), 2049–2066.

[32] Bégovic M. and Copin-Montégut C., 2002. Processes controlling annual variations in the partial pressure of CO2 in surface waters of the central northwestern Mediterranean Sea (Dyfamed site). *Deep Sea Research Part II: Topical Studies in Oceanography*, 49 (11), 2031–2047.

[33] Cantoni C., Luchetta A., Celio M., Cozzi S., Raicich F., and Catalano G., 2012. Carbonate system variability in the Gulf of Trieste (North Adriatic Sea). *Estuarine. Coastal and Shelf Science*, 115, 51–62.

[34] La Ferla R., Azzaro M., Civitarese G., and Ribera d'Alcalà M., 2003. Distribution patterns of carbon oxidation in the eastern Mediterranean Sea: Evidence of changes in the remineralization processes. *Journal of Geophysical Research*, 108 (C9), 8111, doi:10.1029/2002JC001602, C9.

[35] Santana-Casiano J.M., Gonzalez-Davila M. and Laglera L.M., 2002. The carbon dioxide system in the Strait of Gibraltar. *Deep-Sea Research, II. Topical studies in oceanography*, 49, 4145–4161.

[36] Aït-Ameur N. and Goyet C., 2006. Distribution and transport of natural and anthropogenic CO2 in the Gulf of Cadiz. *Deep Sea Research Part II: Topical Studies in Oceanography*, 53 (11-13), 1329–1343.

[37] Huertas I.E., Ríos A.F., García-Lafuente J., Makaoui A., Rodríguez-Gálvez S., Sánchez-Román A., Orbi A., Ruíz J., and Pérez F.F., 2009. Anthropogenic and natural CO2 exchange through the Strait of Gibraltar. Biogeosciences Discussion, 6, 1021–1067.

8

Application of Biosorbent Derived from Cactus Peel for Removal of Colorful Manganese ions from Ground Water

Ayelech Belayneh, Worku Batu

Department of Chemistry, Adigrat University, Adigrat, Tigray, Ethiopia

Email address:

worku.ebo21@gmail.com (A. Belayneh), worku.batu13@gmail.com (W. Batu)

Abstract: Access to ground water is a human right, yet more than 80% people of Ethiopia, especially in rural areas, rely on unimproved sources and the need for finding ways of treating water is crucial. Although the use of natural biosorption coagulant agricultural waste product in drinking water treatment has been discussed for a long time, the method is still not in practice, probably due to availability of material and limited knowledge. In this study, Cactus Peel agricultural waste product made from plant materials found in Adigrat city was applied as coagulation biosorption for treatment of heavy metal in ground water. The Natural Biosorption of Cactus peel were applied based on their size particle as it is without addition of any additive chemicals for treatment of water and at room temperature without any extra cost. Based on particle sizes their efficiencies of treatment, 36.02% for S_1, 17.10% for S_2, 22.80% for S_3 based on Colorful Mn metal/ion concentration were obtained. High treatment efficiencies were recorded at particle sizes of S_1(10mm, 0.5gm).

Keywords: Biosorbent, Cactus Peel, Coagulation, Manganese

1. Introduction

The increase of accessibility of potable water has been lowest in the least developed countries and amongst the poorest people and there are great disparities between the regions of the world. The situation is worst in sub-Saharan Africa, where 40% still do not have improved water and in this area the population growth exceeds the number of people that have gained access to improved water sources. Water from boreholes is often too hard and although most people in the urban area have access to water treated by the municipality through tap or protected pumps; this water is rarely fit for drinking unless treated first [1].

The target of drinking water treatment is to remove colloidal material and impurity suspended as matter in the bodies of water, to achieve the quality drinking water guidelines [1]. Conventional technique is mostly used for surface water treatment. It typically includes chemical coagulation followed by flocculation, sedimentation, filtration and disinfection. There are many steps in water treatment processes, one of which is water treatment processes which can be applied for removal of suspended particles and colloidal material in raw water cause turbidity, synthetic organic polymers and inorganic coagulants can use

as resource in water treatment or can use natural resource as coagulant. Common artificial coagulants are aluminiumsulphate, polyaluminium chlorides, ferric chloride, and synthetic polymers. All of these coagulants have in common the ability of producing charged ions when liquefied in water, which can contribute to charge neutralization [2].Aluminiumsulfate (alum) is a common coagulant generally utilized in water treatment. Alum increases concerns when introduced into the environment towards eco-toxicological impact regarding the application of artificial polymers, the occurrence of remaining monomers is unwanted and have many carcinogenic characteristic [2, 3]. When conventional coagulation is used, chemical coagulants such as iron and alum salts are needed but, natural coagulants such as the seeds from many plants can also be used.

Coagulation is an important step in water treatment processes not only for adsorbing the particles but because it is also for removing the microorganisms that are often attached to the particles. That is, by removing turbidity, coagulants also have the possibility to remove pathogens and to significantly improve water quality and, subsequently, human health.

The aim today is how to give other people access to uncontaminated drinking water by cost effective means, particularly the rural people who can't afford any water treatment chemicals, without affecting the health of their environment. Many countries such as Sudan, Malawi, Zimbabwe and Zambia use plant seeds to cause sedimentation in their own water storage vessels in the home [4].

Water treatment processes exist in many parts of the world to provide safe drinking water and most commonly the municipalities carry out this service through physical and chemical processes.

Currently, the need for better quality water for everyday activities economically, increased the number of treatment units, and the lack of financial investments for the treatment and supply of this product, especially in developing countries, has stimulated research that render the coagulation and flocculation processes more efficient, allowing to increase the production of water without any change in physical body of ETAs and the application of natural polymers as coagulants and coagulant aids primarily or as flocculation [4,5].

In the present work, we described the use of an effective and inexpensive plant waste product of Cactus Peel biosorption material for the removal and treatment of Cadmium, Lead and Manganese species from ground water.

1.1. Application of Agricultural Biosorphation for Treatment of Water

Coagulation is an essential process in the treatment of industrial or ground contaminated water. Examples of chemical based coagulants that are available commercially include lime, alum, ferric chloride and poly-aluminum chloride. While the effectiveness of these chemicals as coagulants are well noted, there are none the less, disadvantages linked with usage of these coagulants such as comparatively high costs, harmful effects on human health as well as the fact that they appreciably affect the pH of treated waters [1-3]. As such, it is desirable to substitute these chemical coagulants with cost-effective natural coagulants to offset the aforesaid disadvantages. Cactus (Opuntia) exhibited high turbidity removal efficiency [1,4].

Coagulation is a vital process in the treatment of both surface water and industrial wastewater. Its application includes removal of dissolved chemical species and turbidity from the aforesaid water via addition of chemical-based coagulants such as alum ($AlCl_3$), ferric chloride ($FeCl_3$) and polyaluminium chloride (PAC). While the effectiveness of these chemicals as coagulants are well noted [6, 7] there are, nonetheless, drawbacks associated with usage of these

coagulants such as relatively high procurement costs, detrimental effects on human health, production of large sludge volumes as well as the fact that they significantly affect the pH of treated water. It is therefore, desirable to replace these chemical coagulants with cost-effective natural coagulants to counteract the aforementioned drawbacks.

Research on natural coagulants have been focused on Moringa oleifera [3-7] for the past two decades but more researchers are studying application of other natural coagulants such as long bean extract and cactus opuntia. It was determined, via two separate studies, that standalone long bean extract [8] was ineffective in removing turbidity while cactus opuntia [9] exhibited high turbidity removal efficiency for sewage and seawater treatment. Hence, the positive outcome of the latter study justifies further research on usage of cactus opuntia as a natural macromolecular coagulant to treat other types of highly turbid wastewater such as landfill Leachate.

The previous literature review indicates [9], Coagulation is an important wastewater treatment process used to reduce water turbidity and normally precedes the more complex secondary and tertiary water treatment process. The effectiveness of a natural coagulant derived from a cactus species for turbidity removal from dye industry effluent and their parameters such as pH as well as colour were also studied. The objectives of this study will be to apply and evaluate the effectiveness of biosorption of cactus peel for Cd, Pb and Mn removal from Ground water and determine the effect of dosage and Particle size of cactus powder on heavy metal removal and treatment efficiencies.

1.2. Conventional Water Treatment Processes

Water treatment processes exists in many parts of the world to provide safe drinking water and most commonly the municipalities carry out this service through physical and chemical processes. Drinking water treatment involves a number of combined processes based on the quality of the water source such as turbidity, amount of microbial load present in water and the others include cost and availability of chemicals in achieving desired level of treatment.

Generally drinking water treatment protocols consist of two major steps: coagulation/flocculation and disinfection (Figure 1). Commonly alum (aluminum sulfate) is used as a coagulation agent, as it is efficient and relatively cost-effective in developed countries; while, disinfection is achieved by the addition of chemical disinfectants like chlorine-based compounds.

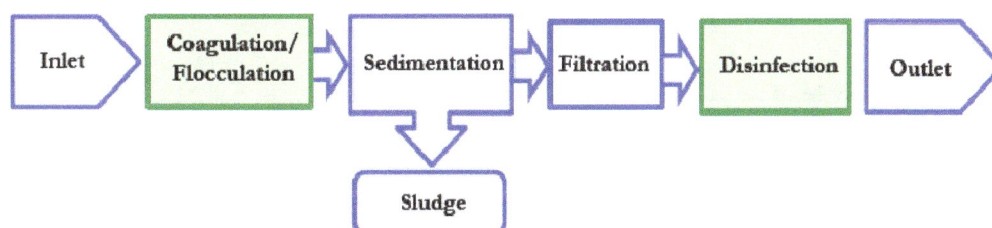

Figure 1. Schematics over conventional water treatment process [9].

1.3. Cactus Wastewater Treatment Plant

Biodredging provides cost-effective and environmentally-friendly residuals removals of unwanted material from water source were practiced by the GWI, 2009, American Water Resources management; awards represent what the industry perceives as being the most deserving of merit during the year 2009.

Figure 2. Cactus plant Leaf and Fruit.

1.4. Chemical Composition of Cactus Opuntia with Moringa Oleifera

Elemental analysis (Table 1) was carried out to provide a preliminary comparison between the elemental compositions of the cactus with that of a conventional natural coagulant, *Moringa oleifera* as determined by

Literature, cactus opuntia contained 2.3% nitrogen, 29.4% carbon and 1.7% hydrogen."*Cactus opuntia* is known to contain 2.3% nitrogen, 29.4% carbon and 1.7% hydrogen (As can be seen in Table 1, the carbon percentage of both the shelled and non-shelled *Moringa* were almost twice the carbon percentage of cactus. This is attributed to the composition of *Moringa* which consisted of more organic matter as compared to cactus.

Table 1. Elemental analysis of natural coagulants.

Content	Cactus opuntia	Shelled Moringa seeds	Non-shelled Moringa seeds
N%	2.3	6.1	5.0
C%	29.4	54.8	53.3
H%	1.7	8.5	7.7

1.5. Effect of Cactus Dosage on Turbidity Removal from Surface Water

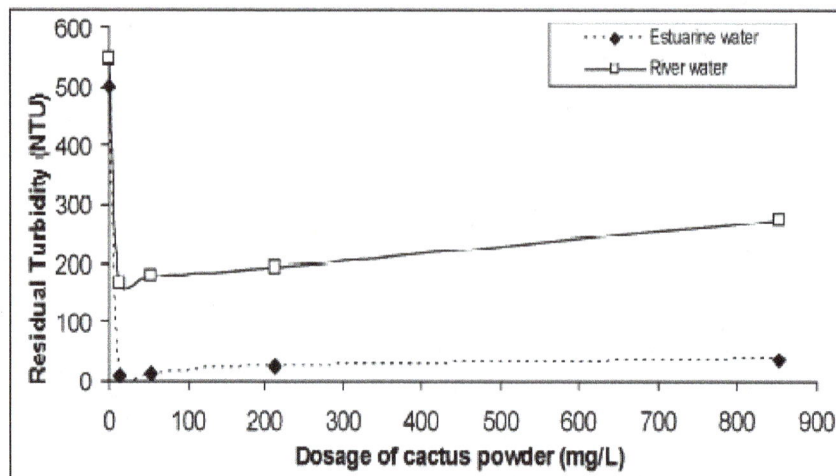

Figure 3. Effect of dosage of cactus powder on turbidity of estuarine and river water.

Literature reviews reveal that the powdered cactus formed large flocs with impurities in the sample which facilitated settling and as a result, clear supernatant was produced. Similar observation was also noted for treated river water. Figure 1 shows the effect of dosage of cactus powder on residual turbidity of surface water.

1.6. Chemistry of Heavy Metal and Their Exposure

By definition, heavy metals are any metallic chemical element that has a relatively high density (superior to 5 g/cm^3); most of them are toxic or carcinogenic even at low concentrations, such as mercury (Hg), cadmium (Cd), arsenic (As) and chromium (Cr).

Poisoning by exposure to heavy metal is well known to affect central nervous function, damage blood composition, lungs, kidneys, liver and other vital organs. Long-term exposure can cause slower progressing physical, muscular, and neurological degenerative processes. Allergies may also occur and repeated long term contact with some metals, or their compounds may become carcinogenic [10].

1.7. Manganese Metals in the Environment

Mn is present in the solid phase and in solution, as free ions, $MnCl_2$, $MnSO_4$, or in salt forms, MnO_2, as a mineral *pyrolusite* or adsorbed to soil colloidal particles. The heavy metal concentration in topsoil is a result of soil-forming processes, as well as agricultural and human activities [10].

Heavy metals are currently of much environmental concern. These metals are dangerous because they tend to bioaccumulation the food chain and they are harmful to humans and animals. Bioaccumulation means an increase in the concentration of a chemical in a biological organism over time, compared to the chemical's concentration in the environment. The threat that heavy metals pose to human and animal health is aggravated by their long-term persistence in the environment [10, 14].

Estimation of the migration ability of any pollutant in the natural environment is considered to be a necessary stage for predicting the ecological situation.

Manganese is a mineral that naturally occurs in rocks and soils and is a normal constituent of the human diet. It exists in well water as a naturally occurring ground water mineral but more also are present due to underground pollution sources. Mn may become noticeable in tap water at concentrations greater than 0.05mg/l by parting the color, odor, or taste of to the water. Mn is seldom found alone in water supply; It is frequently found in iron bearing waters but it is more rare than iron. When Mn is present in water; its every bit is as annoying as iron perhaps even more so. In low concentrations it produces extremely objectionable stains on everything with which it comes in contact. Deposits collect in pipe lines and tap water may contain black sediment and turbidity due to precipitated Mn [10-14].

When fabrics are washed in Mn bearing water dark brown or black stains are formed due to the oxidation of the manganese. The US EPA secondary drinking water regulations recommend a limit of 0.05mg/l Mn because of the staining which may be caused. It is a hard metal and is very brittle. It is hard to melt but easily oxidized. Manganese (Mn) is easily reactive when pure and as a powder it will burn in oxygen. It reacts with water and dissolves in dilute acids.

1.8. Health Effect of Mn

Exposure to high concentration of manganese over the course of years has been associated with toxicity to the nervous system, producing a syndrome that resembles Parkinsonism. This type of effect may be more likely to occur in the elderly. Manganese is unlikely to produce other types of toxicity such as cancer or reproductive damage. Manganese (Mn) is very common compound that can be found everywhere on earth. Manganese (Mn) is one out of three toxic essential trace elements which means that it is not only necessary for human to survive but it is also toxic when too high concentrations are present in a human body [10-14].

2. Methodology

2.1. Description of Study Area

The present study is conducted in Adigrat University, in northern part of Ethiopia. Adigrat is a city of eastern zone Tigray; that found in Tigray regional state of Ethiopia. It is located in the eastern zone around 900 km far from Addis Ababa which is capital city of Ethiopia with longitude 14°16′N 39°27′E coordinates and latitude 14°16′N 39°27′E with an elevation of 2457metersabove sea level [15, 16]. Adigrat is gifted with ground water which is their population used for house hold level activity as a source of water. The largest pharmaceutical manufacturing plant in Ethiopia, 'Addis Pharmaceuticals Factory SC, Walwallo Alcohol Factory SC, includes small scale industries are also located in Adigrat.

2.2. Cactus and Water Sampling and Analysis

2.2.1. Samples Collection

Cactus peel and groundwater samples were collected systematically from the sources during survey times. The cactus samples were collected in plastic bags. The water sample was also collected simultaneously by using plastic bottles from selected site.

2.2.2. Preparation of Samples

Fresh cactus peel sample were collected from agricultural farm lands surrounding the study areas of Adigrat city, as its availability and transportation is easy. Then the leaves of cactus sample collected were first washed by tap water followed by distilled water, to eliminate air borne pollutants. The peels were allowed to dry in shade for 4-6 days until all the moisture content is lost from it and the color change could be observed from yellow to brownish black. Using the domestic mixer the peels were incompletely powdered for later usage in sieve analysis unit to obtain different fractions of peels.

2.3. Mechanical Shaking of Biosorption with Analyte by Centrifuge Stirrer Test

Centrifuge Stirrer test was used to determine the effectiveness of using cactus powder as coagulant. The test was conducted using centrifuge stirrer apparatus using 20ml to 50ml centrifuge test. Raw effluent sample of 20ml were stirred in the stirrer. Cactus powder of dosages of 0.5 and 0.8 grams were added to the effluent samples. This mixing stage was conducted at 150rpm.After a period of time natural coagulation gets started and the samples were allowed to stand for 30 minutes after each treatment was completed (settling stage).Then the filtrates were digested by 3ml of Conc.HNO_3 at $150^{o}C$ for half hour; clear solution contain blue-black like oil was filtered , diluted to 50ml. Then, Uv-visible Absorption Spectrophotometer was applied based on

the color of the solution which was detectable in the visible region of absorbance related with the original analyzed sample. Since manganese ions are colorfully due to the presence of d-electron, absorbance was done at 400nm-480nm.The result of the absorbance also indicated only the Mn ions which were colorful and easily detectable.

3. Results and Discussion

3.1. Heavy Metal Concentration in Ground Water before and after Treatment

Figure 4 shows the heavy metal concentration in ground water before and after treatment with Cactus peel biosorption were recorded by Uv-Vis Spectrophotometer at 400- 480nm for Mn metal concentrations. The potential of biosorbent Cactus peel was determined related with the original sample of ground water.

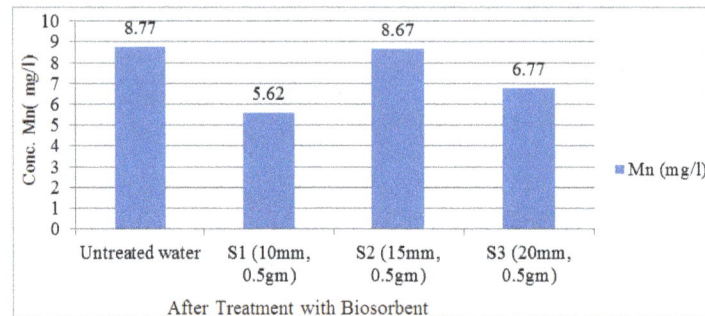

Figure 4. Reduction of Mn concentration by Biosorbent Cactus Peel in Water Samples.

At each different size of biosorption the concentration of metal recorded in each sample were less than the concentrations of metals in original sample. At each size more than 0.1- 3.15mg/l of concentrations Mn were reduced. Biosorption Cactus Peel has high potentials to adsorb Mn ion to it.

3.2. Treatment Efficiency of Bio-Sorption Cactus Peel for Removal of Metal Available in Water

The treatment efficiency of biosorption Cactus peel under this investigation was determined relative to the

concentration of metal in original sample and the results were obtained after several treatment stages. The treatment values of Mn metal/ion were analyzed based on different sizes of biosorbent Cactus peel at normal pHvalue of water or without using additive materials and at room temperature. The removal of Mn concentration in ground water obtained at different particle sizes of biosorption were, 36.02% (S_1), 17.10% (S_2) and 22.80% (S_3). The highest treatment efficiency was determined at S_1 (10mm, 0.5gm), when the lowest obtained at S_2(15mm, 0.5gm).

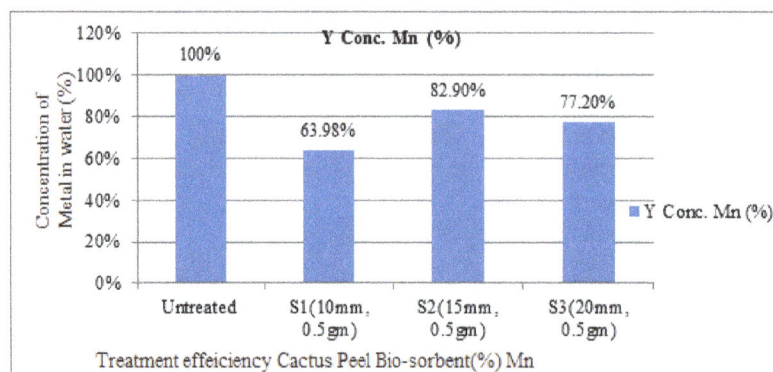

*Key:- 100%Mn = 8.77mg/l,

Figure 5. Treatment efficiency of Biosorbent Cactus peel for removal of Mn metals in water.

Figure 5, recommended that the treatment efficiency determine at S$_1$ (10mm, 0.5gm) was higher for metal absorption in ground water. Therefore also the lowest treatment efficiency of biosorbent Cactus peel was obtained at larger size of biosorbent for both metal/ion. The results show the reduction of treatment as the size of particle biosorbent increases. Therefore the message observed from Figure 5 showed that the biosorbent derived from the Cactus peel for removal of toxic metal from source water can be easily applicable at house hold level without any cost.

4. Conclusion and Recommendation

In the present investigation it was evident that Biosorbent cactus peel can be used as low cost-effective of plants waste product used as biosorbent for the removal of Mn metal from ground water. The treatments values determined were stimulated the potential value of natural biosorbent cactus peel to be highly size dependent. It was observed that maximum percentage removal for each metal was observed at smaller size. This paper give an evidence on biosorbent derived from cactus peel for removal of metals/ions in water can be easily applied as house hold level; without requiring any additional additive chemicals, but by adjusting only their particle size.

Acknowledgement

I would like to thank Mr. Worku Batu (M.Sc) my Advisor for his sincere and valuable suggestion, constructive comments, guidance and advice throughout the project work.

I am grateful to Adigrat University, Department of Chemistry for creating a good educational environment. I would like to express my deep gratitude to my families; all of as for their love, motivation, encouragement and for helping me.

References

[1] DR. T. Kannadasan, M. Thirumarimurugan, K. S. Sowmya, Sukanya karuppannan, Dye Industry Effluent Treatment Using Cactus (Opuntia) and Water Hyacinth (Eichhornia Crassipes)". *Journal of Environmental Science, Toxicology and Food Technology. 2013, pp 41-43.*

[2] Edzwald, J. k "Coagulation in drinking water treatment: particles, organics, and coagulants."Water science technology, (1993); 27(11), 21-35.

[3] Fatoki, O. S and ogunfowokan, A. O. .Effect of coagulant treatment on the metal composition of raw water." Water SA,(2002); 28(3),293-298.

[4] Kang, M. ,kamei, T. and Magara,y. Comparing polyaluminium chloride and ferric chloride for antimony removal."Water research, (2003); 37(17), 4171-4179.

[5] Liew, A. G., Noor, M. J. M. M. and Ng, Y. M."Coagulation of turbid waters using extraction of long bean."Malaysian journal of science, (2004); 23,185-191.

[6] J. K. Edzwald, "Coagulation in Drinking Water Treatment: Particles, Organics and Coagulants" Water Science and Te c h n o l o g y, (1993); Vol. 27, No. 11, pp. 21 – 35.

[7] M. Kang, T. Kamei and Y. Magara, "Comparing Polyaluminium Chloride and Ferric Chloride for Antimony Removal" Water Research, (2003); Vol. 37, No. 17, pp. 4171 – 4179.

[8] T. Okuda, A.U. Baes, W. Nishijima and M. Okada, *"Coagulation Mechanism of Salt Solution Extracted Active Components in Moringa Oleifera Seeds"* Water Research, (2001); Vol. 35, No. 3, pp. 830 – 834.

[9] S. Katayon, et al I. Azni, J. Ahmad, BC. Khor and A.M. Suleyman, *"Effects of Storage Conditions of Moringa Oleifera Seeds on its Performance in Coagulation"* Bioresource Technology, (2006); Vol. 97, No. 13 pp. 1455-1460.

[10] Yin Chun Yang1, Suhaimi Bin Abdul Talib, Turbidity Removal From Surface Water And Landfill Leachate Using Cactus *Opuntia, Journal - The Institution Of Engineers, Malaysia 2007 (Vol. 68, No. 1, March).*

[11] Maria De Lurdes Dinis and António Fiúza. Exposure Assessment To Heavy Metals In The Environment: Measures To Eliminate Or Reduce The Exposure To Critical Receptors.

[12] L. I. Simeonov et al. (eds.), *Environmental Heavy Metal Pollution and Effects on Child Mental. Development: Risk Assessment and Prevention Strategies,* 2011.

[13] Kumari P., Sharma P., Srivastava S. and Srivastava M.M. [2006] Biosorption studies on shelled *Moringa oleifera* Lamarck seed powder: Removal and recovery of arsenic from aqueous system *Int. J.Mineral Proc.* 78: 131-138.

[14] Mondal P., Majumder C.B., Mohanty B. [2008] Treatment of arsenic water in a batch reactor by using ralstonia eutropha MTCC 2487 and granular activated carbon *J. Hazard. Mater.* 153: 588-599.

[15] G. Gebretsadik, G. Animut and F. Tegegne. Assessment of the potential of cactus pear (*Opuntia ficus indica*) as livestock feed in Northern Ethiopia, *Livestock Research for Rural Development, 2013,* 25 (2).

[16] D. Worku Batu and T. Wunesh Solomon, Quantitative Determination of Sugar Levels in Natural Plants of Cactus Pear (Opuntia ficus indica) and Votre-Coach Alimantaire Cultivated in Adigrat, North of Ethiopia, *International Journal of Innovation and Scientific Research,* Vol. 10 No. 1, pp. 125-134, October, 2014.

9

Effects of Water Pipe Leaks on Water Quality and on Non-Revenue Water: Case of Arusha Municipality

Josephat Alexander Saria

The Faculty of Science, Technology and Environmental Studies, The Open University of Tanzania, Dar es Salaam, Tanzania

Email address:

josephat.saria@out.ac.tz

Abstract: Utilities can no longer tolerate inefficiencies in water distribution systems and the resulting loss of revenue associated with underground water system leakage. Increases in pumping, treatment and operational costs make these losses prohibitive. Chronic water losses have been the hallmark of Tanzania especially AUWSA water management over the decades. The aim of this research was to assess effects of water pipe leaks on water quality and on non revenue water. Out of 140 respondents it was found out that unauthorized connections (illegal and by pass) account to 24%, while 30% of respondents identified leakages from water system (transmission mains, distribution mains, utility's reservoirs, service connections); 16% due to stopped/malfunctions water meters (Bulk and customer water meters), whilst 14% respondents enumerated that is caused by other reasons like corruption and bribery among meter readers. The combination of both leakage and low pressure is a source of contamination of pipe water because it allows contaminants to siphon into the water system hence bacterial contamination (TC averaged to 67.5 CFU/100ml and FC averaged to 4.0 CFU/100ml while pH range from 6.5 – 7.7). Community awareness programmes about NRW to all consumers must be conducted to ensure that customer confidence in the utility's services is maintained. A key element in this is open communication like public should be able to easily contact the utility to report burst pipes, leakages, or other concerns.

Keywords: Non Revenue Water, EWURA, Faecal Coliform, AUWSA, Arusha

1. Introduction

Improving access to water supply and sanitation services (WSS) has been an issue on the development agenda to most of developing countries and still these services does not reach a significant proportion of the world's population. Every year, this becomes more of a challenge due to factors such as rapid population growth and increasing urbanization [1]. Water supply and sanitation in Tanzania is characterized by low levels of access, in particular in urban slums and in rural areas, as well as poor service quality in the form of irregular water supply [2,3]. Determining data on access is difficult because different definitions and sources are used, which results in significant discrepancies. According to a report, household surveys regularly return lower rural water supply coverage than estimates by the Ministry of Water (which is collected by district water engineers and urban water and sanitation authorities). For urban areas, survey data are consistently higher because they also include households that are not connected to the

formal water supply network and access water from neighbors, protected wells or boreholes [4].

Rapid population growth and urbanization in Tanzania exerts enormous pressure on the delivery of water supply and sanitation services to her citizens [5]. Since Tanzania is not in a position to meet the costs of maintaining and improving water supply and sanitation services from public revenues, this has led to the introduction of a commercial approach to the provision of these services on which they must be paid for, rather than a free right. Regulation of water supply and sanitation services is particularly important since water entities which provide water supply and sanitation services do not face competitive pressures that would otherwise ensure the quality of service provision to consumers is adequate.

Water quality and the Non-Revenue Water (NRW) are the big challenges to the water industry in general and it openly affects the level of income realized from water billings, service level and consumer satisfaction to meet the water demand from consumers which in turn affects service sustainability in water sector [6]. One of the major issues affecting water utilities in the developing world is the considerable difference between the amount of water put into the distribution system and the amount of water billed to consumers. High levels of NRW reflect huge volumes of water being lost through leaks, not being invoiced to customers, or both [6].

A lack of understanding of the magnitude and sources of NRW is one of the main reasons for insufficient NRW reduction efforts around the world. Only by quantifying NRW and its components and calculating appropriate performance indicators can the NRW situation be properly understood, cost estimates be made, and a fair contract model be developed [7,8,9]. It is also of utmost importance to have good pressure and supply time data, as those have a fundamental impact on leakage levels and its reduction/increase potential.

The existence of NRW is contributed by a number of reasons, depending on the technological development level and its purpose in the industry [7]. The developed countries have managed to control the level of NRW to the lower level like Singapore with 5% and Vitens in the Netherlands with 6% and below, while the developing countries including Tanzania with an average of 35.2% of NRW for regional UWSAs [8]. In Tanzania, for instance the suggested level is 20% or below as per the regulator's (EWURA), but none of the water Authorities has ever met that level [7]. This research aims to examine the factors causing Non-Revenue Water (NRW) in Arusha Urban Water supply and Sewerage Authority (AUWSA) as well as quality of pipe water supplied in Arusha city.

It is the intention of this study to determine the knowledge and understanding of NRW for Arusha Urban Water Supply Authority management in relation with poor management system, equity and quality of potable water supply in rapid growth of city like Arusha Municipality. We intend to develop strategy for management of non revenue water (NRW) to gain a better understanding of the reasons for NRW and the factors which influence its components. Then techniques and procedures can be developed and tailored to the specific characteristics of the network and local influencing factors, to tackle each of the components in order of priority. This diagnostic approach, followed by the practical implementation of solutions which are practicable and achievable, can be applied to any water company, anywhere in the world, to develop a strategy for NRW management.

2. Methodology

2.1. Description of Arusha City

Arusha City is located between 2° and 6° S as well as longitude 34.5° and 38° E. The City is also located on the southern slopes of Mount Meru lying at a height between 1160m to 1450m above sea level (Figure 1). It lies on the Great North road at the center between Cairo and Cape Town. Arusha City is an industrial, international trading, tourism

and conference centre. Arusha City Council has a population of 416,442 (199,524 male and 216,918 female) [5].

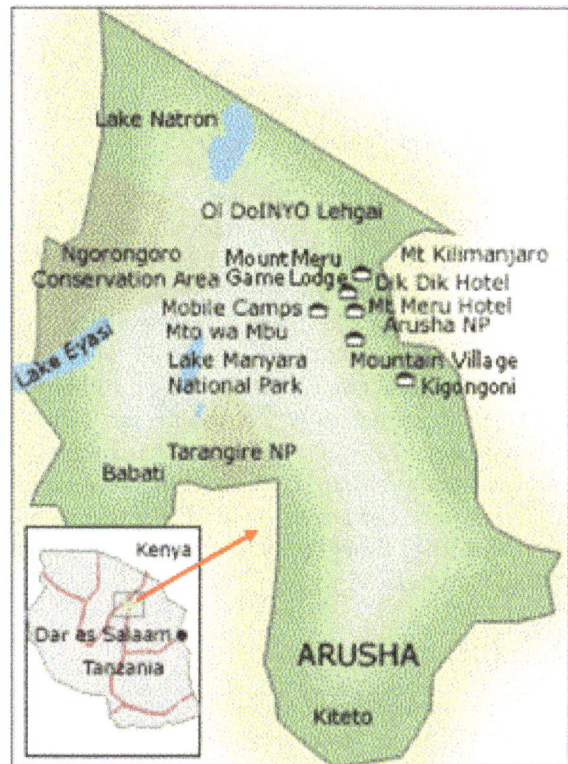

Figure 1. *A Map of Study Site: Tanzania Locating Arusha City.*

2.2. Population and Sampling Procedures

Multistage sampling technique was used to select, one hundred and forty (140) respondents. Questionnaires were distributed to some of AUWSA employees (Management team, Assistants/Supervisors, Technicians/Meter Readers, Normal staff) and customers of different categories (Domestic, Commercial, Industrial and Parastatal). The choice of these units of inquiry is made basing on their involvement, knowledge, experience and number of years stayed in AUWSA. Out of 140 people 21 (15%) were interviewed and 119 (85%) responded through questionnaire as shown in Table 1.

Table 1. *Sampling Frame.*

Unity of Inquiry	Sample Size			Percent (%)
AUWSA Employees	Questionnaires	Intervie wed	Total	
Management team	6	0	6	100
Assistants/Supervis ors	6	0	6	100
Technicians/Meter Readers	10	5	15	100
Normal staff	35	8	43	100
Domestic	46	6	52	87
Parastatal	6	2	8	80
Institutions	5	0	5	100
Industrial	5	0	5	100
Total	119	21	140	89

The main tools for data collection were review of relevant literature, household survey with a structured questionnaire, semi-structured interviews with key informants within the study areas. Data analysis was carried out using the Statistical Package for Social Sciences (SPSS version 16.0) as well as content analysis for qualitative data.

2.3. Microbiological Analysis

Total and Faecal bacteria were analysed using membrane techniques as described in American Health Public Association [10] whereby 100 ml of undiluted and diluted (1 - 10 times) water sample were filtered through 0.45 µm pore size membrane filters. The filters were transferred to two selective media where FC was grown on m-FC broth and TC was grown on m-Endo broth. Plates for Faecal coliform (FC) bacteria were incubated at 44.5 ± 0.5 C° for 24 hrs. FC colonies appeared blue while plates for total coliform (TC) bacteria incubated at 37 ± 0.5 C° for 24 hours were reddish in color. Environmental parameter (pH), were measured *in situ* at each station using a multi-parameter water quality checker (HoribaU-10, Japan).

3. Results and Discussion

In this development water authorities faced a number of challenges like water production which does not meet demand and old infrastructures resulting to water losses that lessen financial capability of the water utilities hence poor services. For many years, the most widely used performance indicators in the World and Tanzania is not exception for Non-Revenue Water and Real (Physical) losses have been percentages loss. Water produced and delivered to the distribution system is intended to be sold to the customer not lost from the distribution system without authorization [11]. When required to identify the main causes of Non-Revenue Water in Arusha Urban Water Supply and Sewerage Authority, the results for respondents are as shown on the figure 2.

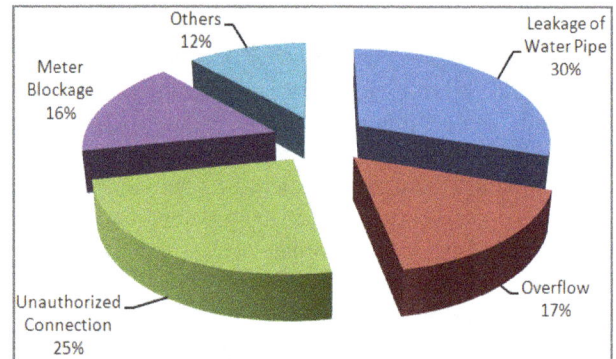

Figure 2. *Reasons for an Increase in Non Revenue Water*

3.1. Water Leakage

Physical losses can occur along the entire distribution system, from storage reservoirs and the primary network to the smallest service connections. Sometimes, when people think about leakage, they normally think of big and spectacular pipe bursts. These often cause a lot of damage but are insignificant in volume compared to all the other leaks that do not come to the surface.

In Arusha municipality leakage of water pipes is estimated to be around 30% [12]. When management were interviewed they identified it is due to different reasons including old and poorly constructed pipelines, inadequate corrosion protection, poorly maintained valves, mechanical break and high pressure (Figure 3).

Figure 3. *Pipe Bust Due to High Pressure in the Water System.*

Figure 2 revealed that, 24% of respondents pointed out that the main causes of NRW in Arusha Water Supply and Sewerage Authority was caused by unauthorized connections (illegal and by pass); 30% of respondents pointed out that NRW in AUWSA was caused by leakages from water system due to age of the water supply system (Figure 4).

Figure 4. Aged Water Pipe Broken.

About 16% of respondents mentioned that the causes of water losses caused by stopped/unfunction water meters (bulk and customer water meters), whilst 14% respondents enumerated that non- revenue water was caused by other reasons like corruption and bribery among meter readers, wrong data capturing, improper customers records, water meter reversal, wrong water meter installations and unread water meters) and 16% of respondents mentioned that NRW was caused by overflow from storage tanks and break pressure tanks.

The most commonly used indicator to measure NRW is the percentage of NRW as a share of water produced. When losses in terms of absolute volume are constant the percentage of NRW varies greatly with total water use, i.e. if water use increases and the volume of losses remains constant the percentage of NRW declines. This problem can be eliminated by measuring NRW not as a share, but in terms of absolute losses per connection per day, as recommended by other researchers [9]. Nevertheless, the use of percentage figures to compare levels of NRW remains common despite its shortcomings. Losses per kilometer of network are more appropriate to benchmark real losses, while losses per connection are more appropriate to benchmark apparent losses. On average, Arusha WSSA continued to have the highest NRW (m^3 lost per km per day) which was 60.2m^3 of water per day in a kilometer length of the distribution network. Previously in 2011/12, Arusha WSSA had the highest volume of 54.48 m^3 of water lost per km per day (Figure 5).

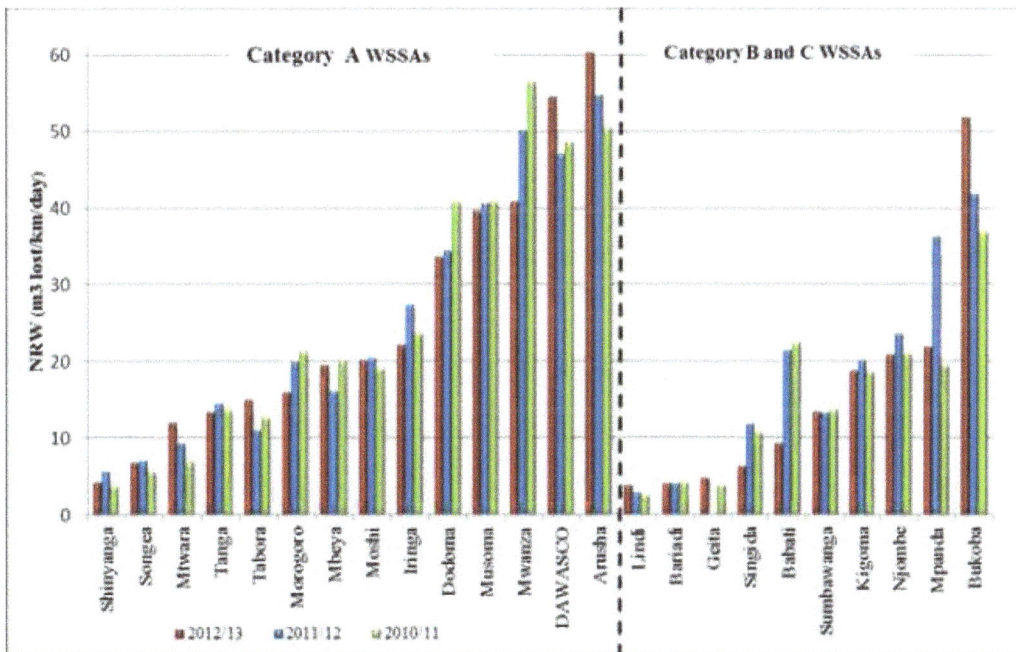

Figure 5. NRW Measured in a Kilometer Length of the Pipe Network in One Day [7].

Apart from Arusha Municipality having high volume of water lost per kilometer in a day, also Arusha WSSA continued to have the highest water connection density in a kilometer of the water distribution network with 132 connections per kilometer of water distribution network. Meanwhile, Lindi WSSA remained with the lowest reported water connections density of 15 connections per kilometer. All these are the big source of NRW.

3.2. Unhygienic Water Supply

Non Revenue Water gives negative impact on the quality of the national water supply services. It reflects the problem of instability of water supply infrastructure. Damaged pipes network, water supply interruption to the consumers and quality of treated water supplied are among the subject that shall be given attention in ensuring high quality of water supply service.

The water runs through AUWSA pipe network is generally of poor quality not only with regards to taste, smell, appearance and temperature but also from a public health point of view. The last is the worst, the consumer can be misled, because the water can look clean and taste pleasant and yet still contain pathogenic microorganisms. In the interview with AUWSA officials they indicated

"...the main pipelines have been installed in phases parallel with development of water works and matched with the past consumer patterns in the different areas of Arusha city, without regard to the development prognosis".

This was inline with other study [13] which indicated many of the main pipelines in the developing countries city today are of wrong dimensions causing many areas to suffer from high pressure, too low or completely lack of water pressure. The combination of both leakage and low pressure is a source of contamination of pipe water because it allows contaminants to siphon into the water system hence bacterial contamination (Table 2).

Table 2. *Analysis of Tape Water Samples (N = 10).*

Sampling Site	T.C. CFU/100ml	F.C. CFU/100ml	pH
Njiro	74	05	6.7- 7.2
Naura	61	03	6.5 – 7.7

The responses showed that water from the taps is seen dirty (muddy) when there is a burst or broken pipe which creates a leakage somewhere in the network allowing mud to get into the network (Figure 6). Sometimes mud is seen in the taps when there is a new water connection made somewhere in the neighbourhood.

The highest number of respondents more than 60% indicated that they have seen rubbishes like charcoal, muddy and sand coming from their pipes.

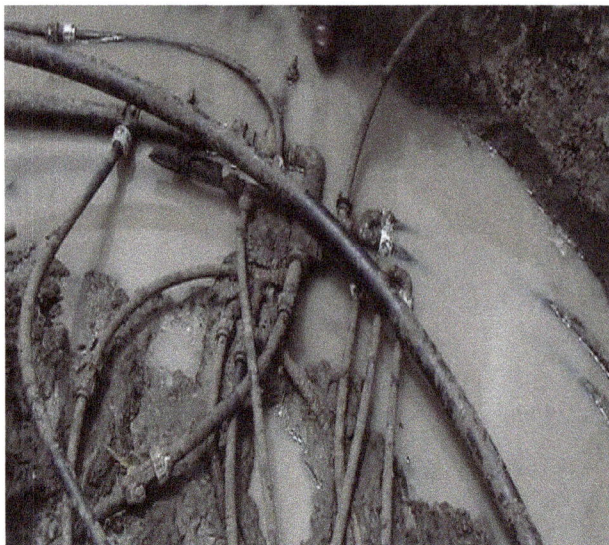

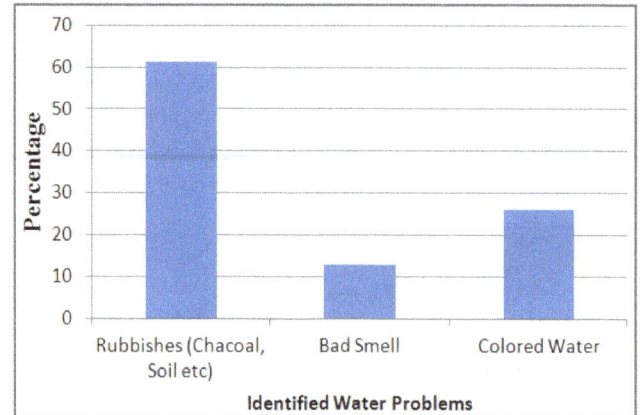

Figure 6. *Issues Identified by Customers on the Quality of Water.*

3.3. Unauthorized Connection and Meter by-Passing

The urban poor are often blamed for high levels of NRW, especially due to illegal connections. On the other hand, the poor are significantly affected by high water losses. While theft of water in low-income communities is certainly a reality in many African cities, its impact must be put in the proper perspective. Poor connections and meter bypassing (Figure 7) has been a major issue of concern and is implicated a significant number of the corporation's staff involved in helping customers to carry out illegal consumption in form of meter bypasses, meter tempering/reversal and under-reading. In addition, the corporation uses a set of forgiveness calls for both its employees and customers to, willingly, disclose illegal consumption. In one of the branches where illegal consumption was uncontrolled, more than 20% of the customer accounts were voluntarily surrendered as suspected illegal connections by staff [11, 14].

Figure 7. *Poor Connections and Meter By-Passing.*

In the discussion with AUWSAs management and the Annual reports the number of unauthorized connections increased from 129 during financial year 2009/2010 to 304. All these seem to be a big issue in having high NRW in Arusha Municipality.

4. Conclusion

Community awareness programmes about NRW to all consumers must be conducted to ensure that customer confidence in the utility's services is maintained. A key element in this is open communication. For example, the public should be able to easily contact the utility to report burst pipes, leakages, or other concerns. Water customers must be sensitized by the AUWSA on the water loss on the supply network. It was discovered that AUWSA conduct very few water customer awareness meeting and the few meetings are conducted to very few customers during the National Annual Water Week starting from 16th to 22nd March of every year which is not enough.

Acknowledge

The author is extending his acknowledgements to the Arusha Water Supply and Sewerage Authority Management team, DAWASCO management as well as Arusha city water customers who participated in this study and all those who participated in one way or another. Also I would like to extend my acknowledgement to Dr G. F. Mhando (MD) who has being constantly encouraging me to study about bacterial contamination due to wide spread of cholera in Tanzania.

References

[1] United Nations (2003), The UN World Water Development Report - Water for People, UNESCO and Berghan Books, Barcelona.

[2] WHO (2003), The Global Water Supply and Sani4tation Assessment 2000 [http://www.who.int/docstore/water_sanitation_health/Globassessment/GlobalTOC.htm], (accessed on 04/08/14).

[3] Thompson, J.; Porras, I.T.; Wood, E.; Tumewine, J.K.; Mujiwahuzi, M.R.; Katui -Katua, M. And Johnstone, N. (2000), Waiting at the Tap: Changes in Urban Water use in East Africa Over Three Decades, *Environment and Urbanization*, 12(2) 37-52.

[4] Mukoyogo, S. M. (1987), Planning and Budgeting for O&M of Rural Water Supplies, Seminar, Regional and District Water Committees.

[5] URT (2012), National Bureau of Statistics, Ministry of Finance, Tanzania in Figures, htt/www.nbs.go.tz .

[6] Farley, M. and Trow, S. (2003), Losses in Water Distribution Networks, IWA Publishing.

[7] URT (2013a), EWURA, Water Review Report 2012/13.

[8] URT (2013b), Arusha Urban Water Supply and Sewerage Authority (AUWSA) (2013), Status of Water Supply and Waste Water Disposal Services in Arusha City, Arusha, Tanzania.

[9] Lambert, A. (2001), What do we know about Pressure: Leakage Relationships in Water Distribution Systems? IWA Conference System Approach to Leakage Control and Water Distribution Systems Management, Brno, Czech Republic.

[10] APHA, (2005), *Standard Methods for the Examination of Water and Wastewater*, 18th Ed., American Public Health Association (APHA), American Water Works Association (AWWA) and Water Pollution Control Federation (WPCF), Washington, D.C.

[11] Farley M., Wyeth G., Ghazali Z.B.M., Istandar A., and Signh S. (2008), The Manager's Non-Revenue Water Handbook, A Guide to Understanding Water Losses.

[12] AUWSA (2013), Non Revenue Water Management Project for AUWSA - Arusha Urban Water Supply and Sewerage Authority.

[13] Thornton, J., Shaw, M., Aguiar, M. and Liemberger, R. (2005), "How Low Can You Go? A Practical Approach to Pressure Control in Low Pressure Systems," Conference Proceedings, IWA Leakage 2005 Conference in Halifax, Nova Scotia, Canada.

[14] Kingdom, B., Liemberger, R. and Marin, P. (2006), The Challenge of Reducing Non-Revenue Water in Developing Countries—How the Private Sector Can Help: A Look at Performance-Based Service Contracting, World Bank, Paper No. 8, Dec 2006.

10

Nitrogen and Phosphorus Dynamics in the Waters of the Great Ruaha River, Tanzania

Matobola Joel Mihale

Department of Physical Sciences, Open University of Tanzania, Dar es Salaam, Tanzania

Email address:
matobola.mihale@gmail.com

Abstract: This study assessed the levels of nitrogenous (ammonia, nitrite and nitrate) and phosphate compounds in the water column in the Great Ruaha River (GRR) in response to natural and human pressures. Water were sampled using Teflon capped plastic containers and analysed using standard methods. High levels of ammonia were observed in the Mtera dam and low in the GRR and its tributaries as well as at Ruaha Mbuyuni. Nitrite levels were observed only in Mswiswi and Luwosi tributaries in the upstream and in the Mtera and Ruaha Mbuyuni in the downstream. High levels of nitrite were observed in the Mtera dam probably due to increased microbial activity. Levels of nitrates were high in the Mtera dam and in the Mswiswi and Lunwa tributaries and low in the Little Ruaha, Ruaha Mbuyuni and in other tributaries. Nitrate levels were increasing from the tributaries to the main river and later decreasing at Ruaha Mbuyuni probably indicative of some changes in land cover, land use, soil type and groundwater level. Observed levels of phosphate followed an irregular pattern and were relatively were high probably due to anthropogenic activities and absence of Ca, Mg, Al and Fe minerals that can precipitate phosphate. Levels of nitrogen and phosphorus in the GRR were lower than the US EPA and Health Canada set standards, indicative of a natural source. Principal Component Analysis (PCA) indicate that the nitrite and ammonia are so closely related to each other than to nitrate, indicating probably that most nitrites originate from the nitrification of ammonia than the denitrification of nitrate. Similarly, nitrate and phosphate were observed in the same principal component clearly indicative of the same source, which could be fertiliser application in agricultural fields. However, increasing agricultural and livestock activities as aggravated by increasing human population pose a threat on the future dynamics of the nitrogen and phosphorus in the area.

Keywords: Nitrate, Ammonia, Nitrite, Phosphate, Great Ruaha River, Tanzania

1. Introduction

Nitrogen and phosphorus are the two key biogenic elements often controlling the overall primary production in aquatic systems. Nitrogen is a key building block in all life forms[1] and phosphorus plays a major role in biological molecules such as hormones, enzymes, proteins and vitamins[2].Nitrogen exist as a gas in the atmosphere (almost 80%) and is not available for use by most organisms. For plants and animals to be able to use nitrogen, N_2 gas must first be converted to more chemically available forms [1] through nitrogen fixation, or chemical manufacture of ammonia. Ammonia then undergoes a series of reactions that consequently produce nitrate. For example, urea is hydrolysed to ammonium (NH_4^+) under natural chemical process. Then, *Nitrosomonas* bacteria convert ammonium to nitrite (NO_2^-) which is rapidly converted to nitrate (NO_3^-) by *Nitrobacter* species. Nitrate and ammonium can then be taken up by plants or used by other soil organisms. The plants are later consumed by animals to build their body tissues (meat, milk, wool). Animal manure, sewage sludge, crop residues, and roots add organic nitrogen materials through mineralisation to give ammonium, which is further transformed to nitrate [1] as shown in Fig. 1.

Phosphorus is released into the environment through weathering of minerals [4]. In aquatic systems, phosphorus originates naturally from the dissolution of phosphate minerals and mineralisation of algae. Anthropogenically, phosphorus originates from sewage discharges, industrial effluents, as well as diffuse inputs from grazing and agricultural land. While phosphorus may exist in either the dissolved, colloidal, or particulate form, the predominant species is orthophosphate in the mono-protonated ($HPO4^{2-}$) or di-protonated ($H_2PO_4^-$) forms [5].

Phosphate is taken up by plants from soils, utilized by animals that consume plants, and returned to soils as organic phosphate residues. This organic phosphate is slowly released either as inorganic phosphate or incorporated into more stable organic materials and become part of the soil organic matter (Fig. 2). This mineralisation process is highly influenced by soil temperature and moisture. How much phosphorus is present in soil, and in what form will depend on the parent material from which the soil has been developed, amount of material that has been weathered, and land use.

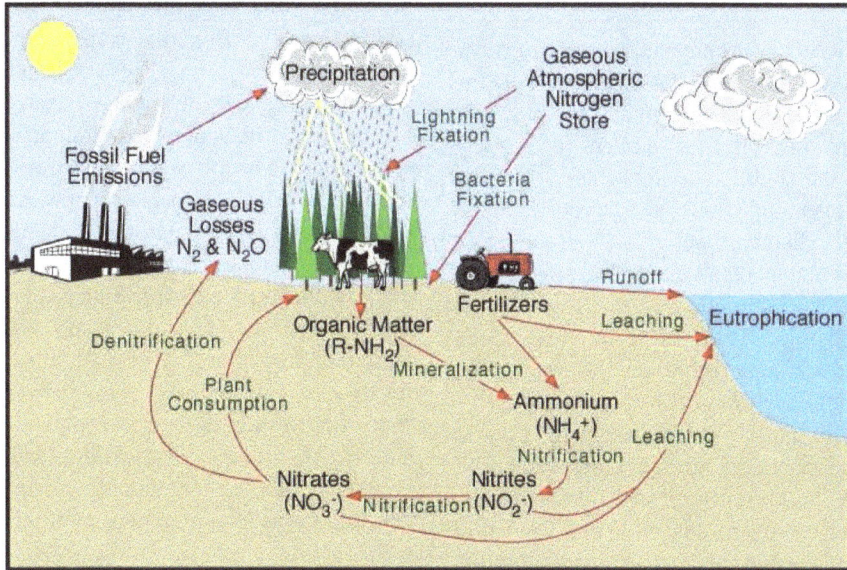

Figure 1. *The Nitrogen Cycle [3].*

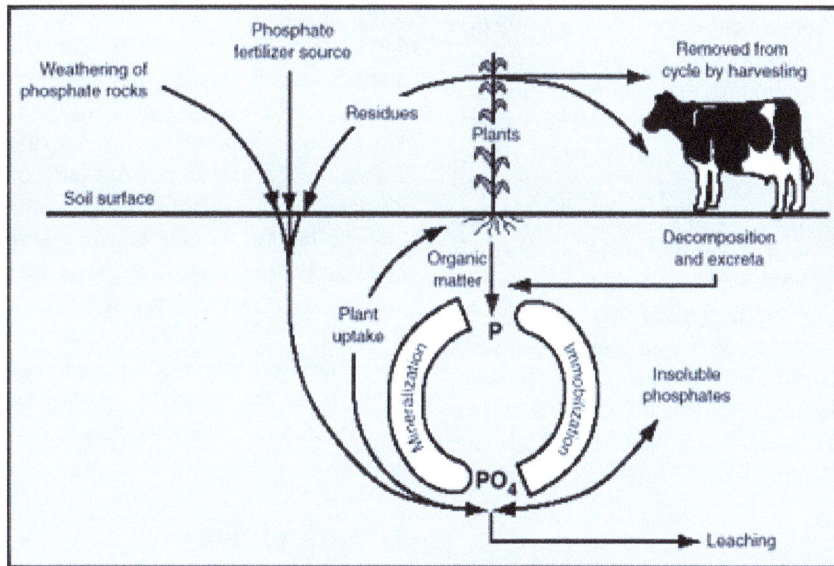

Figure 2. *Phosphorus Cycle [6].*

As fertilizer or manure comes in contact with the soil, moisture from the soil will begin dissolving the particle [9]. The presence of phosphate ions in solution is dependent on pH and is characterized by three equilibrium constants:

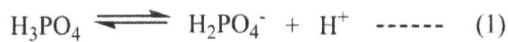

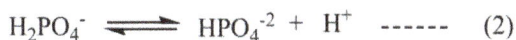

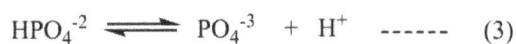

$$H_3PO_4 \rightleftharpoons H_2PO_4^- + H^+ \quad ------ \quad (1)$$

$$H_2PO_4^- \rightleftharpoons HPO_4^{-2} + H^+ \quad ------ \quad (2)$$

$$HPO_4^{-2} \rightleftharpoons PO_4^{-3} + H^+ \quad ------ \quad (3)$$

Whereas the dominant phosphate species is orthophosphoric acid (H_3PO_4) under acidic soil conditions, PO_4^{3-} is the dominant species under alkaline soil conditions. However, HPO_4^{2-} and $H_2PO_4^-$ dominate in the most natural conditions. The relative proportion of each form in a given area depends on the soil, vegetation, and land use characteristics [4].

Phosphorus loads from highlands to many aquatic systems have rapidly increased recently as a result of heavy fertilizer use and the demand to produce more food to meet the demand of the growing population. Converting wetlands to agricultural and urban lands decreased the capacity of existing wetlands to retain phosphorus. This has resulted in phosphorus enrichment of many lakes, rivers, estuaries, and coastal waters [4].

1.1. Environmental Health Effects of Nitrogenous Compounds

The nitrogenous species of environmental concern in water are NH_4^+, NO_2^- and NO_3^-. Ammonia that is released to the atmosphere tends to return to the plant–soil system relatively rapidly, thus increasing the nitrogen content of nearby vegetation. However, the bulk of atmospheric NH_3 is converted to NH_4^+, consuming some of the atmospheric acidity produced by human activities. But, high concentrations of ammonia may result in adverse effects such as leaf burn. Fine-particle ammonium aerosols can travel long distances and cause problems at a regional scale, impacting the water and air quality through wet deposition. NH_3, NO_2^- are usually returned as dry deposition [9].

High concentrations of nitrate and nitrite may lead to eutrophication of the ecosystem due to over enrichment with nutrients [10]. Thus, in the presence of high nitrate levels, aquatic plants and algae are overproduced. This eventually causes settling of dead algae and macrophytes to the bottom of the water system, bacteria proliferation, depletion of dissolved oxygen, sunlight blocking, changes in colour and odours of water and the consequent elimination or reduction of photosynthesis by submerged aquatic vegetation. Exposure to nitrate concentrations in drinking water has been associated with increased incidences of hyperthyroidism (goitre), various cancers, birth defects and increased risk of central nervous system malformations in infants whose mothers consumed drinking water with high nitrate levels [1,11]. Furthermore, nitrate in food and drinking water is reduced to nitrite in the digestive system through bacterial and other reactions [1,11]. The nitrite may then oxidise the iron in the haemoglobin molecule to form methemoglobin [11], reducing the ability of the blood to effectively transport oxygen and carbon dioxide. The nitrite can be further reduced to N-nitroso compounds (NOCs), which are carcinogenic and have been found to induce cancer in animals [1,11].

1.2. Environmental Health Effects of Phosphorus Compounds

Phosphorus, despite being an essential element and non-toxic by itself, has low solubility and as a pollutant has detrimental effects on water quality at quite low concentrations. When soil particles with phosphorus are carried to a river, phosphorus will be contained in the sediment. When soil erosion occurs, more fine particles associated with the phosphorus are removed, causing the soil to be enriched with phosphorus [9]. When rivers are polluted with phosphorus, excessive growth of algae often results. High levels of algae reduce water clarity and can lead to decreased dissolved oxygen as the algae decays [7,10], loss

of biodiversity, decreased recreational uses of waters, and the onset of harmful algal blooms. This is especially true in areas with intensive animal farming, where repeated manure applications have led to excessive accumulation of phosphorus in soils [1]. Once in the river these compounds are transformed through physico-chemical, microbial and biological processes, during their transfer along the aquatic continuum, i.e., from the river spring to estuaries, coastal waters and the ocean. These substances may have several negative effects on the aquatic ecosystem, threaten human health (e.g. poisoning, infection, etc) and impact the economical use of water resources (e.g. drinking water supply, fisheries, hydroelectric power, tourism, etc).

Superimposed by natural variability, the contemporary period is characterised by an ever-increasing utilization of materials, energy and space that emanate from the growing human population, industrial activity and intensive agricultural activities. As a result of severe increase of human impacts on aquatic ecosystems there has been an increasing flux of chemicals in the environment. Nutrient pollution is a significant problem worldwide, and recent reports from national and international organizations show that concentrations of nitrogenous and phosphorus compounds in water are increasing steadily. In developing countries like Tanzania water pollution caused by nutrients is still a serious problem as a result of increased urbanization, poor waste disposal and treatment procedures as well as intensive rural and urban agriculture. In the GRR, the nitrogen and phosphorus substances are brought to the river surface waters mainly due to leaching and erosion of agricultural soils [12]. Since water quality is not monitored systematically in Tanzania, the extent and severity of water pollution from nitrogen and phosphorus fertilisers and the consequences of this pollution is not known. However, data from the increasing use of fertilisers and the expansion of cropped area as well as the increased migration of population and livestock in the basin implies that this is a growing problem. This study was therefore intended to assess the water quality in the Great Ruaha River by determining the nutrient levels in the water column in response to natural and human pressures.

2. Methodology

2.1. Study Area

This study was conducted in the GRR Basin which is located within latitudes 7°41' and 9°25' south, and longitudes 33°40' and 35°40' east (Fig. 3). The Great Ruaha basin is characterised by hills and mountains consisting of granite and volcanic rocks and the plains consisting of alluvial fans and clay soils. The fans were developed by the depositions from the rivers disgorging from the highlands into the plains. The plains are surrounded by alluvial fans [14]. Rainfall distribution is variable and is influenced by topography. The average rainfall in this area is between 400mm to 1,200mm [15], with the most of it received in the highland areas (up to

3,000m above sea level). The water from the Ruaha River and its tributaries forming the GRR flows with a mean flow of 140 m³/s [16]. The mean daily maximum and minimum

temperature range from 28°C to 32°C and 9.5°C to 19.5°C, respectively.

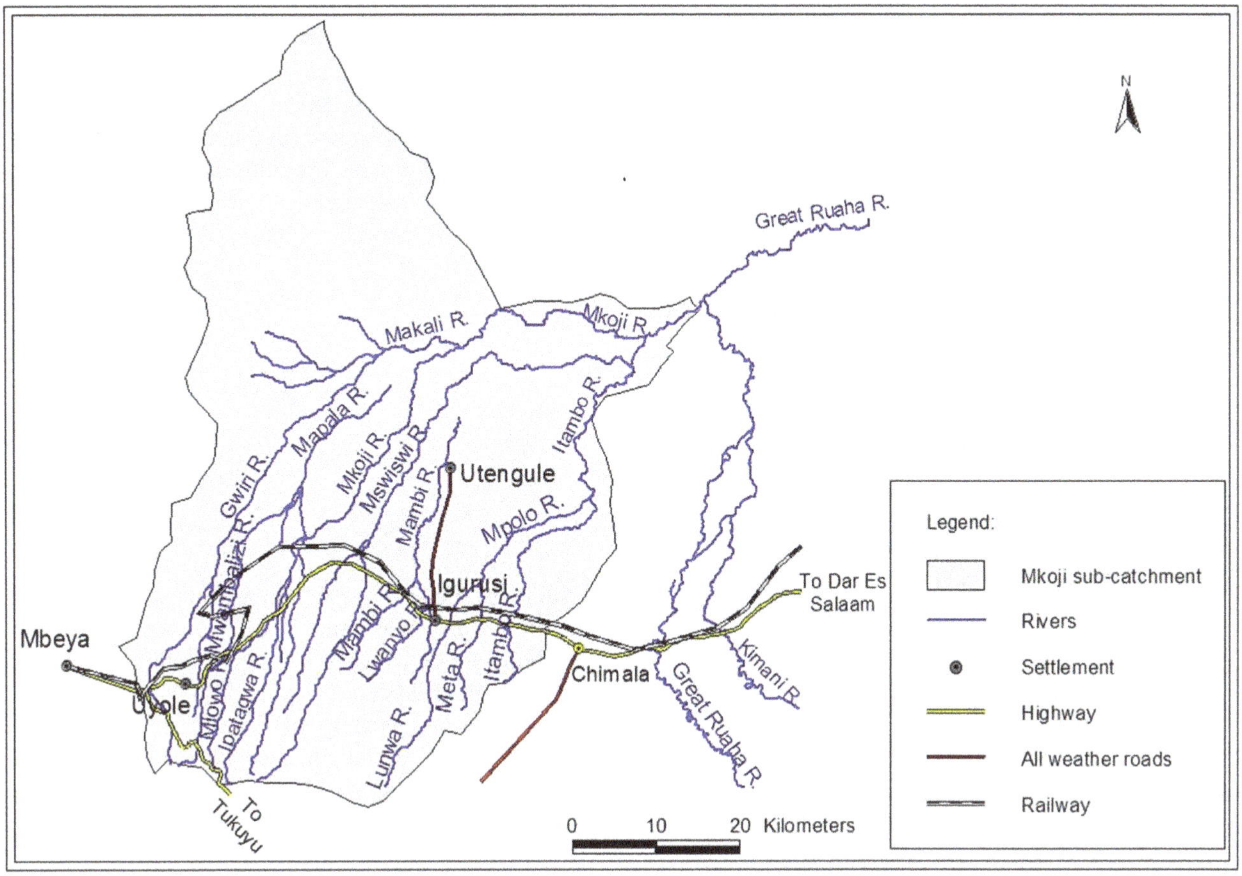

Figure 3. The Map showing the Great Ruaha River and the tributaries as adopted from [13].

The Basin is subjected to continuous change in the population trends as a result of migration and birth. Searching for pastures for livestock keepers from Hanang, Shinyanga, Mwanza, Tabora and Arusha has increased the population influx. The population growth has affected the resource base within the catchment leading to increase demand in the use of resources such as arable land, food and water [14]. The main socio-economic activities in the basin are agriculture and livestock keeping. Agriculture, which is both rain-fed and irrigated has attracted a large population (up to 90%) in the catchment [14] where paddy, maize, beans and irish potatoes are grown as food crops and pyrethrum, vegetables, sunflower, coffee, tea and sugarcane are grown as commercial crops [14,17]. Various fertilisers are applied in the catchment. Inorganic fertilisers applied in the area include Calcium Ammonium Nitrate, CAN, [17,18,20], Nitrogen-Phosphorus-Potassium, NPK, [17,18], Urea [17,18,19,20], Triple super phosphate, TSP [18], Diammonium Phosphate, DAP, [19,20] and Sulphate of Ammonia, SA [20]. Some farmers use organic manure in their farms [19,21]. Livestock keeping is common in the basin and involves local and migrant livestock keepers where cattle, goats, donkeys and sheep are kept [14]. These livestock graze in the riverbanks, in *Ihefu* and *Ifushiro* swamps [16]. There is also mining of

stone aggregates, sand, murram and soil for preparing burnt bricks [14] that take place near the riverbanks. Timber and logging from soft wood (Cyprus wood), pine, and eucalyptus and from the few indigenous species in the area has contributed to environmental degradation [14]. Wastewater from municipalities and industries within the basin are discharged into receiving water bodies with little or no treatment. For example, Iringa urban town discharges raw sewage into the GRR [23]. Nutrients from fertilisers applications during agricultural activities as well as from soil erosion add to the nutrient load reaching the river.

2.2. Data Collection and Sampling

Sampling was done in June 2008 in the Great Ruaha tributary rivers along the Makambako - Mbeya road on the upstream, Mtera dam and at Mbuyuni Bridge along the Morogoro - Iringa road on the downstream (Fig.3). Water samples were collected from pre-selected representative sites within the river basin (Table 1).

Table 1. Description of the water sampling points in the Great Ruaha River.

S/N	Sampling Site	Description of the Sampling Site
1	Mswiswi river	The river is very rock with some few debris from trees

S/N	Sampling Site	Description of the Sampling Site
2	Mambi river	The river is rocky with very little sediments.
3	Lusowi river	The river is rocky with very few debris
4	Lunwa river	The river is rocky and its water are used for drinking
5	Meta river	The river is surrounded with bamboo trees
6	Chimala river	The river is rocky. Whereas the upstream is used for washing clothes and drinking, the downstream is a dumping site for saw dusts. Some sewage effluents from the town are emptied directly into the river
7	The Great Ruaha river	The river is very steep
8	Kimani river	The river has few debris from bamboo trees
9	Igawa river	The river is rocky; people do bath and livestock drink the river water; cow dung and debris from plants were observed
10	Mbarali	Water samples were taken near estate just afterthe water comes out the farms Samples were taken just before water from farm enters the Mbarali river. Water is used for washing clothes and drinking
11	Little Ruaha river	Sample taken at the bridge near Ipogolo; people bath, wash clothes, defecate and urinate on the river banks
12	Mtera dam	Sample taken between the tunnel mouth and the barrier bridge, south of the barrier bridge and at the dam gate on Iringa side. Water for domestic purposes and livestock
13	Ruaha Mbuyuni	Sample taken at the new bridge. Water from the river bank are used for bathing, washing clothes and for livestock use.

Water samples for analysis of ammonia, nitrite, nitrate and phosphate (nutrients) were taken using Teflon capped plastic containers. Since the streams are generally shallow, samples were taken manually from the banks or from shallow spots in the stream by simply immersing the bottle in the water. Due to distance between the field and laboratory, all samples for ammonia analyses were preserved by addition of concentrated H_2SO_4 (2 mL) to maintain a pH <2. All water samples were later stored in cool box at 4°C until brought into the laboratory. In the laboratory, the samples were stored in the refrigerator at 4°C until analysis. The samples were analysed immediately within 7 days (less than 48 hours for nitrates and phosphates). Samples for nitrites and ammonia were analysed without filtration. However, samples for nitrate and phosphate determinations were filtered with 0.45 µm filters prior to analysis.

2.3. Determination of Nutrient Levels in the Samples

Water samples for nutrients were analysed using standard methods as described elsewhere [23], [24]. Identification of the nutrients was done using UV-VIS spectrophotometer as briefly described below. Quantification was done using the Beer-Lambert' equation and expressed in mg/L.

2.3.1. Analysis of Ammonia

Ammonia in water exists in the equilibrium with its ion, NH_4^+, either as a base or an acid. As a base, ammonia exists in equilibrium with ammonia ion by the equation:

$$NH_3 + H_2O \rightleftharpoons NH_4^+ + OH^- \ \text{------ (4)}$$

Similarly, it can exist as an acid by the equation:

$$NH_4^+ + H_2O \rightleftharpoons NH_3 + H_3O^+ \ \text{------ (5)}$$

The amount of ammonia, as indicated by the equations above is influenced by temperature (through its pK_a) and pH.

Ammonia reacts with hypochlorite and salicylate ion in alkaline solution to form green-blue colour (indophenol blue) in presence of nitroprusside catalyst [23].

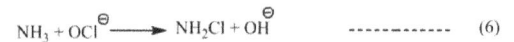

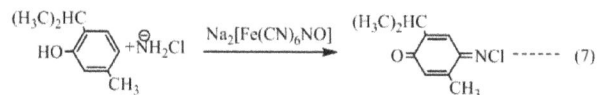

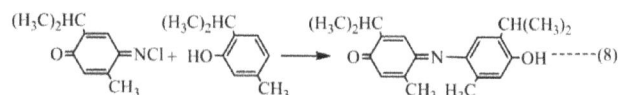

Addition of citrate compound precipitates Mg and Ca and thus eliminates interference. This reaction is conducted at room temperature with pH maintained within pH 4 to 9[23].

Sodium nitroprusside (0.2 g) and Sodium salycilate (18 g) were weighed separately and dissolved together in an Erlenmeyer flask containing distilled water (100 mL). Tri - sodium citrate dihydrate (100 g) and NaOH (10 g) were weighed and dissolved together in distilled water (500 mL). The hypochlorite solution was prepared by dissolving dichloro-sodiumcyanurate-dihydrate(0.58 g)in distilled water (100 mL). The citrate solution (100 mL) and dichloro-sodiumcyanurate-dihydrate solution (25 mL) were mixed immediately before use. Standard ammonia stock solution was prepared by dissolving ammonia chloride (0.3819 g) in distilled water (1L) forming a stock solution of 0.1 g/L NH_4-N. Other working solutions were derived by dilution of the stock solution.

Fifty (50) mL of the sample in triplicate were placed in a 100 ml Erlenmeyer flask. Then, phenolic solution (2 mL) was added and mixed. Later, hypochlorite solution (2 mL) was added and mixed thoroughly. The sample and the reagents were allowed to react for 90 minutes in the dark before its absorbance was measured in the spectrophotometer at 690 nm. Blank samples were prepared using distilled water for zero adjustment of the spectrophotometer. Calibration curve was prepared by making a series of standards with 5 different concentrations in a proper range derived from the ammonia working solution. A plot of concentration versus measured absorbance resulted to a straight line that enabled the calculation of the linear regression factor (f) from calibration curve. Concentrations of ammonia in the water samples were calculated as follows:

$$C \ (\text{mg/L NH4-N}) = f x A \quad (9)$$

where C = Concentration of NH_4-N (in mg/L) in the sample,

A = measured absorbance in water sample, f = factor from linear regression.

Sensitivity was improved by replacing a cuvette of 1 cm with 5 cm light path, and the calculation of concentration became:

$$C \text{ (mg/L NH}_4^-\text{-N)} = \frac{(fxA)}{s} \quad (10)$$

where s = thickness of cuvette (1 or 5 cm)

2.3.2. Analysis of Nitrite

Determination of nitrite in water is based on the reaction of nitrite with an aromatic amine leading to form a diazonium compound, which couples with a second aromatic amine. This azo dye is then quantified by spectrophotometry. In this method, sulphanilamide hydrochloride is used as the amino compound, where nitrite ions react with the sulphanilamide to form a diazo compound.

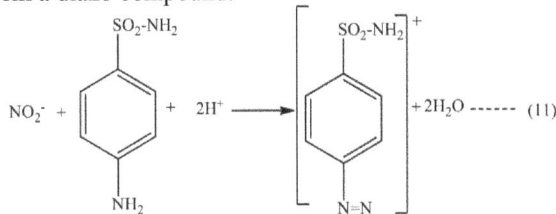

Then, the diazonium compound is coupled with N-(1-naphthyl)-ethylenediaminedihydrochloride(NEDD) in acidic conditions (strictly pH < 2) to produce a final pink-coloured complex that is quantified by spectrophotometry.

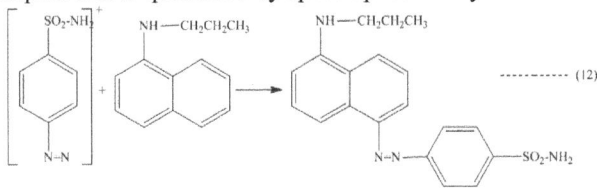

The sulphanilamide reagent was prepared by dissolving crystalline sulphanilamide (10 g) in concentrated hydrochloric acid (100 mL) and then moderately heated to accelerate the dissolution. After cooling, the solution was made up to a 1 L with distilled water and stored in the dark <8 °C. NEDD reagent was prepared by dissolving the amine dihydrochloride (0.5 g) in distilled water (500 mL) and the solution stored in a brown bottle at <8 °C. Standard nitrite stock solution was prepared by dissolving anhydrous sodium nitrite (2.482 g) in distilled water (1 L). Working solutions of NO₃-N were obtained by diluting the stock solution.

Determination of nitrite was done immediately due to high bacterial activity that may rapidly produce nitrite in the sample, hence producing erroneous results. A sample (50 mL), with or without prior filtration, was transferred into a reaction bottle (100 mL volumes) and sulphanilamide regent (1 mL) was added. After mixing and about 1 min of reaction time, NEDD reagent (1 mL) was added. The flask was shaken and the azo dye was allowed to develop for 20 min. The absorbance was measured at 540 nm within 1 h after the addition of the reagents.

Blank samples were prepared using distilled water for zero adjustment of the spectrophotometer. Calibration curve was prepared by making a series of standards with 5 different concentrations in a proper range derived from the nitrite working solution. A plot of concentration versus measured absorbance resulted to a straight line that enabled the calculation of the linear regression factor (f) from calibration curve. Concentrations of nitrite in the water samples were calculated as follows:

$$C \text{ (mg/L NO}_2^-\text{-N)} = fxA \quad (13)$$

where C = Concentration of NO_2^--N (in mg/L) in the sample, A = measured absorbance in water sample, f = factor from linear regression

Sensitivity was improved as in section 2.3.2.

2.3.3. Analysis of Nitrates

Nitrate ions react with 2,6-dimethylphenol to form 4-nitro-2,6-dimethylphenol in presence of sulphuric and phosphoric acids.

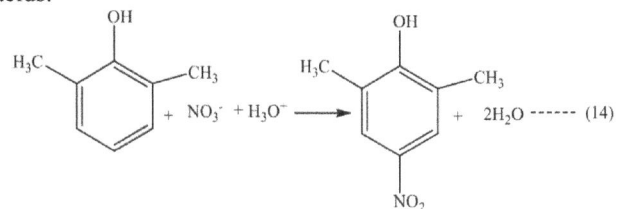

High content of organic compounds, especially aromatic compounds are reduced by sulphuric acids resulting in a yellow colour interfering with analysis. As a result, filtration and sometimes addition of amidosulfonic acid is necessary [23]. The coloured phenolic compound is further quantified by spectrophotometry.

Solutions for the analysis of nitrates were prepared as follows: Solution 1 was prepared by dissolving 2,6-dimethylphenol (1.2 g) in acetic acid (1 L). Solution 2 was prepared by mixing sulphuric and phosphoric acids in a 1:1 ratio. i.e. 1 volume of sulphuric acid is mixed with 1 volume of phosphoric acid. Standard nitrate stock solution was prepared by dissolving dry potassium nitrate (0.722 g) in distilled water (1 L). Working solutions of NO_3-N were obtained by diluting the stock solution.

To a filtered sample (5 mL) in an Erlenmeyer flask (100 ml), amidosulfonic acid (50 mg) was added and left for 10 minutes. Then, solution 2 (40 mL) was added followed by solution 1 (5 mL). The reaction mixture is left to stand for 10 minutes. The absorbance of the resulting solution was measured at 234 nm in a cuvette prior rinsed with solution 2. Blank samples were prepared using distilled water for zero adjustment of the spectrophotometer. Calibration curve was prepared by making a series of standards with 5 different concentrations in a proper range derived from the nitrate working solution. A plot of concentration versus measured absorbance resulted to a straight line that enabled the calculation of the linear regression factor (f) from calibration curve. Concentrations of nitrate in the water samples were calculated as:

$$C \text{ (mg/L NO3--N)} = fxA \qquad (15)$$

where C = Concentration of NO_3^--N in mg/L in the sample, A = measured absorbance in water sample, f = factor from linear regression.

Sensitivity was improved as in section 2.3.2.

2.3.4. Phosphate Analysis

Analysis of phosphorus in water is based on the photometric measurement of 12-phosphormolybdate or the phosphormolybdenum blue species produced when phosphormolybdate is reduced [5]. Under acid conditions, orthophosphate ions react with molybdate ions in presence of antimony catalyst to form antimony-phosphorous-molybdate complex. This complex is then reduced to phosphormolybdate blue by ascorbic acid. The absorbance of the blue compound is easily determined with accuracy using spectrophotometer. Since nitrite concentrations over 1 mg/L generate bleaching of the complex colour, amidosulfonic acid was added to minimise the effect [23].

Solutions for the analysis of phosphates were prepared as follows: solution 1 was prepared by diluting sulphuric acid (192 mL) distilled water (300 mL) while stirring. The solution is then kept cool at 20°C. Ammonium molybdate solution (solution 2) was prepared by dissolving ammonium heptamolybdate-4-hydrate (16.7 g) in distilled water (200 mL). Potassium antimony tartrate (solution 3) was prepared by dissolving potassium antimony (III) oxide tartrate hydrate (0.471 g) in distilled water (100 mL). Ascorbic acid was prepared by dissolving L (+) ascorbic acid (20 g) in distilled water (300 mL). A working mixed reagent was prepared by adding amidosulfonic acid (1.33 g/100 mL) to solution 1 followed by solutions 2 and 3. Standard phosphate stock solution was prepared by dissolving potassium hydrogen phosphate (4.3935 g) in distilled water (1000 mL). Working solutions of the phosphate standard were prepared by diluting the stock solution.

To a filtered water sample (50 mL) in the Erlenmeyer flask (100 mL), mixed reagent (1.5 mL) was added. Then, ascorbic acid solution (0.75 mL) was added and the resulting mixture was left to stand for 6 minutes. The absorbance of the resulting solution was measured at either 800 or 890 nm. Blank samples were prepared using distilled water for zero adjustment of the spectrophotometer. Calibration curve was prepared by making a series of standards with 5 different concentrations in a proper range derived from the phosphate working solution. A plot of concentration versus measured absorbance resulted to a straight line that enabled the calculation of the linear regression factor (f) from calibration curve. Concentrations of phosphates in the water samples were calculated by:

$$C \text{ (mg/L PO4-3)} = fxA \qquad (16)$$

where C = Concentration of PO_4^{-3} in mg/L in the sample, A = measured absorbance in water sample, f = factor from linear regression

Sensitivity was improved as in section 2.3.2.

2.4. Analytical Quality Assurance

The analytical as well as quality assurance and quality control, QA/QC, procedures were used throughout the analytical steps. Blanks and recovery tests were determined to check for the accuracy of the method and reliability of the results obtained. Blanks were subjected to similar treatments like normal samples during the analytical process.

2.5. Data Analysis

Means and standard deviations (STD) of the data obtained were appropriately determined. Further analysis using Pearson Correlation, Principal Component Analysis (PCA) and Student's t-test were performed. PCA was used to evaluate differences and elucidate the relationships between various parameters. Pearson correlation and PCA were performed using Statistical Package for Social Sciences (SPSS, v. 20) for Windows and t-test was determined using Microsoft excel 2007. The findings were presented in form of tables and figures as shown in the next section.

3. Results and Discussion

3.1. Rainfall Trend

The mean annual rainfall data in the Great Ruaha basin are given in Fig.4. The rainfall data showed an irregular trend with most rainfall data being above the average value of 600 mm for the period between 1958 and 1996. However, from 1997 to 2004 there has been a decreasing trend to 320 mm in 2000 and then increasing to 770 mm in 2004 (Fig.4). Reports have shown that the average rainfall for year 2013 from two stations in the Great Ruaha basin was 445.4 mm at Msembe and 741.9 mm at Iringa Maji, indicating a decrease change of 88.4% at Msembe and increase change of 106% at Rufiji Maji [15]. This fluctuating rainfall in recent years and could have an impact on the nutrient dynamics in the GRR.

3.2. Levels of Nutrients in the Great Ruaha River

3.2.1. Ammonia

Observed levels of ammonia were low in the GRR and tributaries and high at the Mtera area, probably due to cumulative effect. The levels at Ruaha Mbuyuni was more or less the same as the levels in the tributaries (Fig. 5). Relatively high levels of ammonia in aqueous environments are not usually observed due to the nitrification process by Nitrosomonas species to form nitrite and then by Nitrobacter species that convert the nitrite to nitrate [1]. The overall nitrification equation is:

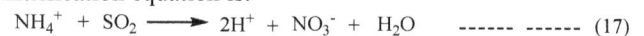

$$NH_4^+ + SO_2 \longrightarrow 2H^+ + NO_3^- + H_2O \qquad (17)$$

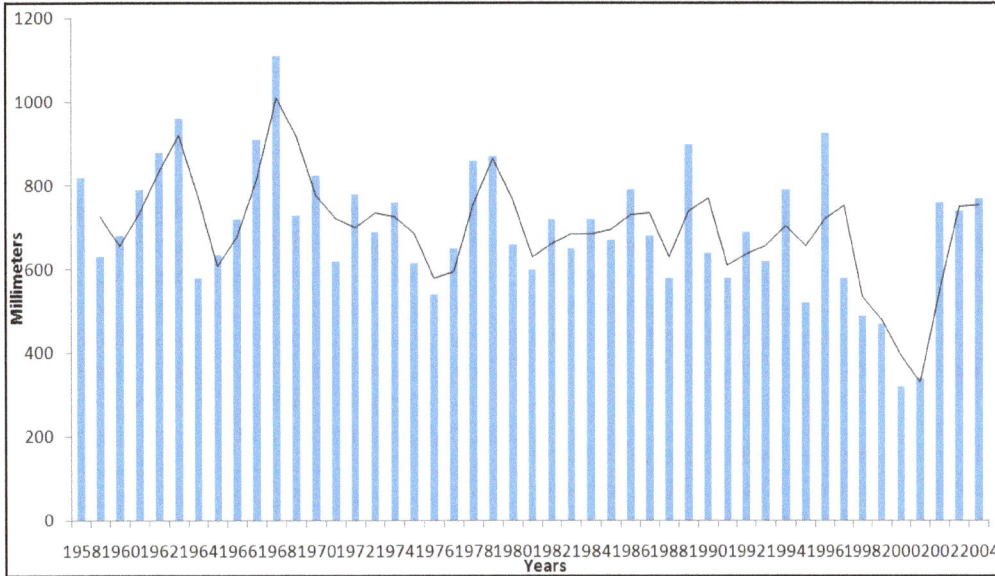

Figure 4. *Mean annual rainfall data for the Great Ruaha Basin (1958 – 2004).*

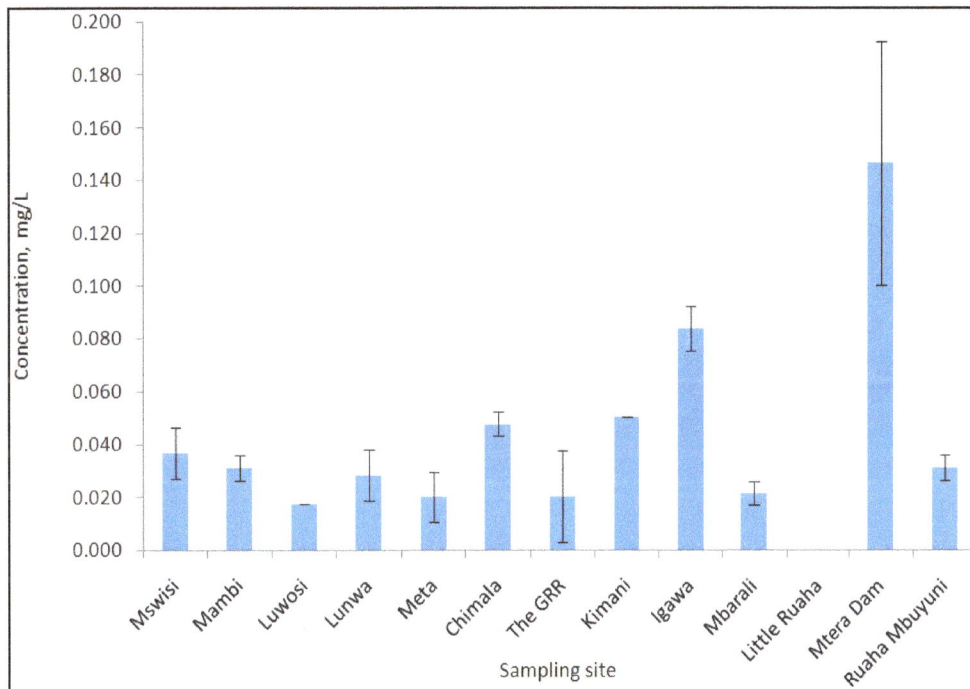

Figure 5. *Concentration of NH$_3$-N (mg/L) in the Great Ruaha River and its tributaries.*

Generally, the ammonia levels in rivers are less than 6 mg/L. In presence of good amount of oxygen, favourable temperature, favourable pH (acidic condition) and microbial population, the levels of ammonia in a given area will always be low [1]. The ammonia levels observed in the study were generally low as expected, despite the fact that levels in Mtera were slightly higher than in other stations. This is probably due to the favourable conditions for microbial growth in the area. Furthermore, volatility of ammonia increases with increasing temperature [9]. Therefore with the high temperatures of the tropical GRR, it is not surprising to observe low levels of ammonia in the area. The observed ammonia levels were generally lower than the set standard

criteria of 0.05mg/L [26] in the upstream and higher than the set value in the Mtera dam. The ammonia levels in the Mtera were even above 1.0 mg/L, which is indicative of suspicious recent pollution from surrounding areas.

3.2.2. Nitrite

Nitrite levels were detected only in Mswiswi and Luwosi tributaries of the GRR in the upstream and in the Mtera and Ruaha Mbuyuni in the downstream (Fig.6). In other areas the levels were below the detectable amount. High levels were observed in the Mtera as compared to other sampling stations.

Nitrite occurs either in the reduction process of nitrate or in the conversion of ammonia to nitrate. Hence, in oxygenated surface water, there is little nitrite present and nitrite concentrations are generally below 0.5 mg/L NO_2^--N because it is readily oxidised to nitrate. High concentrations of nitrite are usually linked to microbial activity, but they may also indicate polluted water [11]. The observed levels of nitrite were higher at the Mtera Dam and lower in river tributaries. This is indicative of the increased biological activities at the Mtera area particularly *Nitrosomonas* bacteria that tend to increase nitrite amount in the water. However, the high concentration levels observed in the Mtera samples were still below the recommended set values of 3 mg/L for nitrite in water, which is equivalent to 1 mg/LNO_2^--N [27].

3.2.3. Nitrate

Slightly high levels of nitrates were observed in the Mswiswi and Lunwa tributaries of the GRR. In other tributaries (Mambi, Meta, Chimala, Kimani, Igawa and Mbarali), the levels were low. The levels were even lower in the Little Ruaha but generally high in the Mtera dam and slightly low at Ruaha Mbuyuni (Fig. 7). This trend indicate that the levels were increasing from the tributaries to the main river as observed by the high concentrations of nitrate in the Mtera dam. However, there is a decrease at Ruaha Mbuyuni indicating some changes in land cover soil type, groundwater level and land use. High concentrations of nitrate are associated with carbonate-type soils, shallow groundwater, and increased use of the land for agriculture [11]. The high levels of nitrate in the GRR could be due to presence of carbonated soils as well as increased anthopogenic activities [28] i.e. agricultural use of the land. The observed levels of nitrate the Mswiswi and Lunwa Rivers and in the Mtera dam could be indicative of

anthropogenic activities on the upstream of the River. Fortunately, the anthropogenic contribution of the pollutant is still not so large to pose any potential environmental impact.

Nitrate is the most common contaminant of water. When present, nitrate does not volatilise, and it is likely to remain in water until consumed by plants or other organisms. Because of its stability and solubility, elevated level of nitrate in water may also indicate the presence of other contaminants, microbial pathogens and pesticides [11]. While nitrate concentration in deep seawater and natural unpolluted groundwater is usually below 0.05 mg/L NO_3^--N, the concentrations in shallow groundwater and surface streams range from less than 0.1 mg/L NO_3^--N to 20 mg/L NO_3^--N [27] depending on soil type, land use practices and depth. In areas of high anthropogenic impacts, the levels can be as high as 30 mg/L NO_3^--N. This implies that although nitrate and nitrite are naturally occurring forms of nitrogen, elevated levels in aquatic systems usually result from human activities [11]. The observed nitrate levels in the GRR ranged from 0.01 mg/L at Meta River to 0.51 mg/L at Mswiswi River. The levels of nitrite and nitrate were both less than 1 mg/L, indicating no agricultural contamination from either fertilisers or seepage [25]. The observed levels in the area are within the ranges of nitrates naturally occurring in surface water, indicative of a natural source. The United States Environmental Protection Agency (US EPA) and Health Canada have set a maximum (allowable) contaminant level (MCL) of 45 mg/L (equivalent to 10 mg/L NO_3^--N) for nitrates in drinking water [1], [26]. The results indicate that NO_3^--N levels observed in the GRR are lower than the US EPA and Health Canada set standards.

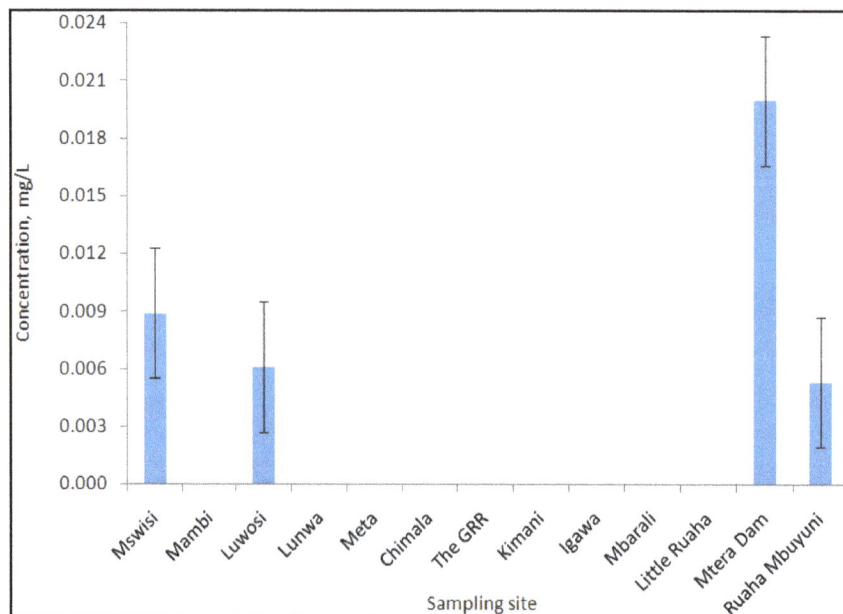

Figure 6. *Concentration of NO_2^--N (mg/L) in the Great Ruaha River and its tributaries.*

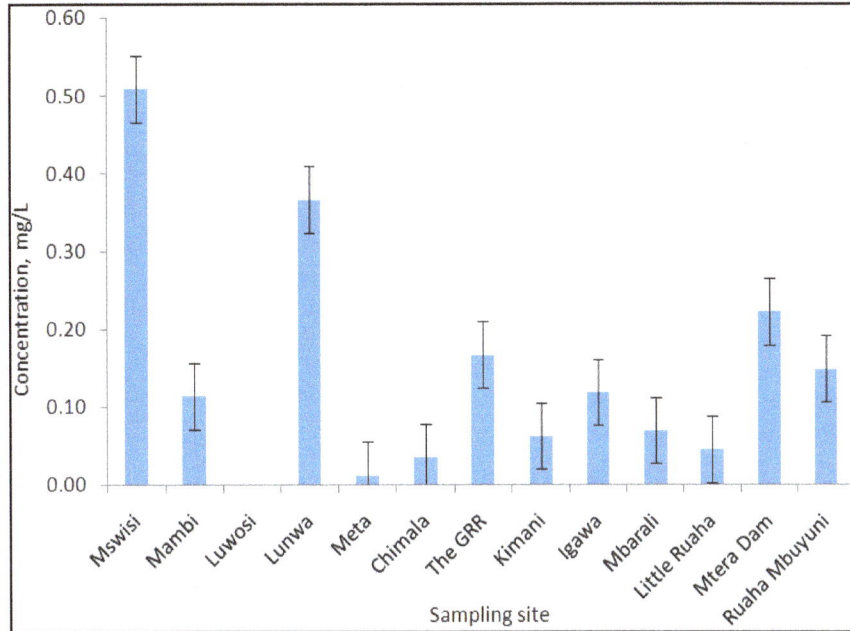

Figure 7. Concentration of NO_3^--N (mg/L) in the Great Ruaha River and its tributaries.

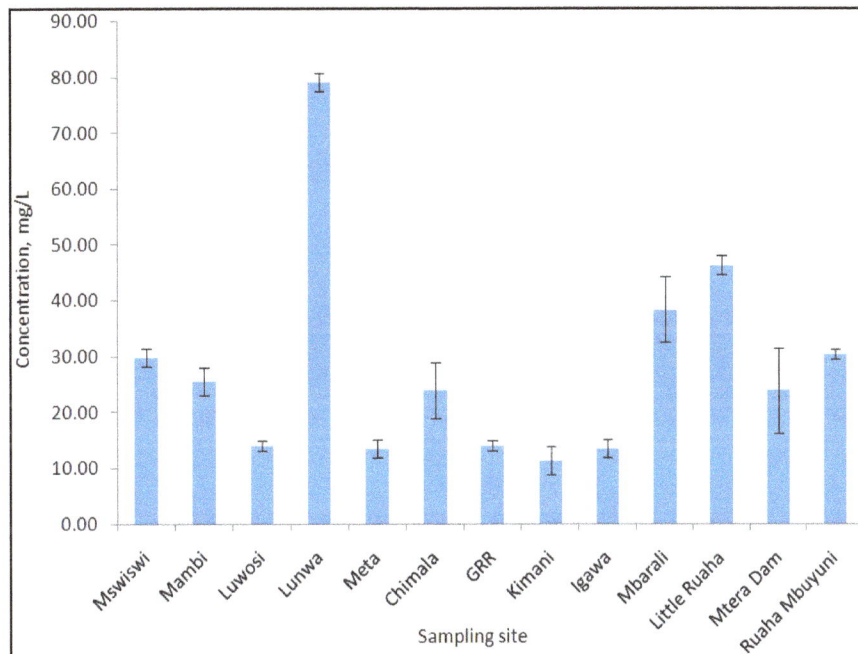

Figure 8. Concentration of phosphate (mg/L) in the Great Ruaha River and its tributaries.

The removal of dissolved forms of the nitrogenous compounds is regulated by various biogeochemical reactions in soil and overlying water column. The relative rates of these processes are affected by physicochemical and biological characteristics of the soil and water column as well as the organic substrates present [4]. Humans have a direct impact on nitrogen inputs to terrestrial systems mainly by increasing the productivity and area sown with addition of fertilisers and growing of N_2-fixing legume crops and pastures. Anthropogenic sources of nitrate and nitrite include intensive use of chemical nitrogenous fertilisers, improper disposal of plant and animal waste, municipal and industrial wastewater discharge, sewage disposal systems and landfills. In the GRR, agriculture is the largest contributor to the pollution of water because of the use of nitrogen fertilisers and also because of the influx of livestock in the area, which produces a great amount of manure as observed in a similar study [11]. As a result of the increased anthropogenic pressures in the GRR, the levels of nitrogenous nutrients in water may pose a threat in the future.

3.2.4. Phosphorus

The observed phosphorus levels in the GRR and its tributaries followed an irregular trend. Highest phosphate levels of 79.11 mg/L were observed at the Lunwa River and lowest levels of 11.31 mg/L were observed at Kimani River (Fig. 8). Slightly higher levels were observed in Mbarali, Little Ruaha, Mtera and Ruaha Mbuyuni.

Many phosphate compounds are not very soluble in water; therefore, most of the phosphate in natural systems exists in solid form. However, soil water and surface water (rivers and lakes) usually contain relatively low concentrations of dissolved (or soluble) phosphorus [10]. Depending on the types of minerals in the area, bodies of water usually contain about 10 µg/mL or more of dissolved phosphorus as orthophosphate [7]. The observed levels in the GRR were higher than the expected values probably due to land use near the tributaries [7] and absence of Ca, Mg, Al and Fe minerals in the soil that can react with the phosphate in water forming solid compounds that are relatively available for plants [8]. In addition, there are probably favourable anaerobic (reducing) conditions that reduce the insoluble Fe^{3+} that can adsorb the phosphate to soluble Fe^{2+} thereby increasing the phosphate levels in the aqueous medium [25]. The observed high levels of phosphate in the waters of the GRR might be due to the reducing conditions that exist in the area.

3.3. Nutrient Source Analysis by Principal Component Analysis (PCA)

Multivariate analysis can be used to identify similarities and differences between nutrient pollutants in samples as a means to detect possible sources. In order to determine the

relationship between nutrients, PCA after varimax rotation was employed. A principal component (PC) was considered significant when its eigenvalue was greater than 1. The measured nutrient values were used as variables (total 4), with the concentrations of the nutrients in the different sampling stations as objects (total 64). The results of the PCA analysis are presented in Table 2. Based on the loading distribution of the variables, the PCA results indicated that the variables can be represented by two principal components that accounted for 79.3 % of the total variance in the original data sets (Fig.9).

Table 2. *Rotated Principal Component (PC) Matrix.*

	Principal Component (83.9%)	
	PC 1 (44.8%)	PC 2 (39.1%)
nitrite	**0.920**	0.139
Ammonia	**0.907**	-0.071
Phophorus	-0.192	**0.891**
nitrate	0.291	**0.863**

As we could expect, ammonia, nitrite and nitrate constituted one related group (PC 1), while the phosphate and nitrate formed the other group (PC 2). This clearly indicates that the sources of the ammonia and nitrite on one hand differ from that of phosphate and nitrate in the other. This is evidenced by the positive correlation between ammonia and nitrite ($r^2 = 0.70$, p = 0.01) and positive correlation between nitrate and phosphate ($r^2 = 0.55$, p = 0.03).

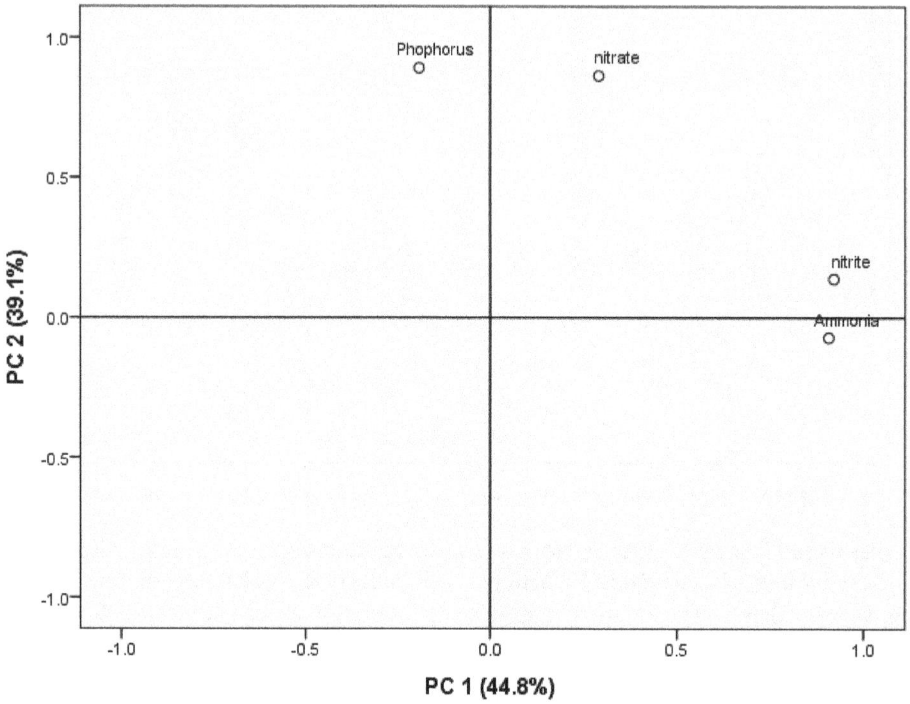

Figure 9. *A two-dimensional score pot of the nutrients in the Great Ruaha River.*

Fig. 9 clearly indicates that ammonia and nitrite have a common source, different from that of phosphorus and nitrate as expected. In addition, PCA indicate that the nitrite and ammonia are so closely related to each other than to nitrate, indicating probably that most nitrite originate from the

nitrification of ammonia than the denitrification of nitrate. In addition, existence of nitrate and phosphate in the same PC clearly indicate that these nutrients have the same source. In the GRR, this could probably be fertiliser application in agricultural fields.

4. Conclusion

Ammonia, nitrite, nitrate and phosphate nutrients in water from the GRR and its tributaries and their source have been determined.The findings have indicated that the nutrients in the basin originate from natural source. However, application of fertilisers in agriculture has been identified as a potential anthropogenic source in the future. As a result, increasing agricultural and livestock activities in the basin as aggravated by the increasing population can pose a threat on the future nutrient dynamics in the GRR basin as a whole.

References

[1] Berkowitz B., Dror, I., Yaron, B (2008), Contaminant geochemistry: transport and fate in the subsurface environment, Springer-Verlag Berlin Heidelberg.

[2] Bünemann, E. K and Condron L. M (2007). Phosphorus and sulphur cycling in terrestrial ecosystems *in* Marschner, P and Rengel, Z (Eds). Nutrient cycling in terrestrial ecosystems, Springer-Verlag Berlin Heldelberg.

[3] Fundamentals of physical geography (2nd Edition) Accessed on 20th November 2014 at http://www.physicalgeography.net/fundamentals/contents.html.

[4] Redd, R, and Delaune R. D., (2008). Biogeochemistry of wetlands, science and applications, Taylor and Francis Group, LLC, CRC Press.

[5] McKelvie, I.D (2000). Phosphates in Nollet L.M.L (Ed). Handbook of water analysis, food science and technology, CRC Press, Marcel Dekker, Inc.

[6] Phosphorus Nutrient Management. Accessed on 20th November 2014 at http://www.ctahr.hawaii.edu/mauisoil/c_nutrients02.aspx.

[7] Adeyemo, O.K., Adedokun, O.A., Yusuf, R.K., Adeleye, E.A., (2008). Seasonal changes in physico-chemical parameters and nutrient load of river sediments in Ibadan city, Nigeria, Global NEST Journal, Volume 10, No 3, pp 326-336.

[8] Busman, L., Lamb, J., Randall, G., Rehm, G and Schmitt, M (2009). The nature of phosphorus in soils: Phosphorus in the agricultural environment, ww-06795-GO, Regents of the University of Minnesota. Accessed at http://www.extension.umn.edu/distribution/cropsystem/DC6795.html on 3rd June 2014.

[9] McNeill, A., and Unkovick, M., (2007). The Nitrogen cycle in the Terrestrial Ecosystems *in* Marschner, P and Rengel, Z (Eds). Nutrient cycling in terrestrial ecosystems, Springer-Verlag Berlin Heldelberg.

[10] Worsfold, P. J., Monbet, P. Tappin A. D., Fitzsimons, M. F., Stiles, D. A., McKelvie, I. D., (2008). Characterisation and quantification of organic phosphorus and organic nitrogen components in aquatic systems: A Review, Analytical ChimicaActa, Volume 624, 37-58.

[11] Oms, M. T., Cerdà, A and Cerdà, V (2000). Analysis of nitrates and nitrites *in* Nollet, L. M. L., (Ed.), Handbook of water analysis, food science and technology, CRC Press, Marcel Dekker, Inc.

[12] Beaudoin, N., Saad, J., Van Laethem, C., Maucorps, J.,

Machet, J.M. and Mary, B., (2005). Nitrate leaching in intensive arable agriculture in Northern France: effect of farming practices, soils and crop rotations. Agriculture, Ecosystem and Environment, 11:292-310.

[13] Mwakalila S (2011) Assessing the hydrological conditions of the Usangu Wetlands in Tanzania, Journal of Water Resource and Protection, Vol.3 (12),DOI:10.4236/jwarp.2011.312097.

[14] Sosovele, H., Ngwale, J.J., (2002). Socio-economic root causes of theloss of biodiversity in the Ruaha catchment area. Report submitted to WWF-TANZANIA. http://assets.panda.org/downloads/rcareportruaha.doc. Accessed on 23rd October 2014.

[15] Rufiji Basin Water Board (2013). Rufiji basin annual hydrological report - 2012/2013, Iringa, Tanzania.

[16] Kashaigili, J.J. and Rajabu, K.R.M (2003). Draft fact sheet on ecohydrology of Great Ruaha River, Raising irrigation productivity and releasing water for intersectoral needs (RIPARWIN): A DFID and IWMI − funded River Basin Management Research Project in Tanzania. http://www.iwmi.cgiar.org/Africa/files/RIPARWIN/05/_downloads/Fact-Factsheet_EcoHyd_GRR.doc. Accessed on 22nd October 2014.

[17] URT -United Republic of Tanzania, Ministry of Agriculture and Food Security (2002). River basin management and smallholder irrigation project, smallholder irrigation improvement component: Initial environmental examination at Mapogoro irrigation scheme, Draft Paper, E968, Volume 11, Environmental cell unit http://www-wds.worldbank.org/external/default/WDSContentServer/WDSP/IB/2004/07/26/000160016_20040726123351/Rendered/INDEX/E9680vol011.txt. Accessed on 22nd October 2014.

[18] URT - United Republic of Tanzania, Ministry of Agriculture and Food-Security (2003). River basin management and smallholder irrigation improvement project (RBMSIIP): Smallholder Irrigation Improvement Component: Ruanda Majenje Irrigation Scheme, Environmental Audit, Draft Report E968, Volume 5. Environmental Resources Consultancy, Consulting environmental conservation. Natural resources management and capacity building. http://www-wds.worldbank.org/external/default/WDSContentServer/WDSP/IB/2004/07/26/000160016_20040726121644/Rendered/PDF/E9680vol05.pdf. Accessed on 24th October 2014.

[19] URT - United Republic of Tanzania, Ministry of Agriculture and Food-Security (2003). River basin management and smallholder irrigation improvement project (RBMSIIP): Smallholder Irrigation Improvement Component: Igomelo Irrigation Scheme, Environmental Audit, Draft Report E968, Volume 4. Environmental resources consultancy, consulting environmental conservation. Natural resources management and capacity building. http://www-wds.worldbank.org/external/default/WDSContentServer/WDSP/IB/2004/07/26/000160016_20040726121315/Rendered/PDF/E9680vol04.pdf. Accessed on 24th October 2014.

[20] URT - United Republic of Tanzania, Ministry of Agriculture and Food Security (2003). River basin management and smallholder irrigation improvement project (RBMSIIP), Small holder irrigation improvement component; Nyamahana irrigation scheme: Environmental Audit, Draft Report E968, Volume 8, Environmental Resources Consultancy. http://www-wds.worldbank.org/external/default/WDSContentServer/WDSP/IB/2004/07/26/000160016_20040726122452/Rendered/INDEX/E9680vol08.txt. Accessed on 22nd October 2014.

[21] URT - United Republic of Tanzania, Ministry of Agriculture and Food-Security (2003). River basin management and smallholder irrigation improvement project (RBMSIIP): Small holder irrigation improvement component: Mangalali irrigation scheme, environmental audit, Draft Report E968, Volume 3. Environmental resources consultancy, consulting environmental conservation. Natural resources management and capacity building. http://www-wds.worldbank.org/servlet/WDSContentServer/WDSP/IB/2004/07/26/000160016_20040726121017/Rendered/INDEX/E9680vol03.txt Accessed on 22nd October 2014.

[22] URT - United Republic of Tanzania, (2006). Water resources assistance strategy: Improving water security for sustaining livelihoods and growth, Water and Urban Unit 1, Africa Region, Report No 35327, The World Bank.

[23] Matsché, N and Kreuzinger, N (2001). Manual on chemical water analysis for the IPGL course / Water chemistry, Institute for Water Quality and Waste Management, Department for Chemistry and Microbiology, Vienna University of Technology, Austria.

[24] Greenberg, A.E., Connors, J.J. and Jenkins, D., Eds (2005). Standard methods for the examination of water and wastewaters, 20th Edition, American Public Health Association(APHA), American Water Works Association (AWWA) and Water Pollution Control Federation (APCF).

[25] Weiner E (2008). Applications of environmental aquatic chemistry: A Practical Guide, 2nd Edition, CRC Press, Taylor & Francis Group.

[26] Health Canada (2013). Guidelines for Canadian drinking water quality: Guideline technical document — Nitrate and Nitrite. Water and Air Quality Bureau, Healthy environments and consumer safety branch, Health Canada, Ottawa, Ontario. (Catalogue No H144-13/2-2013EPDF).

[27] WHO, (2011). Nitrate and nitrite in drinking-water: Background document for development of WHO Guidelines for Drinking-water Quality, Geneva.

[28] Tening, A. S., Chuyong GB., Asongwe G. A.,Beatrice Ambo Fonge B. A., Lifongo, L. L., and Tandia B. K., (2013). Nitrate and ammonium levels of some water bodies and their interaction with some selected properties of soils in Douala metropolis, Cameroon, African Journal of Environmental Science and Technology, Volume 7 Number 7, pp. 648-656.

11

Comparative analysis of heavy metals from groundwater sources situated in Keko and Kigogo residential areas, Dar es Salaam

Isabela Thomas Mkude

Dept. of Environmental Studies, Open University of Tanzania, Dar es Salaam, Tanzania

Email address:

isabela.thomas@yahoo.com

Abstract: The study assessed in comparison the concentrations of heavy metals in water samples (n=12) collected from groundwater sources at Keko and Kigogo areas. Three trace elements; Copper, Zinc and Lead (Cu, Zn and Pb). Analysis was done once in dry season and once in rainy season at Ardhi University laboratory. The water samples were taken in 4oc equipment for transportation. Results show low concentrations of Cu and Zn for both seasons and study areas. Pb concentrations were detected beyond the WHO permissible level (min. 0.01mg/l and max. 0.35mg/l). This presence of high concentrations may be due to urban agriculture activities conducted around the areas surrounding Msimbazi river, and also presence and use of onsite facilities (septic tanks and pit latrines). Discharge and runoff collection from garages and industrial wastes is another reasons thought for this high amount of heavy metals. There is need of adopting cheap water treatment technology for domestic purposes so as to protect people from being affected by the use of untreated water directly.

Keywords: Heavy Metals, Trace Elements, Zinc, Lead, Copper, Groundwater

1. Introduction

Water is essential to human life. In all its use, quality and quantity are the most important terms to consider. Quantity of water differs from place to place due to geographical and climate differences, land uses. On the other hand, the quality of water may differ due to pollution from nutrient pollutant flows, disposal of storm water, sewage and other urban wastes. Besides the quality of water due to climate changes as well as changes in land use (particularly in estuaries, bays and shallow enclosed seas) (11). In Tanzania surface water and ground water are used to supply water in urban and rural areas. The use of surface and ground water in different parts of the country differs caused by differences in topography, rainfall pattern and climate, also in each region. However, ground water is a major source of water for many rural areas and those lacking piped water in Tanzania. It is the most viable alternative supplements especially the central and northern parts of the country in the drier regions. Ground water, these consists of both boreholes and shallow wells a source believed to be safe due to natural protection by the aquifers may be highly polluted by different sources. Pollution from industrial discharges and domestic seepage like the use of latrines and septic tanks or that from improper management of solid waste, lack of sanitation or even open defecation practices may pose a serious pollution to the groundwater (10).

Heavy metals are sometimes called "trace elements", they become of particular interest in recent decades within the framework of environmental investigation. This has without doubt been due to the fact that highly sensitive analytical procedures are available for determining and detecting metal content with high precision (13).

1.1. Problem Statement

Lack of adequate water supply may cause serious health problem and/or even loss of human life. Quality of water used for drinking and other domestic purposes is a fundamental element to be considered during the selection of water sources. The rapid increase of population in the City also increases number of people living in unplanned areas, and this increases also the use of pit latrines and septic tanks as the main sanitation facilities. Improper construction, without consideration of proposed distances to water sources and poor or lack of maintenance of these sanitation facilities causes contamination to the ground water used for domestic purposes and hence poses risk to human health.

1.2. Objective

The main objective of this study was to quantify the selected heavy metals (Cu, Pb and Zn) from groundwater sources at Keko and Kigogo that used for domestic purposes and compare results from two areas.

2. Material and Methods

2.1. Research Design

This research was designed and conducted as a cross sectional multiple case studies, by using qualitative and quantitative data collection approaches. The cases of these studies were Keko and Kigogo areas in different two municipalities of Dar es Salaam City. It was cross sectional design since it referred to a single reference period of study. The phenomenon within the case was the heavy metals in groundwater that used for domestic purposes. The relevant stakeholders involved throughout of the study were the local communities in these two study areas, Keko and Kigogo.

2.2. Study Areas

Keko and Kigogo residential areas were selected due to the fact that, they are located in different municipalities, (Temeke and Kinondoni respectively) but yet they have some common characteristics. They are both unplanned residential areas that people who are living there are in low to middle economic status. Piped water system does not reach to every house; hence most of families find alternative sources of water. Groundwater seems to be the next alternative available since it is cheap and easily available. There are hand dug wells located to their living premises that contain good quantity of water for domestic use, these are normally private owned, while the boreholes (>25m deep) are public owned and located in area that easily accessed by many people through water kiosks. The boreholes are good source of water to these communities since they cost cheaper compared to piped water and its water thought to be cleaner than that from hand dug wells.

2.3. Data Collection Methods

Since it was case study research, multiple data sources were involved to obtain a desirable relevant data. Primary data and secondary data were collected from different data sources and used during the study. Primary quantitative data were obtained by laboratory analysis, while qualitative data were obtained by short interviews with water source owners before water sampling. Secondary data obtained from different previous data sources like research reports, journals, text book, census book, planning reports, and all relevant information to gather all required data mainly through internet. The literature review was conducted to obtain the most relevant information on ground water quality and use in Dar es Salaam.

2.4. Experimental Methods

2.4.1. Sampling and Analysis

Water samples were collected in polyethylene bottles from total of 12 groundwater sources (4 shallow wells and 8 boreholes) from both study areas and analyzed for heavy metals (Cu, Zn and Pb). Duplicate water samples were collected from each source by using grab sampling method and analysis was done in two phases (rainy and dry) seasons. Samples for analysis were stored in sterilized polyethylene bottles at 4 °C and transported for analysis at Ardhi University Laboratory. It is recommended that for better results, analysis need to be performed within 20 to 30 minutes after sampling, and not more than 24 hours after sampling (19). Analysis was done using atomic adsorption instrument. All parameters were analyzed according to the Standard Methods for the Examination of Water and Wastewater (3). Microsoft Excel 2007 program was used for analysis of statistical values; average and their standard deviations.

2.4.2. Sampling Points

Total number of 12 sampling points and their general characteristics are shown in table 1. Boreholes are denoted as BH and Shallow wells as SW.

Table 1. General characteristics of sampling points.

Study Area	Label		Depth(m)	Number of water users/day	Remarks
Kigogo	SW1	Mbuyuni A	5	15	Not lined, used by 2 families
	SW2	Mbuyuni A	8	30	Stones, used for vegetable irrigation
	BH1	Rutihinda	60	1500	Pumped with 10,000lt storage tank. Extended to 8 kiosks
	BH2	Kigogo CCM	60	3000	Pumped with 10,000lt storage tank. Extended to 4 kiosks
	BH3	Police	65	3500	Pumped with 10,000lt storage tank.
	BH4	Mburahati	60	3700	Pumped with 10,000lt storage tank.
Keko	SW3	Keko Mwanga	3	10	Not lined
	SW4	Keko Machungwa	4	80	Cement, used for different domestic purposes
	BH5	Magereza	50	10-15 water tankers	Pumped with 15,000lt storage tank
	BH6	Unubini	35	1000	Pumped with 10,000lt storage tank.
	BH7	Magurumbasi	40	1500	Pumped with 15,000lt storage tank.
	BH8	Toroli	25	1200	Pumped with 10,000lt and 5000lt storage tanks.

3. Results and Discussion

Tables 2 and 3 give summary of mean values of selected trace elements (Cu, Zn and Pb) from all sampling points during dry and rainy seasons respectively.

Table 2. Average values for analysed heavy metals from groundwater on dry season.

	DRY SEASON			
	Sampling Points	Cu (mg/l)	Pb(mg/l)	Zn(mg/l)
Kigogo	SW1	0.010	0.010	0.092
	SW2	0.010	0.010	0.010
	BH1	ND	0.090	0.059
	BH2	ND	0.311	0.054
	BH3	ND	0.081	0.058
	BH4	ND	0.124	0.054
Keko	SW3	0.010	0.270	0.012
	SW4	0.010	0.194	0.010
	BH5	ND	0.309	0.235
	BH6	ND	0.350	0.048
	BH7	ND	0.078	0.046
	BH8	ND	0.232	0.035

ND=Not detected

Table 3. Average values for analysed heavy metals from groundwater on rainy season.

	RAINY SEASON			
	Sampling Points	Cu(mg/l)	Pb(mg/l)	Zn(mg/l)
Kigogo	SW1	0.01	0.02	0.076
	SW2	0.01	0.03	0.01
	BH1	0.013	0.01	0.059
	BH2	0.01	0.034	0.054
	BH3	0.01	0.01	0.058
	BH4	0.01	0.035	0.054
Keko	SW3	0.01	0.32	0.015
	SW4	0.01	0.169	0.01
	BH5	0.06	0.078	0.28
	BH6	0.01	0.28	0.041
	BH7	0.01	0.075	0.048
	BH8	0.01	0.22	0.033

3.1. Copper (Cu)

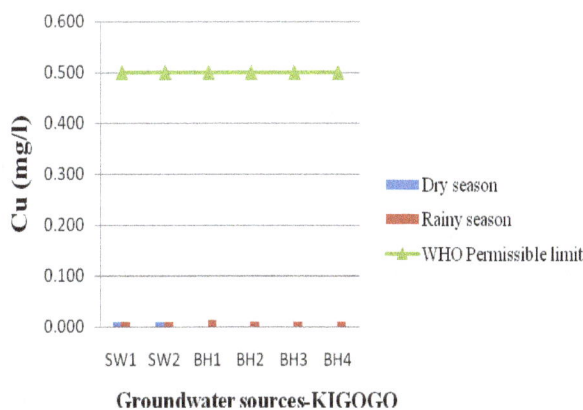

Figure 1. Mean concentration values for Cu (mg/l) during dry and rainy seasons from Kigogo area.

Results of mean concentrations of Cu from analysed samples in seasonal comparison are represented in Figures 1 and 2. All 12 water sources detected with some amount of Cu during rainy season (range from 0.0 to 0.013mg/l) although in very minimal amount. In dry season Cu concentration were not detected in most of samples from boreholes while handdug wells were detected with Cu concentrations in both dry and rainy seasons.

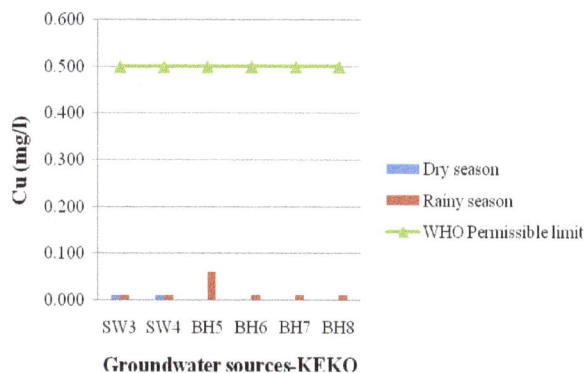

Figure 2. Mean concentration values for Cu (mg/l) during dry and rainy seasons from Keko area.

Source of copper may be due to the intrusion of industrial and domestic wastes (2). This is true since, there are some industries at Temeke Municipality that discharge waste water through Msimbazi River that pass through Keko residential areas. In turn during rainy season water runoff collects some amount of discharged wastewater to the groundwater sources. WHO suggests standards for Cu concentrations in drinking water to be 0.5mg/l (20). All water samples are found to be within the WHO permissible limit.

3.2. Lead (Pb)

Lead concentration was found to range from 0.01 mg/l to 0.35mg/l as shown in Figures 3 and 4. Values were elevated from dry season to that of rainy season. Boreholes detected with high amount of Pb compared to hand dug wells at Kigogo. WHO standards (2008), suggest the permissible range of Pb for drinking water to be 0.01mg/l. At Keko area, the situation is worse, since all groundwater sources found to have large amount of Pb in both seasons. It is reported by different scientist that, high concentration of Pb (> 0.01mg/l) has been implicated in causing anaemia, kidney damage and cerebral oedema to human (18), (4). In this case, high concentrations of lead in the body can cause death or permanent damage to the central nervous system and brain which the effects can be in memory. Other effects are high blood pressure, hearing problems, headaches, slowed growth, reproductive problems in men and women, digestive problems, muscle and joint pain (7), (13). Not only that Lead accumulative effects overtime, but also is considered to be the number one health threat to children. Moreover, high concentration of lead is linked to crime and anti-social behavior in children (19). Source of Lead contamination to ground water may be the result of entry from industrial effluents, old plumbing, household sewages, agricultural run-off containing phosphate fertilizers and human and animal excreta (15).

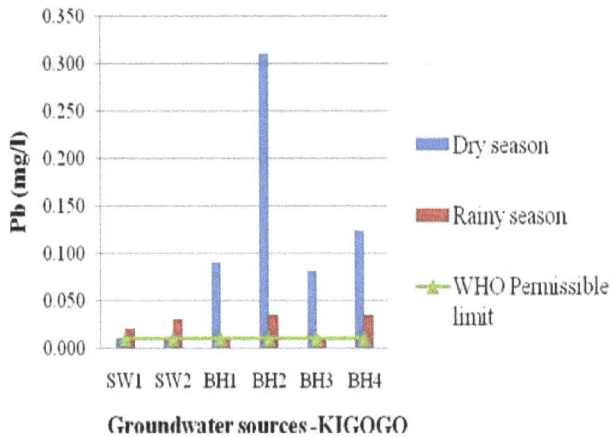

Figure 3. Mean concentration values for Pb (mg/l) during dry and rainy seasons from Kigogo area.

These results in Figures 3 and 4 are high compared to that obtained at Ojota area in Lagos state, where the mean value at dry season was 3.2mg/l and the value obtained in rainy season was 1.5mg/l (12). Similarly, study in Chennai City, Tamil Nadu water was detected with slightly amount of Pb ranging 0.01- 0.07ppm for ground water samples (16).

High elevated amount of Pb at Keko may be due to the reported various anthropogenic activities especially runoff and infiltration water from urban agriculture plots which dominate the area. In this area, there is an informal urban vegetable agriculture activity that uses fertilizers. Unfortunately many of these agricultural plots are near surface and groundwater sources. In addition, the onsite sanitation systems (septic tanks and pit latrines) are the ones used with most of dwellers which may also influence the seepage of human waste to groundwater sources leading the high amount of lead

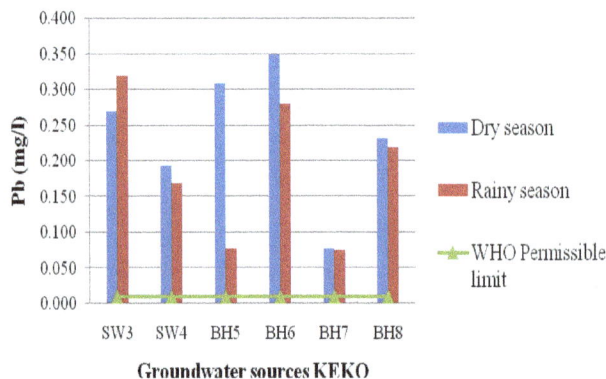

Figure 4. Mean concentration values for Pb (mg/l) during dry and rainy seasons from Keko area.

3.3. Zinc (Zn)

The amount of Zinc (Zn) in the study area is given by Figures 5 and 6. The minimum concentrations of Zn were found to be 0.01mg/l while maximum is 0.28mg/l. The results

show that borehole in this case was detected with show high amount of Zn compared to hand dug wells for both seasons. The high concentrations of zinc may be attributed to the pollution due to persistent leaching along the top layers of the soil and its widespread dispersion could be attributable to the use of liquid manure, compost material and agrochemicals (16).

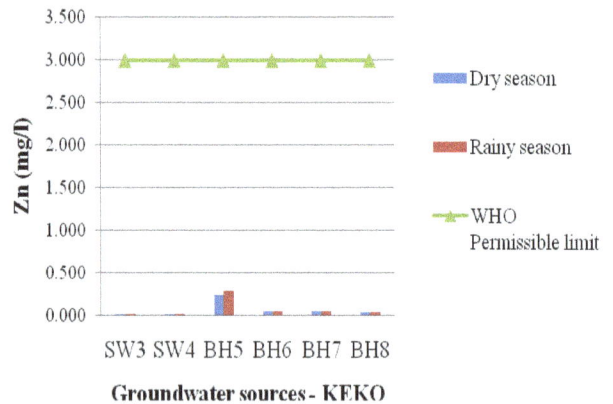

Figure 5. Mean concentration values for Zn (mg/l) during dry and rainy seasons from Kigogo area.

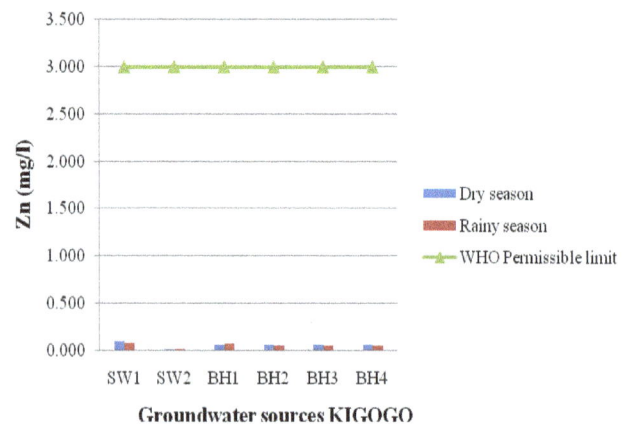

Figure 6. Mean concentration values for Zn (mg/l) during dry and rainy seasons from Keko area.

Results were compared to that obtained by (17) in Tamilnadu area, India which also found to be below WHO permissible level. Results from this study also seems to be lower compared to that obtained from lake Manzala water in Egypt (min 0.32mg/l and max 0.66mg/l) which reported to be due to considerable amounts of zinc leached from protection plates of boats containing the active zinc (14).

The water may however be suitable for other uses such as sanitation and irrigation. All samples met the WHO permissible limit of Zn in drinking water which is 3.0mg/l. However the occurrence of small amount of zinc in the sampled may tend water to be not safe for human consumption particularly drinking and cooking. This shows that Zinc toxicity is absent in the study area. However, source of slight high value (BH5) may due to agricultural inputs, domestic waste discharges and the industrial effluents.

4. Conclusion

Total of 12 water samples were collected from groundwater sources from two residential areas that have common characteristics, Keko and Kigogo in two seasonal conditions; dry and rainy seasons. Analysis of dissolved heavy metals in these water samples was done for Pb, Zn and Cu. Concentrations of Zn and Cu were found to be in permissible range of WHO standards of drinking water. Minimum and maximum concentrations of Pb for both study areas in dry and rainy seasons were detected higher than WHO permissible limit of 0.01mg/l. Human health of Keko residents especially children are at high risk throughout the year to be effected by consumption of high amount of Pb because, water from these sources are used for drinking, cooking and other domestic purposes that lead them to ingest this amount of Pb direct to their bodies. It is recommended to adopt some kind of inexpensive treatment to reduce the levels of trace metals in areas supplying water directly to consumers without any type of treatment.

Acknowledgements

The author extends her acknowledgements to all water sources owners who participated and technical staff at Ardhi University Laboratory, Dar es Salaam and all those who participated in one way or another.

References

[1] Abdul Jameel, A., Sirajudeen, J. And Abdul vahith, R., (2012). Studies on heavy metal pollution of ground water sources between Tamilnadu and Pondicherry, India. Advances in Applied Science Research, 3 (1):424-429

[2] Aggarwal, T. R., Singh, K. N. And Gupta, A. K., (2000). Impact of sewage containing domestic waste and heavy metals on the chemistry of Varuna river water.Poll. Res., 19(3): 491-494

[3] APHA, (1998). Standard Methods for the Examinations of water and wastewater. 18th Edition, American Public Health Association, Washington D.C

[4] Egborge A.B.M, (1991). Industrialization and heavy metal pollution in Warri River, University of Benin Press, Inaugural lecture series 32.

[5] FAO, (2008). Coping with water scarcity; An action framework for agriculture and food security.FAO Water Reports, 38. Food and Agriculture Organization of the UN, Rome.

[6] Hussain, A. Z and K. M. Mohamed Sheriff (2013). Status of heavy metal concentrations in groundwater samples situated in and around on the bank of Cooum river at Chennai City, Tamil Nadu Journal of Chemical and Pharmaceutical Research, 5(3):73-77

[7] Jennings, G., Sneed D., Clair, E. (1996): Metals in drinking water. North Carolina Cooperative Extension service Publication no.: AG-473-1. Electronic version 3

[8] Maganga, P. F., Butterworth, J. A. And Marioty, P., (2002). "Domestic Water Supply, Competition for Water Resources and Iwrm in Tanzania: A Review and Discussion Paper." Phys & Chem. Of the earth 27: 919-926.

[9] Mato, R. R. A. M., (2002). Groundwater Pollution in Urban Dar Es Salaam. Tanzania: Assessing Vulnerability and Protection Priorities. University Press, Eindhoven University of Technology. Eindhoven

[10] Ministry of Water and Iirrigation (mowi), (2009). Water sector status report. Ministry of water and irrigation, Dar es Salaam

[11] Mwakalila, S. (2007). "Residents' Perceptions of Institutional Performance in Water Supply in Dar Es Salaam." Physics and Chemistry of the Earth, Parts A/B/C 32(15-18): 1285-1290.

[12] Oyeku O.T and Eludoyin, A. O. (2010). Heavy metal contamination of groundwater resources in a Nigerian urban settlement. African Journal of Environmental Science and Technology Vol. 4(4), pp. 201-214,

[13] Salem H. M. Et al. (2000). Heavy Metals in Drinking Water and Their Environmental mpact on Human Health. Cairo University, Egypt, September, 2000, page 542- 556

[14] Samir M. S and Ibrahim M. Shaker, (2008). Assessment of heavy metals pollution in water and sediments on orechromis niloticus in the northern delta lakes, and their effect, Egypt. 8th International Symposium on Tilapia in Aquaculture

[15] Sirajudeen J, Abdul Jameel A, (2006). Studies on heavy metal pollution of Groundwater sources between Tamilnadu and Pondicherry India. J. Ecotoxicol. Environ. Monit, ,16(5) 443-446.

[16] Srinivas Gowd S. And Govil, P.K. (2007). Distribution of heavy metals in surface water of Ranipet industrial area in Tamil Nadu, India Environ. Monit Assess, 136, 197-207.

[17] Thambavani, S. D. And Mageswari T.S.R,. (2013). A Comprehensive Geochemical Evaluation of Heavy Metals in Drinking Water. Journal of Chemical, Biological and Physical Sciences. Vol. 3, No. 4; 2942-2956

[18] Townsend, A., (1991). Encyclopedia of Analytical science, Academic Press, London

[19] USEPA (2002). On-Site Wastewater Treatment System Manual. Office of water and office of research development. Washington DC.

[20] WHO 2008, Guidelines for drinking water quality, World Health Organization, Geneva.

12

The role of novel 3D point contact transmission in ship noise pollution reduction

Alexei Popov[1], George Nerubenko[2]

[1]Mechanical Department, National University of Shipbuilding, Nikolaev, Ukraine
[2]NER*MAR Limited, Toronto, Canada, National University of Shipbuilding, Nikolaev, Ukraine

Email address:

optimalproject@hotmail.com (G. Nerubenko)

Abstract: The ship noise causes marine pollution problems. The gear transmission contributes a significant part to ship noise. The replacement of regular gears by novel 3D point contact tooth gears is considered. The reduced noise levels up to 19 dBA are the advantages of transmissions based on invented gears having 3D point system of engaging. The implementation of invented novel transmission gives the opportunity of providing the industry by safe equipment allowing reduce the important component of ocean pollution, i.e. noise pollution.

Keywords: Ship Gear Transmission, 3D Point Contact Tooth Gears, Tooth Contact Stresses, Marine Noise Pollution

1. Introduction

The usage and exploitation of water resources brings the people to problem of providing pollution free environment. It should be achieved when exploiting technology is safe and designed specially for meeting the environment requirements. Ships [1], [2] provide the significant input to the marine noise pollution. Noise from the ship is disrupting marine life where almost every living creature depends on sound as a primary sense [3] – [5] for mating, hunting, and survival. Sound travels more rapidly and over larger distances in the sea than in the atmosphere. One of the ship's noise sources is a main transmission. It is necessary to say that a main transmission is in diesel, steam - and gas – turbine powered ships. Despite of different types of power source , a transmission in these ships has a great impact on overall noise levels. Marine engineers are trying reducing the transmission noise level by using the advanced approaches. Increasing challenges on reducing ocean noise pollution has opened the new directions in ship transmission technologies. The example of such technology implemented in geared transmission is the usage of novel type of gears with 3D point system of mesh engaging which was invented, patented and developed by Alexei P. Popov. The paper identifies the milestones of R&D that were undertaken to create the new type of geared transmission having reduced noise levels. The new shapes of tooth forced the engineers to build ship transmissions with high stress capacity and low levels of noise. The presented results of theoretical analysis combined with experiments illustrate the effectiveness of developed approach. The synthesis of new type of gearing is presented. The implementation of invented novel transmission gives the opportunity of providing the industry by safe equipment allowing reduce the important component of a marine pollution, i.e. noise pollution. An engine's geared transmission is loaded by external and internal forces during operation. The purpose of geared transmission is to deliver engine torque to vehicle wheels within the narrow proper range of engine's rotational speeds. So the transmission uses gears to make the engine's torque more effective, and to keep the engine operating at an appropriate speed. Traditionally gears are designed in an assumption that the gear teeth are subjected to the Hertzian contact compressive stresses, deformations and elasto-hydrodynamic lubrication. The main factors impacting on the character and intensity of teeth wear are the contact compressive stresses and deformations. Thin under-surface layers of tooth are the most stressed areas of gear material. The big value of stress gradients are monitored in these layers. The main aspects of Hertz assumptions [6] – [8] for contact problems would be 1. a surface of contacting bodies are perfectly smooth, isotropic, and homogeneous, 2. the deformations are elastic only, 3. the squeezing press force is normal to contact spot area, 4. the dimensions of contact spot area are very small in comparison to characteristic sizes of contacting bodies. The Hertz theory was improved by hundreds of scientists and engineers during last 100 years, see for instance [11] – [16], [23] – [28], [32] – [33]. Some of the researches deviated from Hertz theory and took for base of their calculations the Non-Hertzian contact approach for the explanation of new phenomena in contacting objects. Such approach could be found, for instance, in [17] – [19] dedicated to Non-Hertzian contact of elastic spheres. Professor Alexei P. Popov using the Winkler hypothesis [20] – [22] had found [9], [10] that the point contact interaction of teeth could be characterized by certain values of a contact stresses which are significantly lower than the values of a contact stresses for teeth having linear contacting touch. This invention provides the opportunity for increasing of the tooth load capacity, the reduction of sizes and weights, and a reduction of noise and vibration. The measured numbers of overall noise reduction in tested prototypes of the marine transmissions is up to 19 dBA.

2. Introduction of Gears with 3D Point System of Mesh Engaging

Nowadays the Hertz formula (with the several correcting coefficients) is developed for an analysis of a linear contact phenomenon in the gearings with a point contact such as the hypoid gear transmissions, the spiral bevel transmissions and the Novikov's transmissions. The Hertz theory does not reflect the physical idea of stressed tooth condition. In reality the engineer is facing the volumetric 3D stressed operation during contact interaction of teeth for such type of gearing. This lack of conformity to experimental values and the disharmony in explanation of test results regarding values of maximum strengths Q_{max} pushes the research efforts for construction of new theory of contact strength and invention of new profiles of teeth. The confirmation of proof is based on new theoretical and experimental studies of point contact in statics and dynamics on prototypes.

New approach is proposed by Professor Alexei P. Popov. The developed theory [9], [10] is based on obtaining of two equally weighted functions of the contact deformations. First function of the contact deformations is determined for the contacting objects taking into consideration their shapes and changes in shape before and after load. The Winkler hypothesis is used for determination of second function. The Winkler hypothesis was improved in [9], [10] and the area of its application was expanded and enlarged, specifically for solution of 3D point contacting systems. The proposed approach [9], [10] introduces the cause - effect relationship between the function of a contact deformations and a function of a contact stresses. The function of contact deformations is the cause or principle. It is reflecting the function of a contact stresses, which is the consequent result or outcome. It is assumed in invented model that patented gear design could lead to new modified shapes of contacting tooth. Under the load in the contact area of a gear pair the observer could see a sort of "point". This "point" looks like microscopic ellipse. Technically this fact is confirmed by experiments. During

proposed modeling the area of point contact is presented by elliptical area. According to above mentioned approach [9], [10] the equations for maximum contact stresses Q_{max} and size of elliptical area of a contact are obtained from the following model of a contact (see Figure 1).

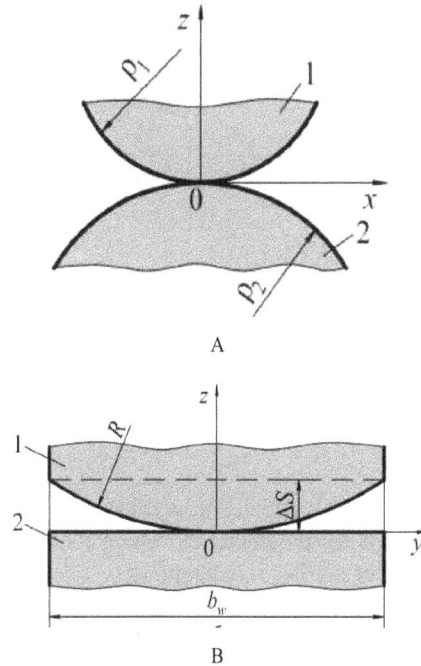

Figure 1. The model for calculations: teeth contact in plane z0x (A) and contact of cylinder with plane in projection on coordinate plane z0y (B); 1 – pinion tooth, 2 – gear tooth.

The introduced model is characterizing by external and internal touching contact of round cylinders having radii in r_1 and r_2 in z0x plane (see Figure 1A), and by interaction of cylinder of radius R with certain plane having width of b_w in mutually perpendicular z0y plane (see Figure 1B). It is necessary to underline that the contact in both planes has point nature despite the fact that there is a contact of round cylinders in z0x plane as shown in Figure 1A, and there is a contact of cylinder and flat plane in z0y projection as shown in Figure 1B.

There is a model of proposed pinion tooth having the side surfaces designed with help of curve lines in Figure 2. There is the basic tooth shape in Figure 2 having radii in r_1 similar to one shown in Figure 1A. The width of tooth is b_w in Figure 2, similar to one shown in Figure 1B. The form radii are R in Figure 2, similar to one shown in Figure 1B.

The solutions for teeth contact problems for linear and nonlinear relationship between elastic deformations of teeth and outcome stresses are obtained. Following the formulas derived in [10], the maximum point contact stress Q_{max} would be determined as

$$Q_{max} = 0.33 \ Z_j \ (K_j \ a(a+v)^2 \ E^2 T_0/(r_w^2 R_p))^{1/3} \qquad (1)$$

where (from Figure1 and Figure2)

$$a = (r/R)^{1/2},$$

$$R = bw^2/8(dS).$$

and according to [10]: Rp – pitch radius, Kj is the product of coefficients including the contact stress coefficient, the coefficient reflecting the influence of type of engine on the contact tooth strength and other coefficients proposed in [10], Zj is the combination of coefficients determined in [10].

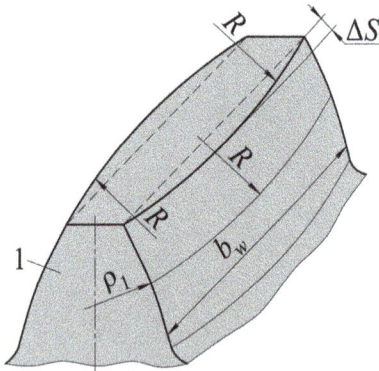

Figure 2. The general view of tooth for modified pinion

Linear or nonlinear contact properties are modeled by the geometrical deviation parameter dS, which could be estimated using the linear or nonlinear assumptions.

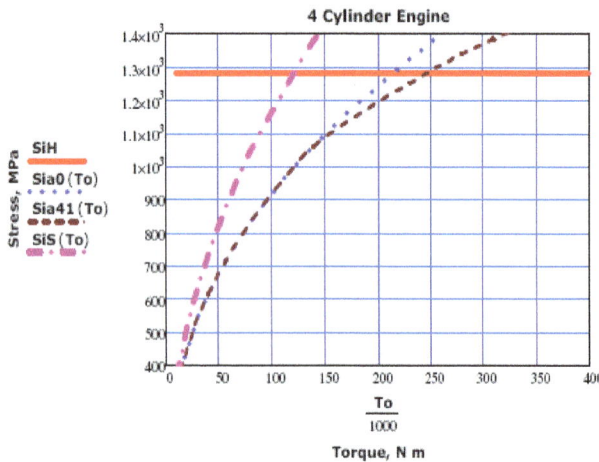

Figure 3. The contact stresses. 4 cylinder engine application.

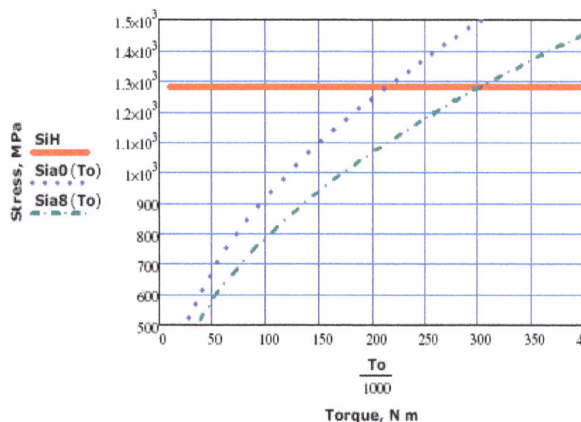

Figure 4. The contact stresses. 4 cylinder engine vs. 8cyliner engine.

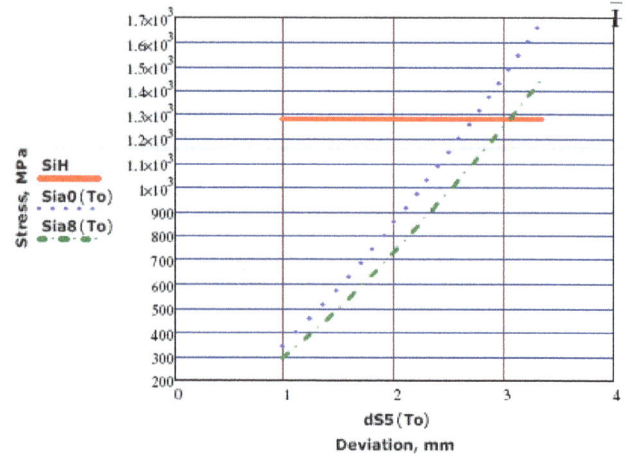

Figure 5. The contact stresses vs. deviation numbers.

The following example of calculations was formed for the illustrative purpose. The prototype of geared reducer has several shifts. The prototype of geared reducer has several shifts. The considered example is for 1st shift [10] of the transmission having point mesh. Parameters of 1st shift are z1 = 33; z2 = 127 (ratio is 3.85); m = 4 mm; aw = 20 degrees; bw = 77 mm; rw = 17,916 mm; v= 0.3; E = 2.1x10^5 MPa. The allowable compressive stress for selected material and surface hardness is calculated based on recommendation of [11], and would be 1282 MPa. The estimates are done for a model of linear or nonlinear contact properties. The direct calculations were used for evaluation of contact stresses (1) for a case of linear contact properties, and the final element technique was used for a case of nonlinear contact properties. The graphical interpretation of analysis is presented in Figures 3, 4, and 5. The ordinate axis is standing for compressive stress numbers in these graphs. The abscissa axis is for torque magnitudes To in Figure 3 and Figure 4. The deviation numbers dS are given in the abscissa axis in Figure 5. The red solid straight line is reflected the value of allowable compressive stress in this gear pair in Figures 3-5. The Figure 3 is the graphical illustration of contact stress calculations for transmission used for 4 cylinder engine.

There are three curves are shown in Figure 3. Brown dashed curve is reflecting the relationship between the applied torque and the contact compressive stresses for linear model, and blue dotted curve is showing the dependence of the contact compressive stresses from applied torque for nonlinear model. The dash – dotted pink curve is built using the standard formulas (for instance see [11], [25], [26]) for stress estimates in traditional contact case, namely in a case with "the line of a contact". The comparison of calculations for "the line of a contact" case (a dash – dotted pink curve) and "the point of a contact" case (a blue dotted curve) demonstrates that for low loads (low values of torque) both curves are practically coincided, but when the value of applied torque is increasing the load capacity of gear is higher for case of point contact. In given example the maximum stress would be reached at torque of 120 N m for "the line of a contact" case, and maximum stress would be reached at torque of 220 N m for "the point of a contact" case. Further analysis shows that for the lower values of torques the both curves reflecting the point contact case are practically coincided up to value of torque of 125 N m.

The nonlinear model analysis demonstrates that the maximum value of stress could be reached at torque of 220 N m, while the maximum value of stress for linear model would be at torque of 250 N m. The received result could help for formulation of the recommendation - use the nonlinear model for having conservative safety factor. The influence of engine type (4 cylinder engine vs. 8 cylinder engine) is shown in Figure 4. The blue dotted curve still is reflecting the relationship between the applied torque and the contact compressive stresses for transmission used with 4 cylinder engine (nonlinear model), and green dotted – dashed curve is reflecting the relationship between the applied torque and the contact compressive stresses for transmission used with 8 cylinder engine (similar nonlinear model). The replacement of 4 cylinder engine to 8 cylinder engine allows increasing of torque (in considered case from 220 N m to 300 N m) keeping the same level of contact stress in gear pair. Figure 5 gives the idea of relationship between parameter dS (torque dependent) and contact stresses (nonlinear model is applied). The blue dotted curve is reflecting the relationship between the applied torque and the contact compressive stresses for transmission used with 4 cylinder engines, and green dotted – dashed curve is reflecting the relationship between the applied torque and the contact compressive stresses for transmission used with 8 cylinder engines. The analysis of curves demonstrates that the magnitude of parameter dS is increasing with torque rise and simultaneously this increase is forcing to increase of the contact stresses. The maximum contact stress (in case of 4 cylinder engine) corresponds to value of deviation dS of 2.7 mm, while the maximum contact stress (in case of 8 cylinder engine) corresponds to value of deviation dS of 3.1 mm.

3. Static Experiment Results

In order to identify the applicability of introduced theory and to make its verification, a series of experimental tests were employed. First of all the static tests were executed [10]. There was contact compressive imitation of two cylinders loaded by force Fn; the circle cylinder with radius $r_1 = 60$ mm was contacting the barrel – shaped cylinder with maximal radius $r_2 = 40$ mm. The form radius of a barrel shape was R = 41.667 m. The length of cylinders was bw = 100 mm. The force Fn was applied several times as step load in range of 10000 N – 30000 N, step equal to 10000 N.

The minor axis 2Aa (conjugate diameter) of elliptical area of a contact could be calculated from formula

$$Aa = 1.202(a\ rw\ Fn\ /\ (a+v)\ E)^{0.33}$$

The major axis 2Bb (transverse diameter) could be calculated from formula

$$Bb = 1.202\ (rw\ Fn\ /\ a^2(a+v)\ E)^{0.33}$$

The summary results of test are presented in Figure 6 (for minor axis Aa) and in Figure 7 (for major axis Bb) as a comparison of calculated values of minor axis 2Aa of elliptical area of contact and major axis 2Bb vs. experimentally obtained data. The theoretical values of minor axis 2Aa and major axis 2Bb were determined using the nonlinear model. The analysis of data presented in Figures 6 and 7 shows very good compliance of theoretical estimated with test results. The obtained maximum relative deviation is 2.8%. The mean value of relative deviations for minor axis 2Aa was 1.41 % in tests, and the mean value of relative deviations for major axis 2Bb was 1.32 %. The obtained maximum relative deviation for value of tested elliptical area is 3.3%, and means value of relative deviations is 2.3%. The results of test confirm the satisfactory coincidence of configurations of modeled and real contact zones. There is a sample of a computer formed image of the stress distribution on a contact point (spot) for pinion tooth in Figure 8. The picture is confirming the theoretical assumption about elliptical area. The stress values in contact point are over 700 MPa in given example, in zones far from center of contact the stress values declined to 400 MPa. The shape of contact zone resembles the ellipse.

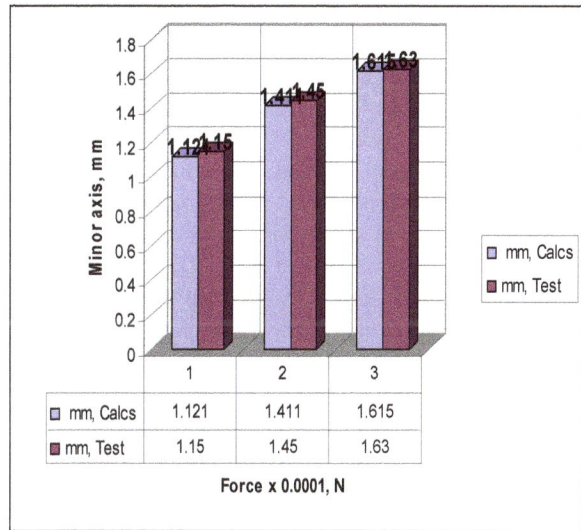

Figure 6. The comparison of calculations and test results for minor axis.

Force x 0.0001, N	1	2	3
mm, Calcs	1.121	1.411	1.615
mm, Test	1.15	1.45	1.63

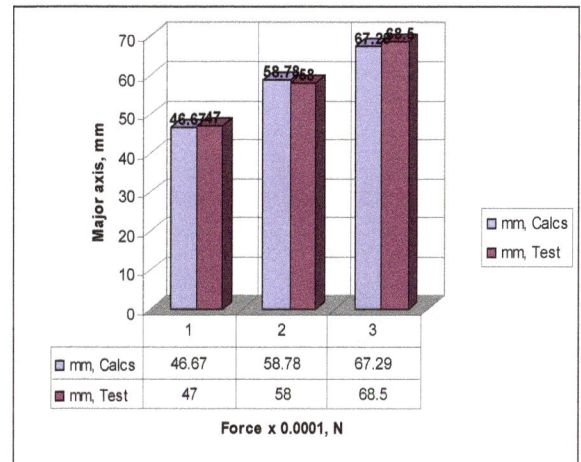

Figure 7. The comparison of calculations and test results for major axis.

Force x 0.0001, N	1	2	3
mm, Calcs	46.67	58.78	67.29
mm, Test	47	58	68.5

4. Ship Transmission Tests

The propeller of ship is driven by Main Engine (see Figure 9). The Main Engine could consist of one or several prime movers. Prime drive could be steam turbine, gas turbine or diesel. Engines must operate at relatively high speeds for maximum efficiency. Propellers must operate at lower speeds for maximum efficiency. The prime drive or engine could have higher operational angular velocity, so it is necessary to put the special device between engine and propeller for the reduction of engine's rotation speed. Therefore, reduction gears are used to allow both the engine and the propeller to operate within their most efficient revolutions per minute (RPM) intervals. The usage of gears is by no means limited to ship propulsion. Other machinery, such as ship's service generators and various pumps, also has reduction gears. Reduction gears are classified by the number of steps used to bring about the speed reduction and the arrangement of the gearing. A gear mechanism consisting of a pair of gears or a small gear (pinion) driven by the engine shaft, which directly drives a large (bull) gear on the propeller shaft, is called a single-reduction gear. In this type of arrangement, the ratio of speed reduction is proportional to the ratio of diameters of the pinion and the gear. In mechanical drives, reversing the direction of rotation of the propeller shaft may be accomplished in one of two ways. It could be reverse of the direction of engine rotation or the usage of the reverse gears. Reverse gears are used on marine engines to reverse the rotation of the propeller shaft during maneuvering without reversing the rotation of the engine.

The device named as a reducer or reduction transmission is shown in Figure 9. Figure 9 reflects the typical equipment configuration for ship propulsion. The reduction mechanism is based on gears with teeth. The noise emitting by gear reducer has a significant impact on marine noise pollution. Hence the improvement of teeth profiles from point of view of reduction of noise levels could be the perspective way of fighting with marine pollution.

It was designed and manufactured the prototype a of ship reducer with two types of gear teeth: first type was based on regular profiles, and second type was based on profiles shown in previous study made by Professor Alexei Popov. The tested reducer had spur or straight-cut gears with point contact. Actually the designed and tested transmission is a prototype of 1st box of heavy ship reduction transmission. The photo of manufactured prototype is presented in Figure 10.

The presented prototype is the 2 stage gear reducer. Applied nominal power Po is 5250 kW. The power is split by three streams on 1st shift (see Figure 11). The nominal speed of rotation of entering first shaft is 12840 RPM. First shaft has a pinion (see Figure 12). Initially the length of teeth of pinion was 135 mm; this length was reduced to 77 mm in modified pinion (see left side and right side photos in Figure 12). In other words the reduction of pinion teeth length was in 1.75 times. The point contact was provided by changing of teeth profile in accordance to [29] – [31]. The straight formation lines of side surfaces were substituted by the curved formation lines having the radius of curvature as R = 61.7 m. The teeth of geared wheels are made with deep profile and their addendum (radial distance from the pitch surface to the outermost point of the tooth) was 1.25 of module. The angle of action was aw = 20o. The profile modification was done for dedendum and addendum of pinion teeth for improving the teeth interaction during entering and exiting. The special program «Aeroflank» was used for the abovementioned modification. The heavy transmission was manufactured in ZORYA Company facility (Nikilaev – city, Ukraine), and modification of pinion teeth was done in Ukrainian company MOTOR – SICH (Zaporoghye – city) on Gear Profile Grinding Machine Gleason-Pfauter 600. The tests were arranged in ZORYA Company facility. The load regimes were selected in interval of 0.4 – 1.2 of applied power Po as shown in Figure 13. In this Figure 13 the red triangles indicate the applied power in each load regime. There are numbers of RPM (which are correspondent to each load) shown in Figure 14.The total amount of cycles of loads was $128 \cdot 10^6$, and as follows from Figure 15 - 90% of total amount of cycles ($115.56 \cdot 10^6$) was measured for nominal regime of load, and 3.6% cycles were measured for overload regime. The number of cycles of loads was in accordance to recommendations for industrial standard test requirements [26]. It was no damages, stops, recorded deviations during transmission test. The transmission was disassembled and inspected after test; all teeth of gears are in good shape, all parts are in operational conditions, no complains or admonitions. The following factors were taking into consideration during design: a. the expected increasing of contact and bending flexibilities of contacting pairs of teeth of 1st shift, b. the increased flexibility of a clamping area of a bottom of dedendum and pinion shaft of teeth having 77 mm width (instead of regular 135 mm width), c. the raised overlap coefficient of gearing caused by deep profile, d. the number of micro-impacts during contacting of pair of gears was radically reduced for the modified pinion in comparison to regular pinion. That fact led to noise level reduction caused by contacting gears.

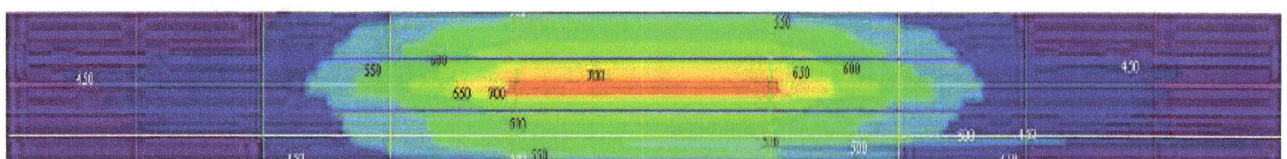

Figure 8. The computer formed image of the tooth contact stresses.

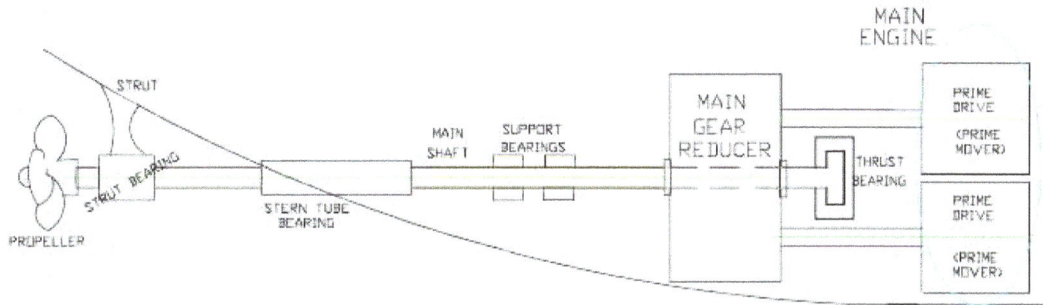

Figure 9. General scheme of ship's prorultion.

Figure 10. Photo of prototype of ship's gear reducer.

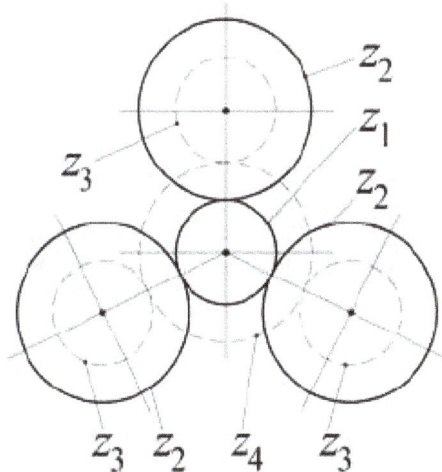

Figure 11. The kinematic scheme of reducer

Figure 12. Photos of 1st stage pinion

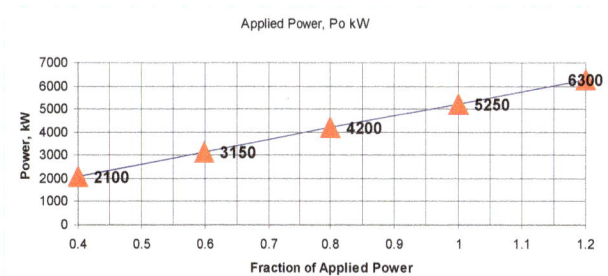

Figure 13. The ship transmission load regimes

The design calculations demonstrated that the predicted reduction of overall noise would be 11.8 dBA. The acoustic engineer recorded the reduction of overall level of noise as 12 dBA during the noise measurement session.

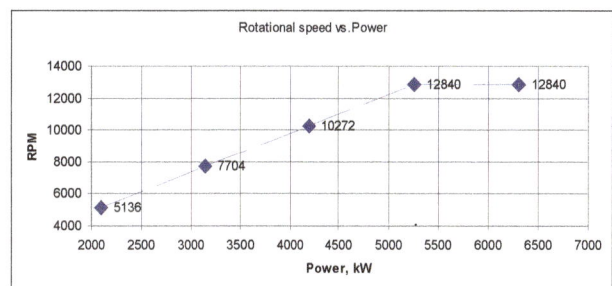

Figure 14. RPM numbers for selected load regimes.

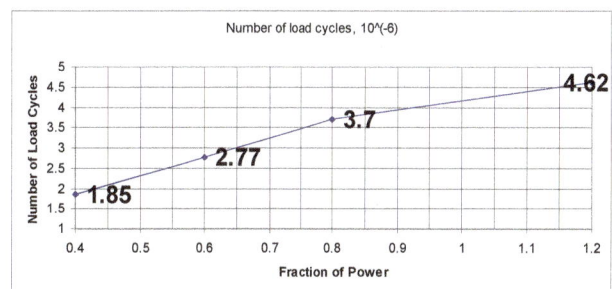

Figure 15. Amount of cycles for selected load regimes.

Now this reducer shown in Figure 10 having the kinematic scheme shown in Figure 11 (with modified pinion shown in right side of Figure 12) is still in operation.

5. Boat Reducer Noise Tests

Several tests were arranged for detailed study of the noise reduction in reducer having proposed modified gears. The 2 stage reducing marine transmission was selected for tests. The gear reducer is installed in boat equipped by IVECO Marine Propulsion 6-cylinders Diesel Engine N67 MNT M28 having 132 kW at 2500 RPM (heavy duty).The Engine performance torque is 560 N m at 1800 RPM. The maximum no load governed speed at max rating is 3150 RPM. The performance graph Torque vs. RPM is shown in Figure 16.

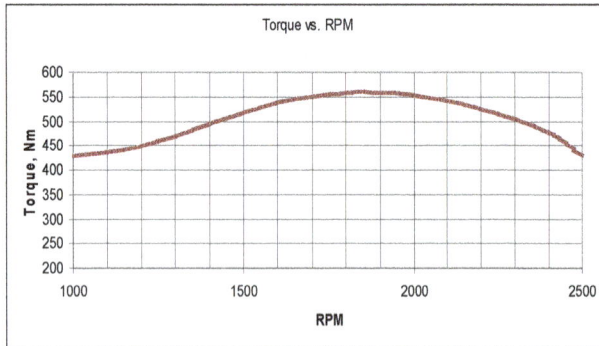

Figure 16. Torque vs. RPM. Engine N67 MNT M28.

The installed gear ratios for selected reducer are:1st stage is 3.00:1, 2nd stage is 1.67:1. Test rig layout was organized in a similar way as for a test stands described in [34], [35]. The direct current electric motor, tested transmission gearbox and electrical brake are residually mounted on rig stand frame. The gearbox was powered by the electric motor and consumes its power on a brake. The transmission noise level was measured on a test stand by sound pressure level (SPL) meter microphone located above the centerline of gearbox surface at a distance of 1 m. Octave Band Spectrum was recorded in each session by that microphone, and the center frequencies for these bands were: 31.5Hz , 63Hz , 125Hz , 250Hz , 500Hz , 1kHz , 2kHz , 4kHz , 8kHz and 16kHz. The measurement methodology and procedure were based mainly on recommendations of [36]. The experiment consisted of two phases and preliminary Background Noise test. For Background Noise test the transmission gearbox was excluded from test and DC electrical motor was connected to brake directly. The tests were done for the following values of motor's rotational speed: 1000 RPM, 2000 RPM, and 3000 RPM. The brake load was provided the torques equivalent to those which were reported for performance curve and the same values of RPM (see Figure 16), i.e. 424 N m at 1000 RPM, 552 N m at 2000 RPM, 483 N m at 3000 RPM The main result of a preliminary test have demonstrated that measured overall noise levels were less than 60 dBA. The original transmission (with regular fabricated gears) was tested during first phase. The tests were done in the equivalent format as for preliminary Background Noise test. The detailed gear inspection after first phase test demonstrated that the gear's teeth were in good operational conditions, there were no visible defects, fatigue spots or damages of surfaces of teeth.

The gear replacement was made for second phase. The modified gears were manufactured using the recommendations and formulas from [9], [10], [29] – [31]. All shafts and other parts had the same geometry and dimensions as the regular ones. Everything was assembled within regular box. It gave the opportunity to compare "apples to apples" during experiment. The modified transmission (with new modified manufactured gears) was tested during second phase. The tests were done in the same manner as for first phase test. The comparison of obtained results during first and second phase tests to levels recorded during preliminary tests brought to conclusion that the corrections for background noise could be ignored. The relationship between the noise levels and rotational speed for different cases of usage of gears (regular vs. modified) were studied and recorded during first and second phases (see Figures 17-19 and 20-22).

The analysis of test measurements data brings to a conclusion that the substitutions the regular gears in transmission by proposed modified gears yields to reduce the overall noise levels in total interval of operating loads and speeds of rotation.

The range of recorded noise reduction is sitting in interval of 12 dBA – 19 dBA (see Figure 23).

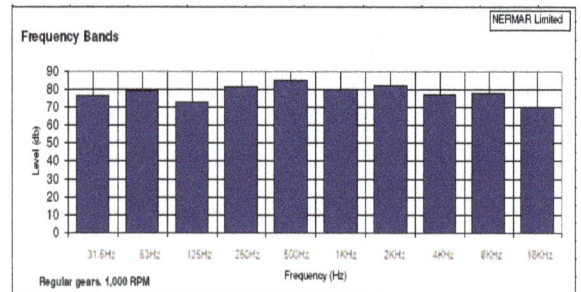

Figure 17. Regular gears. 1000 RPM. Overall noise level is 90 dBA.

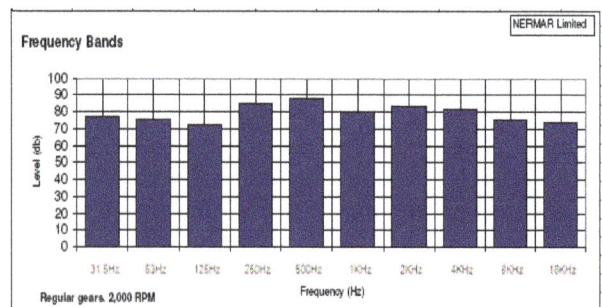

Figure 18. Regular gears. 2000 RPM. Overall noise level is 92 dBA.

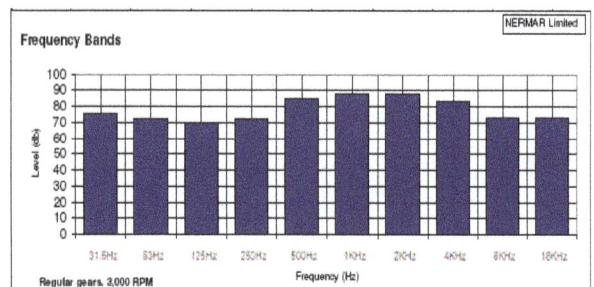

Figure 19. Regular gears. 3000 RPM. Overall noise level is 93 dBA.

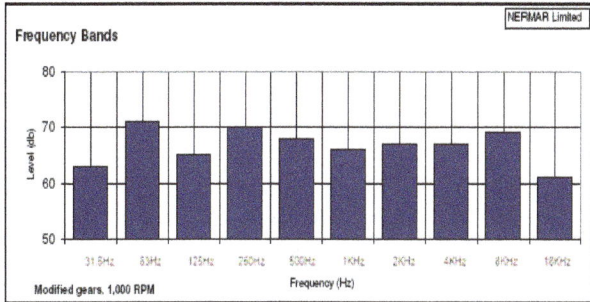

Figure 20. Modified gears. 1000 RPM. Overall noise level is 78 dBA.

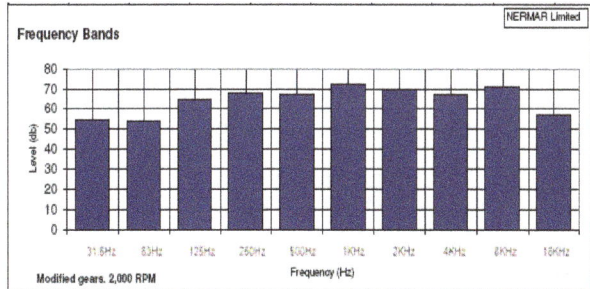

Figure 21. Modified gears. 2000 RPM. Overall noise level is 78 dBA.

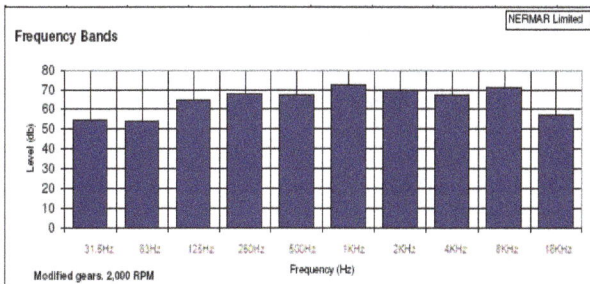

Figure 22. Modified gears. 3000 RPM. Overall noise level is 74 dBA.

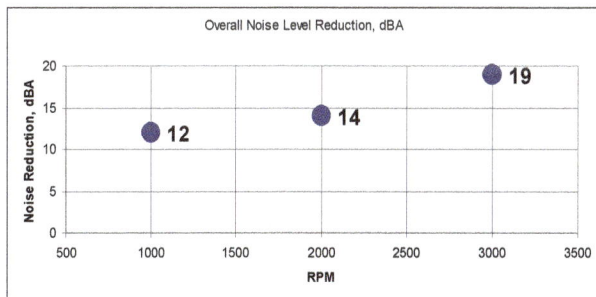

Figure 23. Overall noise level reduction.

The following planned step of implementation of 3D point contact tooth gears is the preparation of documents and paperwork for US Patent application.

6. Conclusion

The proposed transmission having gears with 3-D point contact interaction of teeth provides the effective solution for gearing load capacity optimization and noise level reduction during operation.

Owing to extended properties of new teeth profiles it is possible to increase the value of working contact stresses with parallel reduction of emitting noise levels. The novel type of teeth gears in comparison to conventional gears has a load capacity for contact stresses: higher in 1.7 – 2.4 times, for bending stresses: lower in 1.4 – 1.7 times.

The successful implementation of a transmissions with gears designed with invented point contact interaction of teeth demonstrates the reduction of the sizes and weights of transmissions in 25% - 40%.

The implementation of new type of modified teeth gears in transmissions gives the opportunity improves radically the acoustical performance of transmission operation, and obtains the significantly low levels of noise. The documented overall noise level reductions are 12 dBA – 19 dBA.

The implementation of proposed technology of ship reducer manufacturing gives a qualified chance in the successful fighting with marine noise pollution.

References

[1]   D. Ross, On Ocean Underwater Ambient Noise. Institute of Acoustics Bulletin, St Albans, Herts, UK: Institute of Acoustics, 18. 1993, pages 5-8.

[2]   Megan F. McKenna. Donald Ross, Sean M. Wiggins, John A. Hildebrand. Underwater radiated noise from modern commercial ships. Journal of Acoustic Society of America. 131 (1), January 2012, pages 92-103.

[3]   Mark A. McDonald, John A. Hildebrand, Sean M. Wiggins, Donald Ross. A 50 Year comparison of ambient ocean noise near San Clemente Island: A bathymetrically complex coastal region off Southern California. Journal of Acoustic Society of America. 124 (4), October 2008, pages 1985–1992.

[4]   IFAW. Ocean Noise: Turn it down. A report on ocean noise pollution. June 2008. 42 pages.

[5]   Linda S. Weilgart. The Impact of Ocean Noise Pollution on Marine Biodiversity. Ocean Noise Coalition. Report. Halifax, Canada, 2009. 6 pages.

[6]   Hertz H. Über die Berührug fester elastisher Körper / H. Hertz. // J. für Reine und Angewandte Mathematik. 1881. — Bd. 92.

[7]   Hertz H. Über die Berührug fester elastisher Körper und über die Härte / H. Hertz // Varhandlungen des Vereine zur Beforderung des Geverbefleisses. — Berlin. Nov.1882.— 449 pages.

[8]   Hertz H. Gesammele Werke / H. Hertz. — Leipzig. 1895. — Bd. 1. — S. 155—196.

[9]   Alexei P. Popov. Contact Strength of Geared Mechanisms. Nikolaev. NUK, 2008. 580 pages.

[10] Alexei P. Popov. Geared Mechanisms with point contact of teeth. Nikolaev. ATOLL, 2010. 774 pages.

[11] D. N. Reshetov. Parts of Machines. Moscow. Machine-building. 1963. 723 pages.

[12] Shuting Li. Finite element analyses for contact strength and bending strength of a pair of spur gears with machining errors, assembly errors and tooth modifications. Mechanism and Machine Theory, Vol. 42, 2007, pages 88–114,

[13] A. Kahraman, R. Singh. Interactions between Time-varying Mesh Stiffness and Clearance Non-Linearities in a Geared System. Journal of Sound and Vibration, Vol. 146,135–156, 1991.

[14] L. Walha, T. Fakhfakh, M. Haddar. Nonlinear dynamics of a two-stage gear systems with mesh stiffness fluctuation, bearing flexibility and backlash. Mechanism and Machine Theory, Vol. 44,1058–1069, 2009.

[15] Konstandinos G. Raptis, Theodore N. Costopoulos, Georgios A. Papadopoulos and Andonios D. Tsolakis. Rating of Spur Gear Strength Using Photoelasticity and the Finite Element method, American Journal of Engineering and Applied Sciences, 2010,3(1), pages 222-231.

[16] Faydor L. Litvin, Alfonso Fuentes, Claudio Zanzia, Matteo Pontiggiaa. Design, generation, and stress analysis of two versions of geometry of face-gear drives. Mechanism and Machine Theory. Volume 37, Issue 10, October 2002, Pages 1179–1211.

[17] J.J. Kalker, Transient rolling contact phenomena. Proceedings 25th ASLE meeting, Chicago, Ill. ASLE, New York, 1971, pages 177-184

[18] J. J. Kalker, On the uniqueness of contact problems in the theory of elasticity, Report 387, Laboratory of engineering mechanics,Delft University of Technology. 1971

[19] J.J. Kalker, A survey of the mechanics of contact between solid bodies. Zeitschrift für Angewandte Mathematik und Mechanik 57. 1977, pages T3-T17.

[20] Biot M. A., 1937, "Bending of an infinite beam on an elastic foundation," Journal of Applied Mechanics Transactions of the ASME, Vol. 59, A1-A7.

[21] Bowles, J. E., 1968, Foundation Analysis and Design, McGraw-Hill Book Co., New York.

[22] Cheung, Y. K., and Zienkiewicz, O. C, 1965, "Plates and tanks on elastic foundations- An application of finite-element method," International Journal of Solids and Structures. Vol. 1, pages 451-461.

[23] R. V. Fediakin, V.A. Chesnokov. Calculations for cylindrical Novikov's transmissions and friction transmissions. Proceeding of VVIA. Moscow. 1982. 144 pages.

[24] M. I. Makushin. Stressed condition and strength in contact. Proceedings of Department of Resistance of Materials. MVTU. Moscow. 1947, pages79 -145.

[25] George Nerubenko et al. Atlas of Gear Boxes and Transmissions. Nikolaev, 1990. 32 pages

[26] George Nerubenko et al. The exploitation of ship deck mechanisms. HANDBOOK. Moscow, Transport, 1991. 198 pages.

[27] VNIIMASH. Novikov's transmission with tooth surface hardness $HB > 350$. Strength calculations. Methodical recommendations. Moscow. 1987. 86 pages.

[28] IMASH RAN. Influence of sag on determination of load in gearing and between satellites in planetary transmission. Moscow. 2009. 30 pages.

[29] Alexei P. Popov. Gear Mesh. Ukrainian Patent № 77634. Published 15.12.2006.

[30] Alexei P. Popov. Gear Transmission of Popov A. P. Ukrainian Patent № 81008. Published26.11.2007.

[31] Alexei P. Popov. Gear Transmission with equally strengthen mesh of involute teeth. Ukrainian Patent №84606. Published 10.11.2008.

[32] Davis, J. Gear Materials, Properties and Manufacture. Materials Park OH: ASM International. 2005.

[33] Dudley, D. W. Handbook of Practical Gear Design. Lancaster: Technomic Publishing Company. 1994.

[34] Mats Åkerblom. Gearbox Noise, Correlation with Transmission Error and Influence of Bearing Preload. Doctoral Thesis in Machine Design. Department of Machine Design. Royal institute of Technology. SE100 44 Stockholm, Sweden. 2008.

[35] Shawki S. Abouel-Seoud, Eid S. Mohamed, Ahmed A. Abdel-Hamid, Ahmed S. Abdallah. Analytical Technique for Predicting Passenger Car Gearbox Structure Noise Using Vibration Response Analysis. British Journal of Applied Science & Technology. 3(4): 2013, pages 860-883.

[36] Satya D. Pradhuman. Cirrus Research. Approach and Methodology. 303 S. Broadway, Suite 212, Tarrytown, New York 10591. 2010.

[37] William J. Atherton, Adam Pintz, David G. Lewicki. Automated Acoustic Intensity Measurements and the Effect of Gear Tooth Profile on Noise. Cleveland State University. 1987. Mechanical Engineering Faculty Publications. Paper 4.

13

Effect of Climate Change on Water Resources

Solomon Melaku Melese

Department of Forestry, College of Agriculture, Wollo University, Dessie, Ethiopia

Email address:

Solomon.melaku@wu.edu.et

Abstract: Climate change is one of the global issue that would affect the sustainable development of many regions. The objective of this review is the effect of climate change on water resources. Climate change affects certain components of the hydrological cycle, especially precipitation and runoff. A change in climate can alter the spatial and temporal availability of water resources. The effects of minor levels of climate change are already being felt, with impacts across many economic sectors. These changes will result in increased floods and drought, which will have significant impacts on the availability of freshwater. These impacts on freshwater will be further compounded by rising sea levels, and melting glaciers. In general a warmer climate will accelerate the hydrologic cycle, altering rainfall, magnitude and timing of run-off.

Keywords: Climate, Water, Effect

1. Introduction

Scientific evidence indicates that due to increased concentration of greenhouse gases in the atmosphere, the climate of the Earth is changing; temperature is increasing and the amount and distribution of rainfall is being altered (Houghton et al. 1996). The IPCC Scientific Assessment suggests that global average temperature may increase between 1.5 and 4.5°C, with a 'best estimate' of 2.0°C, in the next century with a doubling of the CO_2 concentration in the atmosphere (Houghton et al. 1996). The United Nations Framework Convention on Climate Change (UNFCCC) defines climate change as, "a change of climate which is attributed directly or indirectly to human activity that alters the composition of the global atmosphere and which is in addition to natural climate variability observed over comparable time periods". Climate change affects certain components of the hydrological cycle, especially precipitation and runoff. A change in climate can alter the spatial and temporal availability of water resources. It reduces either the overall quantity of water or the timing of when water is available for use will have important effects on agriculture, industrial and urban development.

The climatic impact on the water regime may also make worse other environmental and social effects of water management. For instance, reduced river runoff can concentrate the effects of pollutants or exacerbate the spread of water-borne disease. Climate fluctuations can also affect the use of agricultural land associated with irrigation systems. Climate change will greatly complicate the design, operation, and management of water-use systems (Gleick and Shiklomanov, 1989). Scientists around the world now agree that the climatic changes occurring internationally are the result of human activity. Although responsibility for the causes of climate change rests primarily with the developed and industrialized nations, the expenses of climate change will be bear most directly by the poor. This is for a number of reasons including: - many of the region's most likely to be adversely affected fall in the developing world; the poor are disproportionately dependent on occupations, such as farming, that are adversely affected by climate change and because the poor have very limited resources, they do not have the ability to adapt to climate change in the way that wealthier households can. In particular, changes to water quality, quantity and availability will be an impact of ongoing climate change in many areas.

As (Kinfe, 1999)reported on Awash River the general warming simulated by General circulation models (GCMs) under CO_2 doubling would result in a substantial decrease in annual runoff over the Awash River Basin. In addition to this, the total average annual inflow volume into Lake Zuway might decline significantly. This is likely to drop the lake level up to two third of a meter and shrink the water surface area up to 25 km2, which is about 6% of the base period water surface area. This combined with the unbalanced supply-demand equation in the watershed is expected to have significant impact on the lake water balance. Therefore, in Lake Zuway Watershed, runoff is likely to decrease in the future and be insufficient to meet future demands for water of the ever increasing population (Lijalem, *et al.* 2007). Some Current scientific research shows that climate change will have major effects on precipitation, evapotranspiration, and runoff and ultimately on the nations water supply. Climate-induced changes in the water cycle likely will affect the magnitude, frequency, and costs of extreme weather events as well as the availability of water to meet growing demand (Frederick and Gleick, 1999).The objective of this graduate seminar is to review the effect of climate change on water resources.

2. The Impacts of Climate Change on Critical Areas

The effects of minor levels of climate change are already being felt, with impacts across many economic sectors. While there will clearly be some gains from climate change (for example, agriculture in some northern regions should increase in productivity due to a rise in temperatures). But, most of the impacts will be negative, and gains and losses will not be evenly distributed. Of the major critical areas:-

Water: Rising global temperatures will lead to an intensification of the hydrological cycle, resulting in dryer dry seasons and wetter rainy seasons, and subsequently sensitive risks of more extreme and frequent floods and drought occurrence. Changing climate will also have significant impacts on the availability of water, as well as the quality and quantity of water that is available and accessible. Melting glaciers will increase flood risk during the rainy season, and strongly reduce dry-season water supplies to one-sixth of the World's population.

Agriculture: Declining crop yields are likely to leave hundreds of millions without the ability to produce or purchase sufficient food supplies, especially in Africa. At mid to high latitudes, crop yields may increase for low levels of change in temperature, but will decline at higher levels of temperature change.

Ecosystems: Ecosystems are fundamentally dependent on water resources: healthy ecosystems depend on receiving appropriate amounts of water of a certain quality at certain times. Humans, in turn, are dependent on ecosystem processes. For example, primary productivity and inputs from watersheds support food webs, yielding fish for commercial and recreational purposes, while decomposition and biological uptakes remove organic materials and nutrients, purifying water. Ecosystem processes are affected by temperature and flow regimes will be affected by changes in climatic conditions. Changing temperatures will cause ecosystems to shift. Forests, land types and plant species will dieback in some areas as temperatures rise, but increase in other areas. However, in many cases, the speed of change in temperature may be too fast for ecosystems to adjust, resulting in the loss of forests and species. Work across the United States suggests a wide range of serious concerns for ecosystems, with possible extinction of endemic fish species already close to their thermal limits, declining area of wetlands with reductions in waterfowl populations, concerns about stream health, and major habitat loss (Eaton et al., 1996; Covich et al., 1997; Hauer et al., 1997; Meyer, 1997; Schindler, 1997). Researchers also express concern not only about the actual impacts of climate change, but also about the limited ability of natural ecosystems to adapt to or cope with those changes over the short time frame in which the impacts are likely to occur. This limited ability to adapt may lead to irreversible impacts such as species extinction.

Health: Higher temperatures expand the range of some dangerous vector-borne diseases, such as malaria, which already kills one million people annually, most of whom are children in the developing world. Further, heat waves associated with climate change, and increases in water borne diseases, will result in increased health problems

3. Climate Change and the Hydrological Cycle

Land use change is directly related with hydrological cycle (Sreenivasulu and Bhaskar 2010) and, any change in land use and shifting cultivation is the major cause for large-scale changes in associated ecosystem (Chakraborty 2009). Since, much of the solar energy received by the Earth is used to drive the hydrological cycle, higher levels of solar energy trapped in the atmosphere will lead to an intensification of this cycle which resulting in changes in precipitation patters. These changes will result in increased floods and drought, which will have significant impacts on the availability of freshwater. These impacts on freshwater will be further compounded by rising sea levels, and melting glaciers. In general a warmer climate will accelerate the hydrologic cycle, altering rainfall, magnitude and timing of run-off. Warm air holds more moisture and increase evaporation of surface moisture. With more moisture in the atmosphere, rainfall and snowfall events tend to be more intense, increasing the potential for floods. However, if there is little or no moisture in the soil to evaporate, the incident solar radiation goes into raising the temperature, which could contribute to longer and more severe droughts (Mall R. K. *et al*, 2006). Therefore, fig 1 showed that change in climate will affect the soil moisture, groundwater recharge and frequency of flood or drought occurrence and finally groundwater level in different areas. Changes in the water cycle, which are consistent with the warming observed over the past several decades, include:

- Changes in precipitation patterns and intensity
- Changes in the occurrence of drought
- Widespread melting of snow and ice
- Increasing atmospheric water vapor
- Increasing evaporation
- Increasing water temperatures
- Changes in soil moisture and runoff and so on.

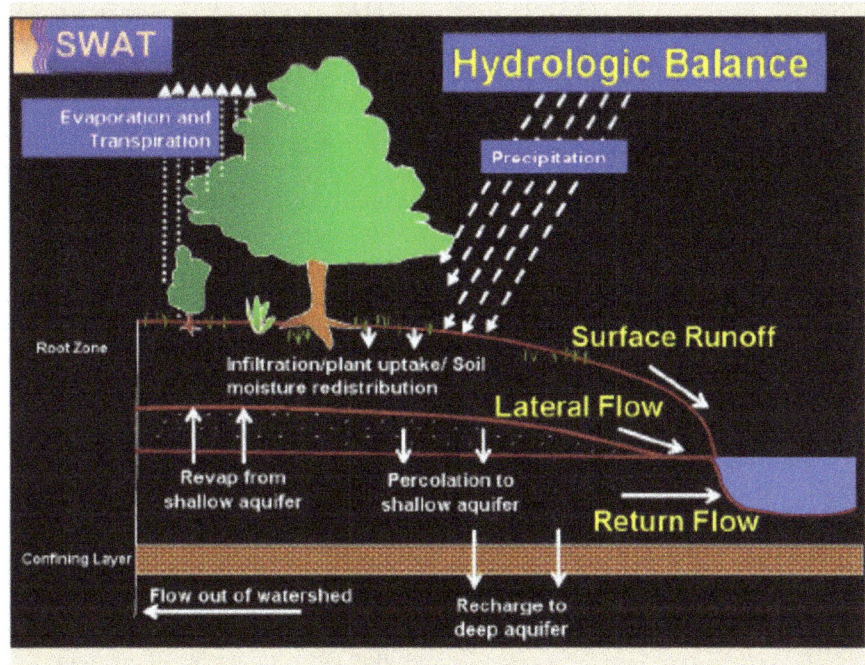

Fig. 1. Soil and water assessment tool.

3.1. Floods and Drought

Floods and droughts are likely to become more common and more intense as regional and seasonal precipitation patterns change, and rainfall becomes more concentrated into heavy events. While it sounds when, a warmer world produces both wetter and drier conditions. Even though total global precipitation increases, the regional and seasonal distribution of precipitation changes and more precipitation comes in heavier rains (which can cause flooding) rather than light events (Frederick and Major, 1997). In the past century, averaged over the United States, total precipitation has increased by about 7 percent, while the heaviest 1 percent of rain events increased by nearly 20 percent. Flooding often occurs when heavy precipitation persists for weeks to months in large river basins. Observations also show that over the past several decades, extended dry periods have become more frequent in parts of the United States, especially the Southwest and the eastern United States. Longer periods between rainfalls, combined with higher air temperatures, dry out soils and vegetation, causing drought. For the future, precipitation intensity is projected to increase everywhere, with the largest increases occurring in areas in which average precipitation increases the most (Matalas and Fiering, 1977). The number of dry days between precipitation events is also projected to increase, especially in the more arid areas. Mid-continental areas and the Southwest are particularly threatened by future drought. So the combination of more intense and frequent storms with land use changes is already proving to be deadly for the world's vulnerable populations. In drier regions, even a slight rise in temperatures will lead to greater loss of moisture, exacerbating drought and desertification (Frederick

and Major, 1997). Drought will lead to decreased water availability and water quality for populations in many water-scarce regions, particularly in southern Africa, North Africa, Central America and central Asia. When less precipitation and higher temperatures occur simultaneously, the availability of water resources is reduced even further while evaporation is increased, leading to a vicious cycle. In Sub-Saharan Africa, for example, long periods of drought are becoming more common and are predicted to become more widespread. While some farmers have been able to survive by selecting seed varieties based on changing conditions, poorer farmers have not been as able to adapt.

3.2. Rising Sea Levels

Sea levels rising in response to thermal expansion of the oceans and increased melting of glaciers and land ice will also affect water availability. The global sea level increased about 18 cm during the past century. The most recent IPCC results suggest average sea level might rise another 15 to 95 cm by the year 2100. Changes in sea level will increase coastal erosion, and flooding of wetlands and lowlands. It will also have a significant impact on the availability of freshwater: Higher sea levels and increased storm surges could adversely affect freshwater supplies in coastal areas. Salt water in river mouths and deltas would advance inland and coastal aquifers would face an increased threat of saltwater intrusion, jeopardizing the quality of water for many domestic, industrial, and agricultural users. For example, sea level rise would aggravate water-supply problems in several coastal areas in the United States, including Long Island, Cape Cod, New Jersey shore communities, and the Florida cities of Miami, Tampa, and

Jacksonville (Frederick and Major, 1997). As the level of the sea rises, additional scarce freshwater supplies are required to prevent saltwater interruption into the delta or farm land. In general, sea level rise end with salinity problems and makes the productive land unproductive or useless.

3.3. Melting Glaciers

Scientists are especially concerned about the Arctic, where the effects of climate change are being felt more quickly and severely than anywhere else on the planet. Arctic temperatures increased by about 5 degrees during the 20th Century – 10 times faster than the global average. Snow cover has declined by some 10 percent in the mid-high latitudes of the Northern Hemisphere since the late 1960s and, in the Russian Arctic, buildings are collapsing because of melting permafrost under the foundations. The dramatic changes occurring in the Arctic are being felt around the world: melting glaciers contribute to rising sea levels, and also increased flash floods as river basins fill more quickly and with a greater volume of water, with the resulting impacts on freshwater availability.

4. Effect of Climate Change on Ground Water

Groundwater will be less directly and more slowly impacted by climate change, as compared to surface water (e.g. rivers). So, ground water is less expected to be affected by climate change. This is because rivers get replenished on a shorter time scale, and drought and floods are quickly reflected in river water levels since, river water temperature ere in close equilibrium with air temperature and as air temperature rise so will river temperatures (Hammond and pryce,2007; Hassan et al.,1998). Groundwater, on the other hand, will be affected much slower. Only after prolonged droughts groundwater levels will show declining trends. Global warming as part of climate change will affect ground water indirectly. Groundwater is the intricate, but often overlooked, link between surface waters and many freshwater and terrestrial ecosystems. As many groundwater systems both discharge into and are recharged from surface water, impacts on surface water flow regimes are expected to affect groundwater. Thus neglecting the consideration of groundwater in the process of IWRM can result in the mismanagement of surface water with severe effects on the population and the environment. Increased variability in rainfall may decrease groundwater recharge in humid areas because more frequent heavy rain will affects the infiltration capacity of the soil, thereby increasing surface runoff. In semi-arid and arid areas, however, increased rainfall variability may increase groundwater recharge, because only high-intensity rainfalls are able to infiltrate fast enough before evaporating, and alluvial aquifers are recharged mainly by inundations during floods.

5. Impact of Climate Change on Water Resources for Human Development

Access to water plays a key role in development. It sustains human life, both through direct consumption and use in agriculture and industrial activities. Availability of water for drinking purposes is also essential and it cannot be separated from wider water resource management issues. The competition between households, agriculture and industry on water can cause conflict over water availability and use. Today, more than one billion people still lack access to safe water, while over two billion lack safe sanitation.

5.1. Changes in the Quantity, Quality and Accessibility of Water Supplies

5.1.1. Quantity of Water

Many of the world's countries already struggle under existing water stress from pressures such as irrigation demands, industrial pollution and water borne sewerage. These pressures will be significantly exacerbated by climate change, which for many regions will result in reduced rainfall and increasing temperatures, further reducing the availability of water for drinking, household use, agriculture and industry. The Stockholm Environment Institute estimates that, based on only a moderate climate change, by 2025 the proportion of the world's population living in countries of significant water stress will increase from approximately 34 percent (in 1995) to 63 percent For example, in Africa's large catchment basins of Niger, Lake Chad and Senegal, the total available water has already decreased by 40-60 percent, and desertification has been aggravated by lower than average annual rainfall, runoff and soil moisture, especially in Northern, Southern and Western Africa. The consequences for water supply include smaller flows in springs and rivers, and decreasing groundwater levels. In case of Ethiopia also some rivers and lake size are shrinking for example, lake Zuway drops to level up to two third of a meter and shrink the water surface area up to 25 km2, which is about 6% of the base period water surface area. This declining in quantity of water affects local user including farmer and investor. As these competing demands intensify under climate change, effective governance for balancing water demands will become essential, particularly in the face of strong pressures to prioritize industrial uses over other uses such as drinking supplies.

5.1.2. Quality of Water

Climate change is expected to affect water quality in both inland and coastal areas. Specifically, precipitation is expected to occur more frequently via high-intensity rainfall events, causing increased runoff and erosion. More sediments and chemical runoff will therefore be transported into streams and groundwater systems, impairing water quality. Water quality may be further impaired if decreases in water supply cause nutrients and contaminants to become more concentrated. Rising air and water temperatures will also impact water quality by increasing primary production, organic matter decomposition, and nutrient cycling rates in lakes and streams, resulting in lower dissolved oxygen levels. Lakes and wetlands associated with return flows from irrigated agriculture are of particular concern. This suite of water quality effects will increase the number of water bodies in violation of today's water quality standards, worsen the quality of water bodies that are currently in violation, and ultimately increase the cost of meeting current water quality goals for both consumptive and environmental purposes (Adams R. M. *et al*, 2008). The quality of existing water supplies will become a further concern in some regions of the globe. Water acquires most of its geochemical and biochemical substance during its cycle from clouds to rivers, through the biosphere, soils and geological layers. Changes in the amounts or patterns of precipitation will change the route/ residence time of water in the watershed, thereby affecting its quality. For example, in areas with relatively high water tables, or under intensive irrigation, increased evaporation due to higher temperatures will raise the concentration of dissolved salts. Further, increased flooding could raise water tables to the point where agrochemicals/ industrial wastes from soil leach into the groundwater supply. Similarly, higher ocean levels will lead to salt water intrusion in coastal groundwater supplies, threatening the quality and quantity of freshwater access to large populations.

5.1.3. Accessibility of Water

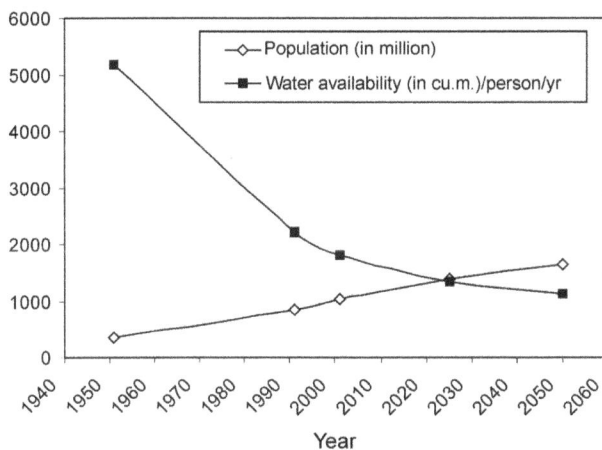

Source: Ministry of Water Resources, Govt of India (2003).

Fig. 2. Average annual freshwater availability and growth of population from 1951 to 2050.

As water quantities and quality decrease as a result of intensification of the hydrological cycle, competition for available resources will intensify. Demand for agricultural and domestic water in particular increases significantly at hotter and drier times of the year. Agriculture has always been the dominant end-use of diverted water; this will only intensify with increasing needs for irrigation brought on by higher temperatures and reduced precipitation, coupled with increasing populations. Meanwhile, demands of industry are expected to become a greater issue in the competition for dwindling resources; in the event of decreasing water

tables as a result of climate change, industrial needs will be forced to compete with agricultural and domestic water supply sources, and could lead to conflict. In general, in this century population numbers are increasing from time to time with declining water resources. This figure 2 shows the observed and projected decline in per capita average annual freshwater availability and growth of population from 1951 to 2050 in India.

5.2. Natural Disasters

The increase in natural disasters, primarily floods and droughts, will further exacerbate issues over water availability and water quality. The Intergovernmental Panel on Climate Change (IPCC) has projected that flooding and landslides pose the most widespread direct risk to human settlements from climate change. The UNFCCC predicts that "a future of more severe storms and floods along the world's increasingly crowded coastlines is likely, and will be a bad combination even under the minimum scenarios forecast". Beyond the immediate and apparent destruction caused by flooding, it also includes loss of life and livelihoods. Flooding has major impacts on water resources, and so humans through:-

- Overburdening of wastewater and drain systems, leading to contamination of water supplies with subsequent outbreaks of dysentery and cholera.
- Disruption of safe water supplies through damage to infrastructure.
- Flood water in low land area creates breeding grounds for mosquitoes with increased risk of malaria, yellow fever and dengue
- Floods can also increase the incidence of diseases such as skin diseases as a result of constant contact with water; and
- Inadequate nutrition following disruption of income and food distribution systems.

6. Consequences for Human Populations

Effect of climate change on water resource is not affects human population only through quantity, quality and accessibility but also through affecting agriculture, health, economic activity, and conflicts over water resources. The potential of Landsat image data is to provide an accurate classification for land cover changes over time (Tsarouchi and Buytaert 2013). Normalized Difference Vegetation Index (NDVI) method is better suitable where analysis is carried out using either past or present images with no ground truth data (Chandrabose et al. 2012).

6.1. Impacts to Agriculture and Food Security

Agriculture is the sector that dominantly affected by climate changes, spatially in developing countries since their agriculture is dependent on rain fails. In addition to this, population pressures are also another problem that needs intensified agricultures. Warmer temperatures will lead to increased water evaporation, intensifying the need for irrigation precisely as water becomes even less available. (Shiklomonov, 2003) predicts that water withdrawal for agriculture will rise from 2600 km^3 in 2000 to 3200 km^3 by 2025. Increasing supply for irrigation will simply not be feasible in many regions, particularly where irrigation capacity is not sufficiently developed to accommodate changing precipitation patterns. In sub-Saharan Africa, for example, where up to 90 percent of agriculture is rain fed, the sector accounts for 70 percent of employment and 35 percent of GNP, and changes in rainfall will havea significant social and economic impact. Meanwhile, it is estimated that a temperature increase of 2-3.5 degrees Celsius in India could result in a decline in farm revenues of between 9 and 25 percent. The International Rice Research Institute, for example, estimates that for every degree Celsius of night time temperature increase, there is at least a 10 percent decrease in rice production for the African region. While some areas will benefit from longer growing seasons (such as northern Asia), changes in water regimes will render other areas unsuitable for traditionally-grown products, and others areas will become susceptible to new forms of crop and livestock diseases. In regions already affected by food shortage and famine, this could cause further disruptions in food supply. The United Nations Development Programme (UNDP) warns that the progress in human development achieved over the last decade may be slowed down or even reversed by climate change, as new threats emerge to water and food security, agricultural production and access, and nutrition and public health. The contribution of climate change for sea level rise, droughts, heat waves, floods and rainfall variation could, by 2080, push another 600 million people into malnutrition and increase the number of people facing water scarcity by 1.8 billion (UNDP, 2008). Agriculture constitutes the backbone of most African economies. It is the largest contributor to GDP; the biggest source of foreign exchange, accounting for about 40% of the continent's foreign currency earnings; and the main generator of savings and tax revenues. In addition, about two-thirds of manufacturing value-added is based on agricultural raw materials. Agriculture remains crucial for pro-poor economic growth in most African countries, as rural areas support 70-80% of the total population. More than in any other sector, improvements in agricultural performance have the potential to increase rural incomes and purchasing power for large numbers of people to lift them out of poverty (NEPAD, 2002; Wiggins, 2006). Climate change, however, is considered as posing the greatest threat to agriculture and food security in the 21st century, particularly in many of the poor, agriculture-based countries of sub-Saharan Africa (SSA) with their low capacity to effectively cope (Shah et al., 2008; Nellemann et al., 2009).

6.2. Health Impacts

Clearly, the health implications of changes to water supply are far-reaching. Currently, more than

3 million people die each year from avoidable water-related disease; most of them are in developing countries. The effects of climate change on water will exacerbate the existing implications of water shortages on human health, as follows:

- *Water-borne diseases*: result from the contamination of water by human/ animal faeces, or by urine infected with pathogenic viruses/ bacteria, both of which are more likely to occur during periods of flood and therefore intensify with the projected increases in natural disasters under climate change. Diseases are transmitted directly when the water is drunk or used in food preparation.
- *Water-washed diseases*: those resulting from inadequate personal hygiene as a result of scarcity or inaccessibility of water (including many water-borne diseases and typhus).
- *Water-based diseases*: those caused by parasites that use intermediate hosts living in/ near water (e.g. guinea worm).
- *Water-related diseases*: borne by insect vectors that has habitats in/near water (such as malaria).
- *Water-dispersed diseases*: infections for which the agents proliferate in fresh water and enter the human body through the respiratory tract (e.g. legionella). Further indirect health challenges include malnutrition arising from agricultural disruption and food insecurity.

7. Possible Soulution on Climate Change

Climate is just one of many factors challenging future water planners and managers. Indeed, changes in population, economic conditions, technology, policies, and the relative value society places on alternative water uses may be more important determinants of future supply and demand conditions than those attributable to climate change (IPCC, 1996b). Vulnerability, or the sensitivity and potential magnitude of damage associated with climate changes, may also be greatest for regions where current stress on water resources is high. There are two ways to manage the risks posed by climate change: mitigation of GHGs to slow or reverse the pace of climate change, and adaptation to climate impacts to minimise their effects.

7.1. Mitigation

Mitigation implies the human measures, structural and non-structural, undertaken to limit the adverse impacts of climate change by reducing the levels of GHGs in the atmosphere. This is accomplished through the development of appropriate technology for reducing emissions and/ or capturing them at their source. Examples of mitigation include measures such as energy efficiency, promotion of renewable energy sources, and carbon trading. Mitigation of GHGs is essential to slow the impact of climate change and it is the most cost-effective and least risky strategy for reducing the future effects of climate change. This is mostly the issue of developed world being they release mere GHGs from their industries.

7.2. Adaptation

Climate change is already occurred; even if mitigation efforts were immediately able to reduce global carbon emissions to zero, there will be certain and ongoing impacts from climate change that need to be addressed. Adaptation describes a set of responses to the actual and potential impacts of climate change to moderate the harm or take advantage of the opportunities that climate change may bring. In countries where the majority of poor people depend on agricultural income, proposed climate change adaptation strategies centre on increasing agricultural productivity and make their agriculture more diversified; including livestock, fishery and forestry that less vulnerable to climate stress and shocks. Beside to this Water management for agricultural production is a critical component that needs to adapt in the face of both climate and socio-economic pressures in the coming decades (Bates et al., 2008). With regard to agricultural production and water, climate change adaptation may include:

- Adoption of varieties and species of crops with increased resistance to heat stress, shock and drought. For example, a private-public partnership under the leadership of the African Agricultural Technology Foundation called Water Efficient Maize for Africa (WEMA) intends to develop drought-tolerant African maize.
- Modification of irrigation techniques, including amount, timing or technology (e.g. drip irrigation systems);
- Adoption of water-efficient technologies to 'harvest' water, conserve soil moisture (e.g. crop residue retention, zero-tillage), and reduce siltation and saltwater intrusion;
- Improved water management to prevent water logging, erosion and nutrient leaching;
- Modification of crop calendars, i.e., timing or location of cropping activities;
- Integration of the crop, livestock, forestry and fishery sectors at farm and catchment levels;
- Implementation of seasonal climate forecasting.
- Enhancing stakeholder participation in water development and climate change adaptation

8. Conclusion

About two-thirds of the population of the world will become more vulnerable to availability and use of water due to climate changes. It is one of the global issue that would affect the sustainable development of many regions. Regional acidification from industrial activity, water pollution (indirectly through air pollution and directly through discharge of pollutants from industrial activity and sewage disposal), desertification, and soil erosion may also be major threats to water resources.

One of the most significant impacts of climate change is likely to be on the hydrological system, and hence on river flows and regional water resources. This will be particularly true in arid and semiarid areas of Africa where water resources are very sensitive to climate variability, particularly rainfall. Climate change is predicted to have serious implications for the hydrology of the sub-basin, affecting the magnitude and seasonality of surface flows, and increasing the frequency of extreme events such as drought and floods. 34Even if Ethiopia is also known as a tower of east Africa, today many rivers and lakes become shrinking in size due to the decrease in river flow and some small streams dry up completely, and finally the magnitude of flow of the medium to large rivers will decrease significantly.

References

[1] Adams, R.M and D. E. Peck. 2008. "Effects of Climate Change on Drought Frequency: Impacts and Mitigation Opportunities"; Chapter 7 in *Mountains, Valleys, and Flood Plains: Managing Water Resources in a Time of Climate Change*. A. Dinar and A. Garrido, eds. Routledge Publishing.

[2] Bates, B.C., Kundzewicz, Z.W., Wu, S. and Palutikof, J.P. (eds) 2008. 'Climate Change and Water'. Technical Paper of the Intergovernmental Panel on Climate Change. Geneva: IPCC Secretariat.

[3] Chandrabose AS, Viswanath GK, Giridhar MVSS, Sridhar P (2012) Assessment of land use land cover changes in middle Godavari (G-5) sub-basin of river Godavari using RS and GIS.

[4] Covich, A.P., S.C. Fritz, P.J. Lamb, R.D. Marzolf, W.J. Matthews, K.A. Poiani, E.E. Prepas, Ecosystems of the Great Plains of North America." *Hydrological Processes* 11: 993–1021.

[5] Eaton, J.G., and R.M. Scheller. 1996. "Effects of Climate Warming on Fish Thermal Habitats in Streams of the United States." *Limnology and Oceanography* 41(5): 1109–1115.

[6] Fredeick. K. D. and Gleick. P. H. 1999. Water and global climate change: Potential Impacts on U.S. Water Resources; Prepared for the Pew Center on Global Climate Change.

[7] Frederick, K.D., and D.C. Major. 1997. "Climate Change and Water Resources." *Climatic Change* 37: 7 – 23.

[8] Gleick P, Shiklomanov IA. 1989. The impact of climate change for water resources. Second meeting of IPCC WG-2, WMO/UNEP, Geneva.

[9] Hammond, D. and Pryce, A.R., 2007. Climate change impacts and water temperature. Environment Agency Science Report SC060017SR. 111PP.

[10] Hassan, H., Aramaki, T., Hanaki, K., Matsuo, T. and Wilby, R.L., 1998. Lake stratification and temperature profiles simulated using downscale GCM output. J. Water Sci. Technol., 38, 217-226.

[11] Hauer, F.R., J.S. Baron, D.H. Campbell, K.D. Fausch, S.W. Hostetler, G.H. Leavesley, P.R. Leavitt, D.M. McKnight, and J.A. Stanford. 1997. "Assessment of Climate Change and Fresh water Ecosystems of the Rocky Mountains, USA and Canada." *Hydrological Processes* 11: 903–924.

[12] Houghton JT, MeiraFilho LG, Callander BA, Harris N, Kattenberg A, Maskell K 1996. Climate change 1995: the science of climate change. Contribution of WGI to the second assessment report of the Intergovernmental Panel on Climate Change. Cambridge University Press, Cambridge.

[13] IPCC. 1996b. Climate Change 1995: Impacts, Adaptations and Mitigation of Climate Change: Scientific - Technical Analyses. Contribution of Working Group II to the Second Assessment Report of the Intergovernmental Panel on Climate Change. Cambridge University Press, NY.

[14] Kinfe Hailemariam. 1999. Impact of climate change on the water resources of Awash River Basin, Ethiopia. Clim Res 12: 91–96.

[15] Lijalem Zeray Abraham, Jackson Roehrig, and Dilnesaw Alamirew Chekol 2007. Climate Change Impact on Lake Ziway Watershed Water Availability, Ethiopia. Catchment and Lake Research.

[16] Mall .R. K. Akhilesh Gupta, Ranjeet Singh, R. S. Singh3 and L. S. Rathore 2006. Water resources and climate change: An Indian perspective: current science, vol. 90, NO. 12.

[17] Matalas, N.C., and M.B. Fiering. 1977. "Water Resource Systems Planning." In *Climate, Climatic Change, and Water Supply.* National Research Council.

[18] Meyer, J.L. 1997. "Stream Health: Incorporating the Human Dimension to Advance Stream Ecology." *Journal of the North American Benthological Society* 16: 439–447.

[19] Nellemann, C., MacDevette, M., Manders, T., Eickhout, B., Svihus, B., Prins, A. and Kaltenborn, B. (eds) .2009. *The Environmental Food Crisis. The environment's role in averting future food crises. A UNEP rapid response assessment.* Arendal, UNDP.

[20] NEPAD. 2002. Comprehensive Africa Agriculture Development Programme.

[21] OECD .2003. *Poverty and Climate Change. Reducing vulnerability of the poor through adaptation.* Paris: OECD.

[22] Schindler, D.W. 1997. "Wide spread Effects of Climatic Warming on Freshwater Ecosystems in North America." *Hydrological Processes* 11: 1043–1067.

[23] Shah, M., Fischer, G. and van Velthuizen, H. 2008. *Food Security and Sustainable Agriculture. The Challenges of Climate Change in Sub-Saharan Africa.* Laxenburg: International Institute for Applied Systems Analysis.

[24] Tsarouchi GT, Buytaert W (2013) Monitoring land use changes in the Upper Ganga Basin, India by using Remote Sensing and GIS techniques on Landsat 5 TM data. Geophys Res Abstr 15: 229.

[25] UNDP. 2008. *Fighting Climate Change - Human Solidarity in a Divided World.* New York: UNDP.

[26] Wiggins, S. 2006. *Agricultural growth and poverty reduction: A scoping study.* Working Paper No. 2 on Globalization, Growth and Poverty. Ottawa: IDRC.

14

Research Method of Silting the Mountain Reservoirs Under the Current Climate Change

Giorgi Metreveli[1], Lia Matchavariani[2, *]

[1]Institute of Applied Ecology, FENS, Ivane Javakhishvili Tbilisi State University, Tbilisi, Georgia

[2]Department of Geography, Faculty of Exact & Natural Sciences, Ivane Javakhishvili Tbilisi State University, Tbilisi, Georgia

Email address:

giorgi.metreveli@tsu.ge (G. Metreveli), lia.matchavariani@tsu.ge (L. Matchavariani)

*Corresponding author

Abstract: The mountain reservoirs are an accumulator of the processing products of river banks and sediments. These elements form the silting prism and its bench, which are located in the reservoir and the riverbeds of the tributaries. As a result, there is increased riverbeds and there are a number of catastrophic flooding risk and deficit of the beach-forming sediment on the sea coast. The aim of our research is the estimation of accumulation of solid material in the reservoirs and tributaries, as well as a change of channel processes. In order to study the sedimentation prism-formation and equilibrium bad processes, the method of field experiment is proposed, which was used in the mountain reservoirs of Georgia. For analysis of obtained results were used the methods of mathematical statistics - least square method and differential calculus. The experiment showed that the final result of the reservoirs' silting is an accumulative terrace, which produced an equilibrium alluvial bed. Fractional distribution of sediments in silting prism is determined by reservoir control type and intra-annual distribution of fluvial sediments. The network of sediment extraction quarries has to be located along the reservoir taking into account this principle.

Keywords: Accumulate Terrace, Caucasus, Mountain Reservoir, Silting

1. Introduction

The current climate change has strengthened processes of sea shore abrasions and mountain reservoir sedimentation. High values of mountain reservoirs condition the current boom of construction of these facilities worldwide. A selection of their location occurs according to their characteristics, and due to this fact is ignored the threat that these reservoirs create to headrace population and infrastructure as a result of possible catastrophic floods, while in seaside countries it causes heavy deficit along the beachmaker sediments. Mentioned negative aspects are the result of the lack of corresponding knowledge. During designing project developers don't take into account rising of water level in beds of tributaries located above the reservoir.

Studies of reservoir silting worldwide, including Georgia are carried out starting from the 30's of the last century. The basic goal of these studies was the determination of operating time for the reservoirs, creation of calculation methodology of accumulated sediment volume and description of processes taking place in tailrace [1]. In 80's was executed global generalization of reservoirs, their integrated use and regional distribution [2]. In the same period reservoirs of the South Caucasus were studied [3]. A stochastic and probabilistic assessment of the flows and sediments from the viewpoint of their forecasting was also made in Georgia [4]. Fundamental studies on issues of reservoirs silting producing energy were carried out in Georgia [5], where the goal was an elaboration of the method of sediments' washout from reservoirs. Interesting studies were also conducted regarding reservoir ecology [6].

Nowadays, in many countries worldwide are especially intensified the studies from the viewpoint of silting, management and modeling of reservoirs; furthermore, monitoring, evaluation, sustainability, balancing, comparative analysis, as well as avoiding the effect on the Environment, catastrophic floods, provision of population security [7, 8, 9, 10, 11, 12, 13, 14, 15, 16, 17, 18, 19, 20, 21, 22], etc.

Today the main attention is focused on accumulative processes taking place in headrace of water reservoirs, since this issue was not examined before. Especial interest in accumulative processes running in headrace is caused by the fact that without study of these processes is impossible to identify settlements and communication lines caught in the zone of catastrophic water floods and assess appropriate risks. The second important reason is that mentioned negatives appear in the middle phase of operation, while dramatic and sometimes tragic nature they take only after expiration of reservoirs' lifetime.

Sediments accumulated in water reservoirs, i.e. silting prism (SP) consists of two growing parts. First of them is a part formed right in the reservoir, and the second part is an accumulated body, so-called bench. Growth of the latter causes rising of water level in the river bed that lasts until SP reaches its maximum limit, i.e. the volume after which accumulative processes in the headrace completely cease. In these conditions river has already a developed equilibrium bed (EB) on the surface of SP, the parameters of which (length L and slope I) are so big that river is able to completely move sediments to the tailrace.

Because of the fact that the study of SP and EB formation processes and their parameters for acting water reservoirs is complicated, expensive and longstanding process, expected negatives are completely ignored during the selection of location for reservoirs. This problem is not highlighted in sufficient volumes, neither in Georgian, nor in foreign scientific-technical literature. Respectively, interests and safety of population domiciled below the reservoirs aren't taken into account when selecting their location, and doesn't occur determination of volumes and fractional composition of accumulated sediments and assessment of risks of seacoast abrasion caused by lack of sediments. Due to the same reason are not elaborated the methods of avoiding or adaptation of these negatives.

Natural experiments are the highly effective means for assessment of solid material accumulation and variations of bed processes in water reservoirs and tributaries and for filling the lack of knowledge. For these purposes must be created artificial reservoirs, where will be possible to carry out an integrated study of SP and EB formation processes for less than two years, to determine functional relations between their parameters and hydrological-hydraulic characteristics of the river. Obtained results allow us to elaborate recommendations on avoidance and adaptation of mentioned negatives.

2. Study Area and Methods of Research

For a study of SP and EB formation in time and space were selected three mountain rivers from the Southern slope of the Caucasus mountains with a different regime (mountain regions of Western and Eastern Georgia). Their relatively straight segments were obstructed by 1-meter height dam and were covered by the network of longitudinal and cross sections of stationary observation points. This network

contains water body and river bed above it, the length of which was at least two-times bigger than water body length. Planning of SP and its bench was made taking into account the frequency of floods and freshet of tributaries. Field studies were carried out using a differential GPS-receiver Leica GS08 (fig. 1), which is connected to a network of the National Agency of Public Registry and provides a high precision (±5 cm) of geodetic measurements throughout the country. For analysis of obtained results were used the methods of mathematical statistics (least square method) and differential calculus (bounds method). Approbation of calculation results were implemented at Gumati reservoir, which is constructed at the Rioni River in 1953 (West Georgia), and Sioni reservoir, created at the Iori River in 1963 (East Georgia).

Figure 1. Differential GPS-receiver Leica GS0.

3. Results and Discussion

As natural experiments (fig. 2) show, reservoir's silting is the most intensive in the first phase of operation. At this time the most part ($r \geq 70\%$) of tributary sediments and solid materials formed due to riverbank deformation experiences sedimentation. SP formation occurs simultaneously, but with different rate in water reservoirs and tributaries (fig. 3). Sediment fractions (clay – fine sand), the hydraulic size of which in turbulent medium $U \leq 1.0$ m/sec, is transported by streams in the initial phase of operation through the whole water reservoir and one part of them forms the layer with size corresponding to reservoir control type. The remaining part of such fractions ($\geq 30\%$) doesn't take part in prism formation, since it is transported by letting out water to the tailrace. Transportation of detritus (coarse sediments) in this phase to the tailrace occurs sporadically, during emergency water outlets.

In following phases the share and diameter of sediments transported from the reservoir gradually increase and reaches its maximum in the third phase of the operation. At the end of this phase SP reaches limit values of development and is formed as accumulation terrace, on which EB is developed by tributaries.

Development of SP and its bench becomes especially active, when coarse-fraction sediment ($d \geq 10,0$ mm), which is precipitated in the zone of flood curve forms crests. This accumulative formation has the form of beach ridge (fulls), which slope exposed towards dam is sharper than the bench – slope formed in tributaries (fig. 3d). Such crest gradually increases in height, comes close to the dam and when reaches it, SP volume and EB length will become equal to their limit values.

The experiment showed that the rate of SP growth is diminishing – average annual volume (r_s) of materials precipitated in water reservoir and tributaries is the biggest in the initial phase of operation, afterwards it decreases in time in second and third phase and finally becomes equal to zero, when SP reaches its limit value.

$$\lim_{n \to T} r_s = 0 \tag{1}$$

Time distribution (r_u) of sediment transported from water reservoir is opposite to this process, it gradually increases and at the end of reservoir operation, i.e. after T time (in years) will become equal to the average annual volume (R) of sediments ($r_u = R$) and this sediment along with letting out water completely moves to tailrace:

$$\lim_{n \to T} r_u = 0 \tag{2}$$

Here, r_s and r_u are average annual volumes (m³/year) of sediments precipitated in reservoirs and tributaries and sediments transported by the stream to tailrace, correspondingly, R is annual volume (m³/year) of sediments brought by tributaries to the water reservoir, T – duration of water reservoir operation time (years), while $n=1,2,\ldots,T-1,T$.

EB starts above the initial river bed, at height (h, m) of dam outlet (fig. 3c). It follows SP surface and ends at the imaginary cross section, above which the river permanently keeps its natural ability of sediment transportation. EB length (L) and slope (I‰) determine the value of the SP surface area (F) and by means of this parameter is possible to calculate F value in any phase of EB development. Processes of SP and EB formation proceeded with the rate corresponding to the ratio (W/V) of their useful capacity (W) and annual amount of sediments (V).

SP surface in limit state is represented by the plain inclined towards the dam (fig. 3d), which begins from dam outlet. Its area significantly (sometimes $f \geq 60\%$) exceeds reservoir surface (fig. 3a). It is extended in tributaries up to the cross section, to which reservoir's water flood curve reaches during floods and freshets. SP bench sizes basically are depended on hydrological and hydraulic characteristics of tributaries: bench length (L), dam outlet height (h), sediment diameter (d), maximum discharge of water and sediment (Q_m, R_m) and inclination of the bed located above the water reservoir (I‰):

$$L = f(h, Q_m, R_m, d, I^{-1}) \tag{3}$$

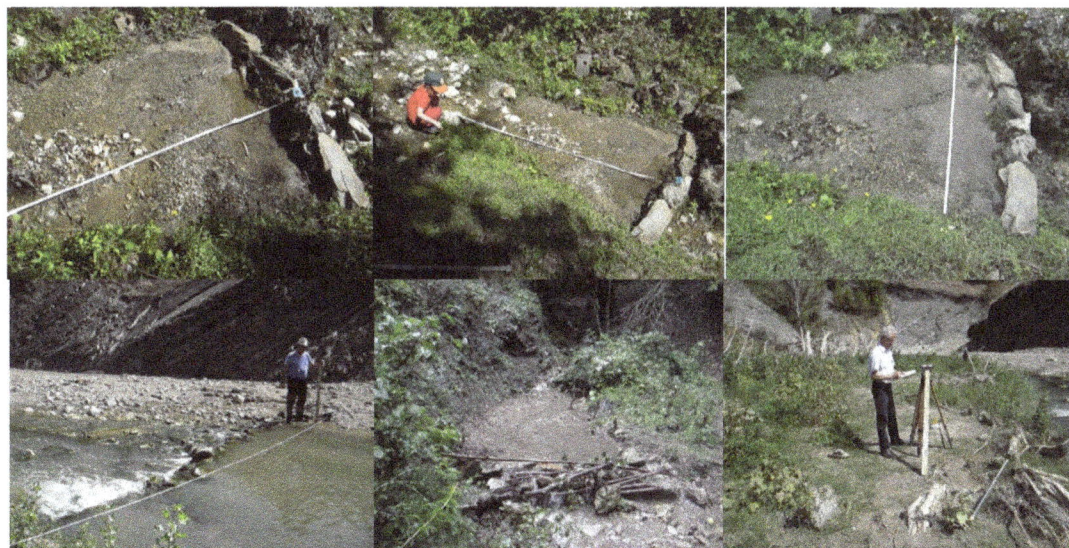

Figure 2. Natural experiment at mountain debris flows.

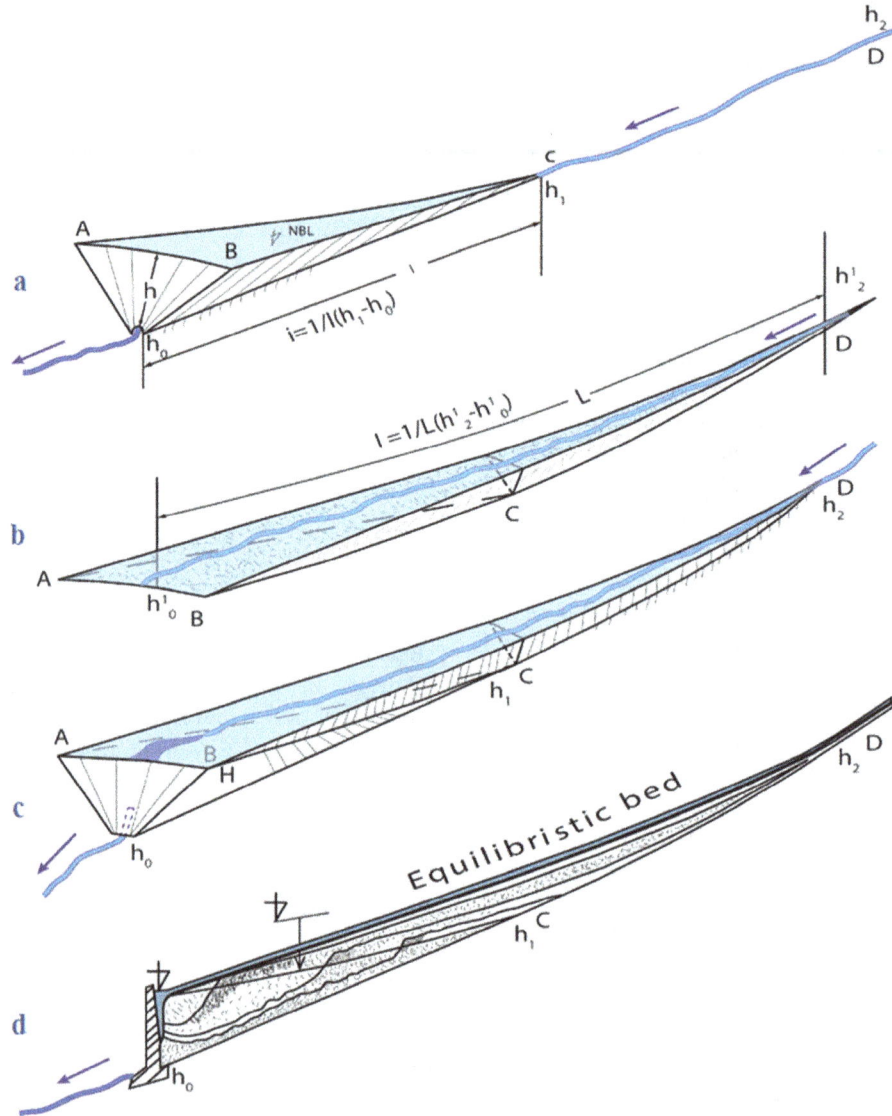

Figure 3. Transformation of silting prism and equilibrium bed.

At the same time EB length is inversely proportional to the inclination of that river section (I‰), which is located above the reservoir. This fact means that the more river bed is inclined, the shorter is EB. According to experimental results, a length of the latter (L) doesn't exceed a double length of the reservoir (2S).

$$L \leq 2S \qquad (4)$$

The experiment showed that the water flood risks for the population and infrastructure located above water reservoir increase proportionally to the height of SP and its bench. The more place is occupied by a bench in the river bed, the more high is a probability of flowage of the river.

A straight line is the simplest approximated form of EB. When the river crosses tectonic fault line its form comes closer to convex or concave curve. In this case it may be described by coordinates of the parabola, which begins from dam outlet and lasts up to above mentioned imaginary cross section.

Results of natural experiments are tested on Gumati and Sioni water reservoirs, since SP and ES are nearest to limited values. Sediments accumulated during many decades in Gumati reservoir and in river bed located above have raised the Rioni river bed level so much that it every year overflows its banks and submerges adjacent settlements and infrastructure. Territory adjacent to Sioni reservoir is in a more dangerous situation, since sediments precipitated above the water reservoir already have raised river bed by 3-5 meters. Due to this fact Iori River several times overflows protective structures and heavily damaged riverine area, while it would seem that settlements are reliably protected by the dam.

98% of Gumati reservoir are occupied by sediments, while Sioni reservoir already lost approx. 75% of volume. Both reservoirs are located in the active tectonic zone of the Southern slope of the Greater Caucasus. As of 2015 SP bench has changed river bed parameters in these reservoirs (approx. by 5-6 meters at Gumati and 2-3 meters at Sioni reservoir) so much that during floods and freshets these rivers every year overflow their banks and do damage to the settlements, infrastructure and environment.

Prediction, based on calculations, made according to results of natural experiments in the near future (2025-2030 years) SP and river bed parameters will change by another 15-20% that creates a real risk of disaster for neighboring settlements. As a consequence the probability of catastrophic floods and related risks are so high that floods, which usually are repeated once in two decades (P≥5-10%), become a serious threat for population and environment.

4. Conclusions

- Natural experiment is a highly effective means for monitoring the formation process of water reservoir SR and EB and study of their functional relations with tributaries' parameters, especially under conditions of time shortage. Thanks to these experiments is possible to gather in short span of time cognitive and application information on designed water reservoirs that will allow population to avoid the risks of catastrophic floods, will reduce a lack of knowledge on mountain reservoirs and thereby the science will get a new necessary and effective tool for designing.
- Creation of SP begins from time of putting a reservoir in the operation and ends with its formation as accumulation terrace. During this process coarse-fraction sediments precipitate in the movement zone of flood curve – near the river heads of tributaries and above them. One part of fine-grained sediments distributed by the flow across the whole water reservoir, forms accumulative layer, while the rest of sediments follow let out water to the tailrace;
- Annual volume of material participating in the formation of SP and its bench is diminishing (waning) in time. Respectively, the process of SP and EB creation is diminishing, i.e. it is the most intense in the initial phase of reservoir operation and is negligible when SP and EB reach limit values.
- SP with limit value of volume completely covers reservoir and a part of a gorge above it up to the imaginary cross-section, above which a river permanently keeps a natural ability of sediment transportation.
- SP surface represents an inclined surface, which begins from dam outlet and lasts up to above mentioned imaginary cross section;
- EB is a river bed developed at the surface of SP, the parameters of which correspond to the natural ability of sediment transportation. After the formation of EB erosion-accumulation processes in the headrace are virtually ceased and these processes are managed by other (anthropological, tectonic, sedimentation) factors.
- In general, fractional distribution of sediments in SP is determined by water reservoir control type and intra-annual distribution of fluvial sediments. Respectively, sediment diameter in SP increases from the movement area of reservoir flood curve (from the headwaters of tributaries) to the above mentioned

imaginary cross-section and reduces starting from this area towards the dam. SP is characterized by sediment foliation and drastic variation of its diameter both in vertical and horizontal directions, due to which the network of sediment extraction quarries has to be located along the reservoir taking into account this principle.

Acknowledgements

This study is funded by Shota Rustaveli National Science Foundation within the grant "Modern Methods of the Joint Problem Realization for Shore Protection and Hydropower" (#AR/220/9-120/14).

References

[1]   Shamov G. I. (1959). River sediments. Gidrometeoizdat, 378 p. (in Russian)

[2]   Avakian A. B., Saltankin V. P. & Sharapov V. A. Reservoirs. Moscow, Misl, 1987, 325 p. (in Russian)

[3]   Metreveli G. S. (1985). Reservoirs of South Caucasus. L., Gidrometeoizdat, 130, 296 p. (in Russian)

[4]   Svanidze G. G. (1977). Mathematical modeling of hydrological series. L., Gidrometeoizdat, 296 p. (in Russian)

[5]   Gvelesiani L. G., Shmaltsel N. P. (1973). Mode of mountainous rivers sediment and reservoirs siltation. Hydrology of lakes and reservoirs. Part 2. Reservoir, 5th session, IHHC UNESCO, Moscow, 38-48

[6]   Iordanishvili I. K. & Iordanishvili K. D. (2012). Issues of mountain reservoirs eco-evolution of Georgia. Institute of Water Industry, GTU, Tbilisi, 186 c. (in Russian)

[7]   Metreveli G. S., Kereselidze D. N. & Rehviashvili Sh.D. (2004). The dynamic of silting prism of mountain reservoirs. Dynamics and Thermic of rivers, reservoirs and sea coastal zone. Proceedings of the VI Conference,. IWP RAS, Moscow, 70-73 (in Russian)

[8]   Amitrano D., Di Martino G., Iodice A., Riccio D., Ruello G., Papa M. N., Ciervo F. & Koussoube Y. (2013, July). High resolution SAR for monitoring of reservoirs sedimentation and soil erosion in semi-arid regions. Geoscience and Remote Sensing Symposium (IGARSS), 2013 IEEE International, 911-914

[9]   Andredaki M., Georgoulas A., Hrissanthou V., & Kotsovinos N. (2014). Assessment of reservoir sedimentation effect on coastal erosion in the case of Nestos River, Greece. International Journal of Sediment Research, 29(1), 34-48

[10]   Bennett S. J., Dunbar J. A., Rhoton F. E., Allen P. M., Bigham J. M., Davidson G. R., & Wren D. G. (2013). Assessing sedimentation issues within aging flood-control reservoirs. Reviews in Engineering Geology, 21, 25-44

[11]   Caputo M., & Carcione J. M. (2013). A memory model of sedimentation in water reservoirs. Journal of Hydrology, Elsevier, 476, 426-432

[12]   Detering M., & Schuettrumpf H. (2014). Reservoir Siltation and Ecological Life Span of Dams. Wasser Wirtschaft, Springer, 104 (1-2), 30-33

[13] Dewals B., Rulot F., Erpicum S., Archambeau P., & Pirotton M. (2012). Long-term sediment management for sustainable hydropower. Comprehensive Renewable Energy. Vol. 6, 355-376

[14] Garg V., & Jothiprakash V. (2013). Evaluation of reservoir sedimentation using data driven techniques. Applied Soft Computing, Elsevier, 13(8), 3567-3581

[15] Gopinath G., Ashitha M. K., & Jayakumar K. V. (2014). Sedimentation assessment in a multipurpose reservoir in Central Kerala, India. Environmental Earth Sciences, 72(11), 4441-4449

[16] Hajji O., Abidi S., Habaieb H. & Mahjoub M. R. (2014). Regionalization and contribution to the study of reservoirs sedimentation: Lakes of Cape Bon and the Tunisia Central. 11th International Conference on Hydroscience & Engineering "Hydro-Engineering for Environmental Challenges", Germany, Proceeding 2014 (pp. 575-582)

[17] Hosseinjanzadeh H., Hosseini K., Kaveh K., & Mousavi S. F. (2015). New proposed method for prediction of reservoir sedimentation distribution. International Journal of Sediment Research, 30(3), 235-240

[18] Mansikkamäki H. (2013). Monthly sedimentation in some reservoirs of hydroelectric stations in Finland. Fennia - International Journal of Geography, 143 (1)

[19] Mattheus C. R., & Norton M. S. (2013). Comparison of pond-sedimentation data with a GIS-based USLE model of sediment yield for a small forested urban watershed. Anthropocene, Elsevier, 2, 89-101

[20] Ran L., Lu X. X., Xin Z., & Yang X. (2013). Cumulative sediment trapping by reservoirs in large river basins: A case study of the Yellow River basin. Global and Planetary Change, 100, 308-319

[21] Wisser D., Frolking S., Hagen S., & Bierkens M. F. (2013). Beyond peak reservoir storage? A global estimate of declining water storage capacity in large reservoirs. Water Resources Research, 49(9), 5732-5739

[22] Yasir S. A., Crosato A., Mohamed Y. A., Abdalla S. H., & Wright N. G. (2014). Sediment balances in the Blue Nile River basin. International Journal of Sediment Research, 29(3), 316-328

15

Underwater technique for prevention of oil pollution

George Nerubenko

NER*MAR Limited, Toronto, Canada – National University of Shipbuilding, Nikolaev, Ukraine

Email address:

optimalproject@hotmail.com

Abstract: The underwater tools, means, apparatus, vehicles and submarines are considered in the paper. It is stated that the majority of initial sources of petroleum pollution are underwater located. The novel concept of composing the emergency rescue complex is provided. The complex consists of best industrial technical components which could effectively solve the problem of underwater oil spillage caused by damaged oil tankers and offshore platforms.

Keywords: Emergency Rescue Operation, Offshore Oil Platform, Oil Spills, Submarine, Underwater Transportation

1. Introduction

The usage and exploitation of water resources brings the people to problem of providing pollution free environment. It could be achieved when exploiting technology is safe and specially designed meeting environment requirement. Some examples illustrate the necessity of such approach. Evaluation of sea polluting by damaged oil tankers had the following numbers: Exxon Valdez Tanker disaster (1989) was $ 6.0 Billion, and Prestige Tanker was $ 4.0 Billion in 2002. Sea polluting by damaged BP Deepwater Platform 2010 in Mexican Gulf was over $ 42.0 Billion. The advanced technologies are presented and the ways of implementation into practice are shown. Marine sector is considered in this paper, specifically paying attention to oil spillage from tankers and from tank barges, and oil spills from offshore platforms. Only the sources of oil spills such as damages, holes, cracks, etc of tankers and oil platforms are targeting. While with transporting by tanker the Company is dealing with an object having a definite "oil spill size" knowing that the total amount of leaked oil will be limited (however the Company doesn't know where the spill might happen or what type of damage might be), with an offshore platform, the situation is the exact reverse. The location of incident is certain but the potential release quantity and duration are unknown. The selection of sources of oil spills as a theme of current paper has a simple explanation. It is cheaper, faster and much more effective to kill the problem in embryo source instead of creating several fences of defense: engineering, technological, biological, chemical, etc. The majority of oil spillage sources are located underwater. The scope of interest of this paper is the oil spills running from sinking damaged oil tankers and crushed elements of offshore platforms. Starting from 1985 the author was involved [1], [2] in design, R&D of underwater tools, means, apparatuses, devices and submarines solving two problems: emergency rescue operations in marine petroleum industry and underwater transporting of oil and petroleum products.

2. Examples of Underwater Petroleum Catastrophes

The operation of technique in marine oil industry is accompanied by some natural malfunctions, accidents, wrecks, breakages, damages which occur from time to time. The amount and scope of damages could be different and should have random probability. But it is obvious that the bigger number of industrial objects would have the higher probability of non proper functioning, damages, breakages, and so on. Only for the Gulf of Mexico the approximate estimates indicate that there are 3600 production platforms, 300 drilling rigs, and 2500 vessels on a daily basis. EPA confirms that there are 20000 oil spills reported in the Gulf of Mexico yearly.

Unfortunately the huge disasters and catastrophes happen in marine oil industry. The Exxon Valdez oil spill [3], [4] occurred in Prince William Sound, Alaska, on March 24, 1989, when Exxon Valdez, an oil tanker struck Prince William Sound's Bligh Reef at 12:04 a.m. local time and spilled at least 41000 m3 of crude oil over the next few days, fouling about 2100 kilometers of coastline. The remote location of the spill and a delayed and inadequate response from Exxon and

Trans-Alaska Pipeline operator ALYESKA made matters even worse. The oil slick had spread over 8000 square kilometers. The authors of [3] stated: "Exxon was not prepared for a spill of this magnitude- nor were Alyeska, the State of Alaska, or the federal government. It is clear that the planning for and response to the Exxon Valdez incident was unequal to the task. Contingency planning in the future needs to incorporate realistic worst-cast scenarios and to include adequate equipment and personnel to handle major spills."

On November 13, 2002, while the Greek-operated, single-hulled oil tanker Prestige [5] was carrying 77000 metric tons of cargo of heavy fuel oil, one of its twelve tanks burst during a storm in northwestern Spain. Thinking that the ship would sink, the captain called for help from Spanish rescue team, with the expectation that the vessel would be brought into harbor. However, pressure from local authorities forced the captain to steer the embattled ship away from the coast and head to French coast. Reportedly after pressure from the French government, the vessel was once again forced to change its course and head south into Portuguese waters. Fearing for its own shore, the Portuguese authorities promptly ordered its navy to prevent it from approaching further. The integrity of the oil tanker was quickly deteriorating, and during the storm it was reported that a 12 - meter section of the starboard hull had broken off, releasing a substantial amount of oil. At around 8:00 a.m. on November 19, the ship split in half. It sank the same afternoon, releasing over 76000 m^3 of oil into the sea. After the sinking, the wreck continued leaking oil. It leaked approximately 125 tons of oil a day, polluting the seabed and contaminating the coastline. The affected area was a very prominent ecological region, supporting coral reefs and many species of sharks, birds and fishes. The heavy coastal pollution forced the local government to suspend offshore fishing for six months.

The *BP Deepwater Horizon* oil pollution disaster [6], [7] was caused by an explosion on the Deepwater Horizon offshore oil platform about 80 kilometers southeast of the Mississippi River delta on April 20, 2010. This resulted in 11 worker fatalities and a massive petroleum release. The *Deepwater Horizon* sank in about 1500 meters of water on April 22, 2010. After a series of failed efforts to plug the leak, BP said on July 15 that it had capped the well, stopping the flow of oil into the Gulf of Mexico for the first time in 86 days. Totally this subsea drilling system discharges petroleum of approximately 780000 m3 to U.S. waters. The oil slick produced by the Deepwater Horizon oil spill covered as much 75000 square kilometers. Such surface slick threatened the ecological systems and the economy of the entire Gulf Coast region. The underwater damages following by oil spills were the common characteristics of all these abovementioned catastrophes.

3. Statistics of oil Spills

There are several societies, institutions, organizations collecting the informational data about oil spills. It is very difficult to accumulate all statistical data in one body because of different nature of spills, different objects, different ways of recording and reporting, variety of areas, different tasks and addresses of reporting and so on. There are data of largest tanker spills collected in [8]. The data are based on reports of oil spills from one tanker and the volume of spilled oil from one tanker had to be over 37854 m^3. Data were collected from 1960 to 1999. Figure 1 shows the treated oil spill volume numbers (in m^3) plotted for interval of years 1960 – 1996. The reported data are presented by squared points interconnected with solid blue curved line. The trend-line approximation is plotted as solid red straight line. The graphical analysis indicates that the trend-line has the slight declining tendency. The absolute numbers of largest tanker spills are slightly reduced in 1960 – 1999 period from 86000 m^3 to 85000 m^3.

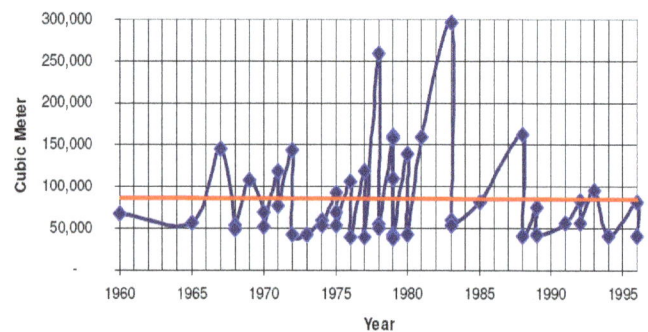

Figure 1. Largest tanker spills in 1960-1996. All spills of 37854 cubic meters or more.

ITOPF [9] gathered, maintained and analyzed a database of oil spills from tankers, combined carriers and barges. The collected information is about the accidental spillages since 1970, except those resulting from acts of war and pirates.

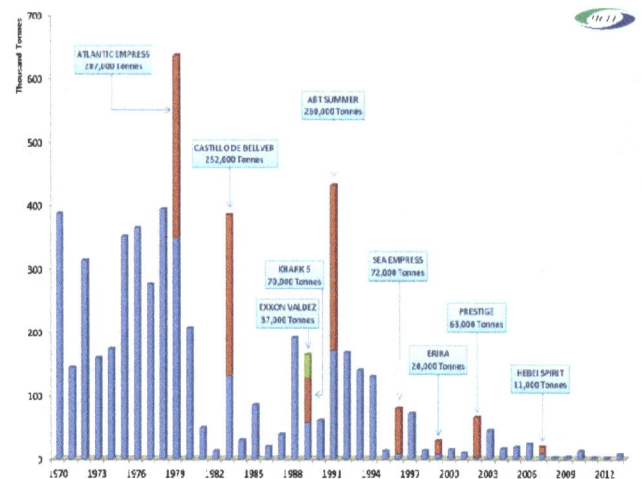

Figure 2. Quantities of oil spilt over 7 tones (rounded to nearest thousand), 1970 to 2013. (Copied from [9])

Quantities of tanker's oil spilt over 7 tones (rounded to nearest thousand), for time window 1970 – 2013 are shown in Figure 2[9]. In [9] it was underlined that usually a few very large spills were responsible for a high percentage of oil spilt. For instance, in four year period 2010-2013 there have been 28 spills of 7 tones and over, resulting in 22000 tones of petroleum lost. 90% of this volume was spilt in 8 incidents. Looking at the chart presented in Figure 2, it could be stated that it is clear evidence of sharp declining of tanker's oil spill tendency line starting from 2000.

It should be noted also that there has been a steady reduction in the number of offshore accidents reported by several studies. According to [10] (if to exclude the BP Deepwater well spill) in the average and median spill sizes, the size of spills from platforms tend to be smaller than those from tankers. For example, assuming that for 1964 -2013 interval the relative volume of oil spills from offshore platforms was 100%, and for last 15 years this number was reduced to 29.5%.Unfortunately the oil spill estimates are subjected by selected method of study (for example, see [11], [12]). There is no 100% proof that the situation with oil spills in the offshore platform industry is radically improved. As pointed out in [13], the most typical causes of accidents "include equipment failure, personnel mistakes, and extreme natural impacts (seismic activity, ice fields, hurricanes, and so on). Their main hazard is connected with the spills and blowouts of oil, gas, and numerous other chemical substances and compounds. The environmental consequences of accidental episodes are especially severe, sometimes dramatic, when they happen near the shore, in shallow waters, or in areas with slow water circulation. Broadly speaking, two major categories of drilling accidents should be distinguished. One of them covers catastrophic situations involving intense and prolonged hydrocarbon gushing. These occur when the pressure in the drilling zone is so high that usual technological methods of well muffling do not help. Drilling accidents are usually associated with unexpected blowouts of liquid and gaseous hydrocarbons from a well as a result of encountering zones with abnormally high pressure. No other situations but tanker oil spills can compete with drilling accidents in frequency and severity." Environmental impacts may arise at all stages of offshore platform activities, including initial exploration, production and final decommissioning. However the total amounts of underwater spills are very high and numbers of accidents are still significant.

One of the goals of petroleum engineers is a reduction of the environmental impacts of the petroleum industry on the marine environment, specifically through implementation of advanced underwater technique.

4. Selection of Advanced Technology.

There was the concept of universal marine transporting and emergency rescue salvage developed in 1980's [1] and [2].

The necessity of proposed concept was formulated during development of the underwater complex covering two tasks: emergency rescue combating the oil spills during salvage operation and underwater transportation of petroleum and its products during time period with no salvage operations. The ideas presented in [1] gave the opportunity propose the Marine Unit [2] of safe underwater transportation of oil and petroleum products combined with guaranteed prevention of oil spills. The concept is illustrated in Figure 3 schematically. The main

schematic members of designed Salvage Unit are (see Figure 3) 1 – two rescue submarines, and 4 – salvage ship. Each of submarines is equipped by 2 - sucking pipe (hose) (for pumping out the oil from sunk damaged tanker or accepting the crude oil from underwater drill or damaged well 5), and 3 - delivery pipe (hose) (for pumping in the collected oil to oil storage tank of ship 4). The ship 4 is equipped by Emergency – Rescuer systems and devices as well.

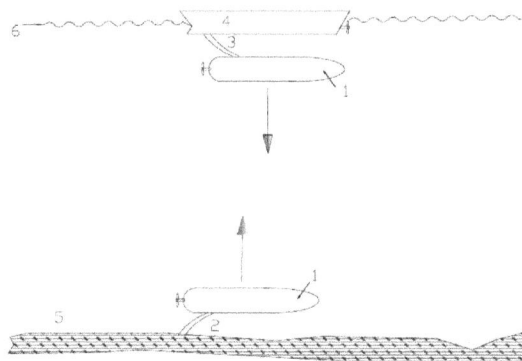

Figure 3. Proposed combined Salvage Unit.

The proposed configuration of Salvage Unit allows empty the damaged tanker (including the sinking tanker) in nonstop mode. One cycle of nonstop mode is as follows: while one submarine is involved in operation of discharging and pumping out the oil from damaged object, the other submarine is delivering the collected oil to ship's storage tank. The roles of submarines would be changed after finishing the cycle. When the emergency situation is over the submarines would be used as regular underwater tankers for transporting the petroleum products. Figure 3 shows the schematic skeleton of developed concept. It was proposed that submarines could have different shape configurations [1], [2], as shown in a Figure 4 as an example.

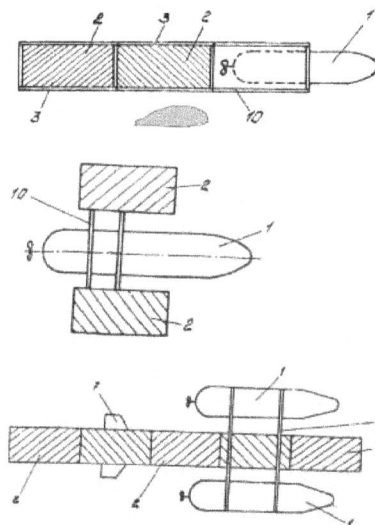

Figure 4. Versions of submarine equipped with additional tanks.

As per [1] the submarine 1 or underwater tugboat (see Figure 4) could carry several additional tanks 2 (or barges) allowing increase the volume of oil transport. Sometimes there are two submarines in one train (as presented in bottom part of Figure 4). Some tanks could have double-shelled 3 body structure. Tanks 2 and submarine 1 are interconnected by stiff framed self-adjusted carrier 10. Framed carrier 10 is affixed to tugboat 1. The submarine 1 and framed carrier 10 are supplied by horizontal stabilizing rudders 7 for providing the maneuverability to whole train.

The bottom part of hose 2 (presented in Figure 3) should be connected to output flange of damaged sinking oil tanker, to specially installed temporary flange on the crashed part of tanker, or to properly organized point (usually also flanged) of damaged offshore platform. The submarine has the remote underwater robot (for instance, as shown in Figure 5 [14]) in its structure for these cases. The robot could have one or several manipulating arms for servicing the troubled object.

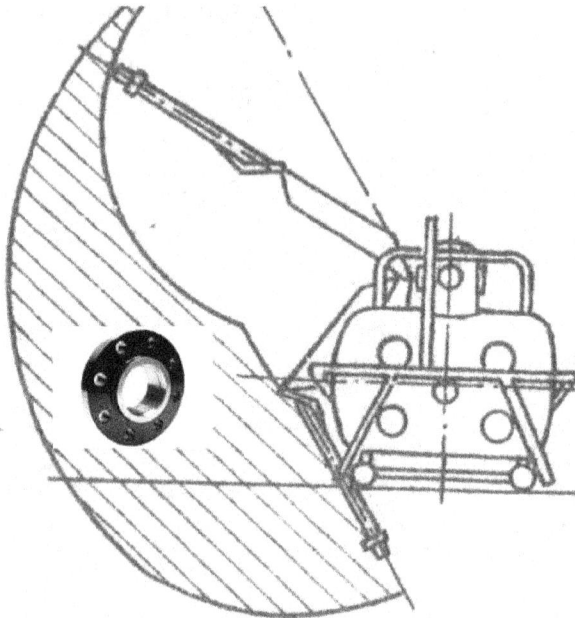

Figure 5. Underwater robot with manipulating arms.

Figure 6. Underwater self propelling connector-acceptor.

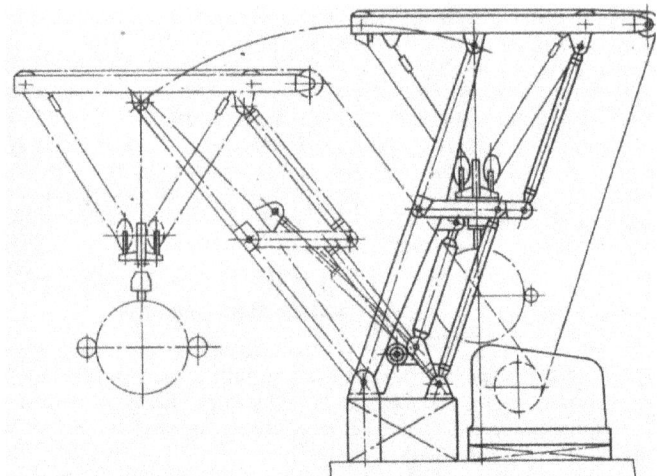

The bottom part of submarine's hose 2 (shown in Figure 3) could be equipped by special connector – receiver as shown in Figure 6 [2]. This device is a remotely controlled self moving tuning apparatus. The main purpose of this device is providing the fast and spill free connection. The connector – receiver (see Figure 6) consists of 1 – the rigid strong housing, 2 - protectors, 3 – fast acting connecting unit, 4 – a cap for connection to a flexible hose, 5 – a flexible hose, 6 – water-jets for maneuvering, 7 – the water-jet drives, 8 - TV cameras.

The presented in Figure 6 device has transfer capacity of 1000 m3 / hr of diesel fuel.

The Emergency – Rescuer ship 4(shown in Figure 3) contains the various types of systems and devices, however the main elements are [15] the Lifting Mechanism presented schematically as an example in Figure 7 in two positions, transit winch (see photo taken from [15] in Figure 8), and winch adjuster (see photo taken from [15] in Figure 9).

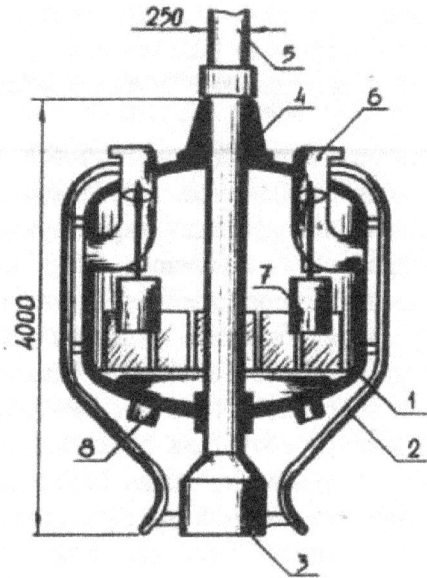

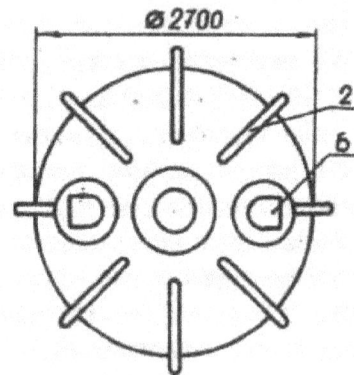

Figure 7. Lifting Mechanism. Max. Load 16 ton.

Professor David Tsagareli [2] had reported about project proving the idea and suggested method. The diesel propelled Russian navy submarine was converted into underwater rescuer.

Figure 8. Transit winch 135 kW. Drum Diameter 1000 mm

This submarine had been equipped by special remotely controlled robot, and pumping system for sucking out the oil from sunk tankers (underwater position). The rescuer could accept around 9,000 MT of oil and oil products during one phase of pumping out operational session.

Figure 9. Winch adjuster 50 kW. Drum Diameter 1200 mm

There were two pumps in operation, and the one phase time was 8 hours. The timing of experiment was as follows: 1 hour - descending underwater and search for "damaged" tanker, 2 hours - robot operation, 1 hour – connection / disconnection, 8 hours - working phase, 1 hour - coming back to base, 8 hours - pump out the tanker oil to base tanks. Actually one session took around 21-22 hours. "Sunken" tanker with 20,000 MT of crude oil was cleaned up in two days.

It is obvious that it should be the free time for the Salvage Unit when no accident happens. The exploitation of submarines could be planned as a transporting vehicle for this particular time interval. Figure 10 demonstrates the scheme of the underwater oil transportation applicable for Arctic regions covered by ice 26 [1]. The submarine – tanker 1 carries the oil tanks 2 interconnected by framed self adjusted carrier 10. The tanks 2 are sitting on technological bed 14. For instance, the delivery station 11 includes the shore oil storage 12 and the

shore ballast water storage 13. The oil pipeline 15 is connecting the shore oil storage 12 to technological bed 14. The ballast water pipeline 16 is connecting the shore ballast water storage 13 to technological bed 14. The customer facility (for example receiving station 17) includes the shore oil storage 19 and the shore ballast water storage 18 similar to 13 and 12.

For increasing of the effectiveness of proposed concept and for providing the flexibility in operation some elements of system could be changed and improved. The content below would be dedicated to descriptions and presentation of the parts and components of discussed concept.

5. Methods and Tools Solving the Underwater Oil Spill Problems

In first phases of emergency rescue operation detailed observation of troubled site could be done using the equipment configuration presented in Figure 11 (taken from [16]). Herein the rescue ship 1 is connected to deep water garage 3 by means of wired cable 2. The deep water garage 3 is connected to 5 - TV-controlled unmanned underwater apparatus by floating cable 4. The considered complex allows executing the necessary jobs at depths down to 6000 m with consumed power of 30 – 60 kW. The TV-controlled unmanned underwater apparatus could do a variety of operations and solve different tasks.

When the operation depth is to 500 m, it is possible usage of other scheme - "a submarine connected to the unmanned apparatus by floating cable" as shown in Figure 12. In cases of dealing with narrow paths and constrained holes on depths to 600 meters the TV-controlled autonomous unmanned underwater apparatus could be used. The example of such apparatus is shown in Figure 13 (copied from [17]). This device could operate with loads up to 20 kg.

A lot of attentions are dedicated to design of underwater apparatus and their components (see for example [14], [18], [19], [21]-[23], [26]). The German Lloyd has published rules for classification and construction [20] of unmanned submersibles (UUV); the operating and monitoring systems are also included in these rules. The rules [20] pointed out that unmanned submersibles may be Remotely Operated Vehicles (ROV) or Autonomous Underwater Vehicles (AUV). Well equipped an autonomous marine vehicle is proposed in [23]. The vehicle's configuration is including a rigid hull having an interior and a periphery, a deck joining the rigid hull at the periphery. Configured element is pivotally attached to the deck, and this element housing a plurality of sensors capable of effecting communication to and from the vehicle. In addition the vehicle could comprise various sensors and mission-specific hardware. Sensors include vehicle-mounted audio/video devices, radar, GPS and RF antennas, and other positioning and collision avoidance devices. Mission specific hardware includes various probes, protection systems, task and operational assemblies.

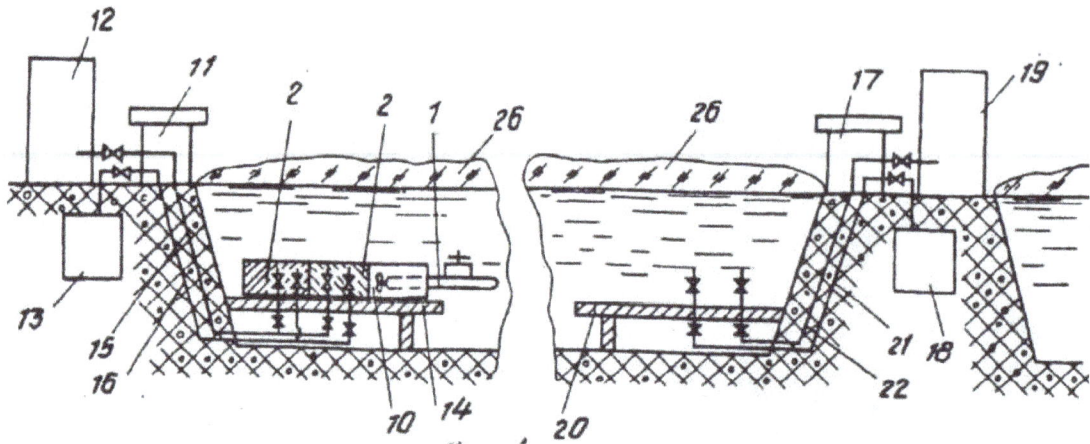

Figure 10. *Technological scheme of underwater under-ice transportation.[1]*

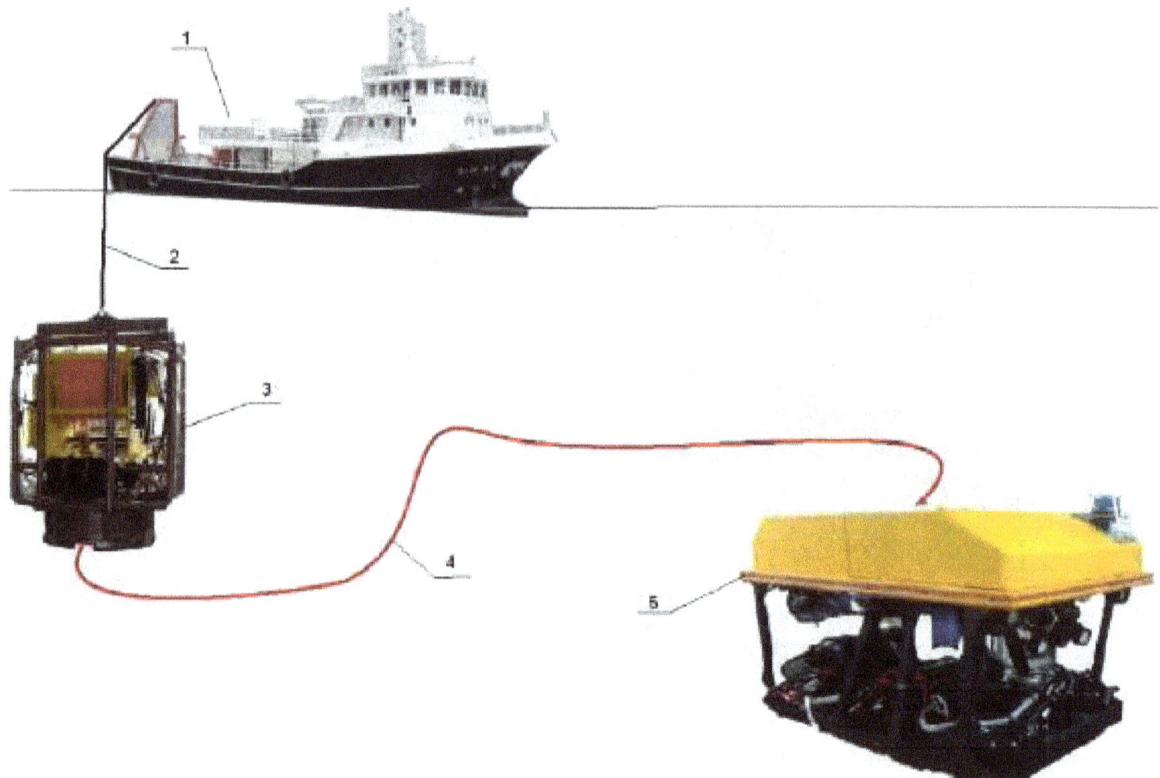

Figure 11. *Scheme of underwater technical unit [16].*

The laser-guided underwater robot is patented in [24]. The initial application of this robot was for reactor pressure vessel inspection, but underwater specifics could make its functionality and effectiveness much broader.

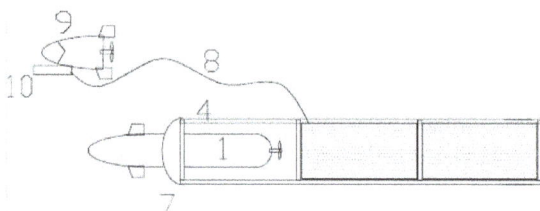

Figure 12. *Underwater tanker with unmanned robot. 1-submarine, 4 – framed carrier, 8 – floating cable, 9 – unmanned device, 10 – Lights and TV camera.*

Figure 13. *Underwater apparatus OBSOR 600 ([17])*

An apparatus in [24] uses a laser driven type underwater robot guided by a laser pointer for the quick and adequate inspection of underwater object. The position control methods are provided for the stable guidance of the robot to desired three dimensional (3D) inspection positions on the various types of underwater objects.

The different tools arrangement for an autonomous unmanned underwater apparatus is presented in [25] (see Figure 14).

Figure 14. Underwater apparatus MMT 2012

Such apparatus could target the following tasks like search, observation and inspection of underwater objects on depths to 3000 meters, and provide a geological research as well.

All apparatus of above mentioned types and configurations would solve the problem of finding of underwater source of oil spillage, and (in some cases) handle the installation of flange for receiving line.

There are a lot of proved engineering solutions for cases similar to BP Deepwater well breakage. First of all it is necessary to cover the place of accident with a sort of cap or sarcophagus if such situations happen. The idea of that device was proposed in 1995 in [27] and presented in Figure 15.

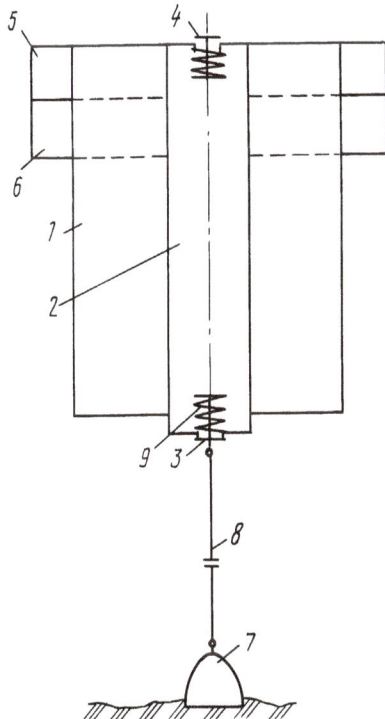

Figure 15. Device for subsurface oil gathering

Floating underwater vessel 1 (see Figure 15) is a ballast tank (unloaded hydrostatically from outer pressure). There is an equalizing tank 2 located in central part of 1. The equalizing tank 2 contains valves 3 (fill up) and 4 (ventilation).There are tanks of permanent buoyancy 5 and of variable buoyancy 6 in the upper part of 1. The special unit 7 is designed as an anchored collector covering the damaged well. The anchored collector is connected to valve 3 by pipe 8. The forced drive 9 is inserted in valve 3 structure. Such devices could fulfill the coverage of crashed well and protect the environment from oil pollution. Another version of design is demonstrated in [28]. The analysis of designed and developed solutions provides the effective means for tackling the problems similar to BP Deepwater well disaster. Figure 16 illustrates the special coverage or sarcophagus which is solving two problems simultaneously- collection of oil running out from damaged unit and pumping the cement for sealing and capping the opening of hole. The presented selected device is a sort of combination of ideas developed in [2], [27] and [28]. Generally speaking the covering vessel is the sarcophagus mounted on ocean bottom by means anchors, keeping the sarcophagus inside volume from penetration of the outer water. There is a pipe on side upper part of the sarcophagus. This pipe is servicing the oil collecting mechanism (not shown in picture).There is a special pipe line mounted on top of the sarcophagus. That line is helping transport and pump-in the cement for capping purposes. It is necessary to note that the designs, drawings and related engineering packages and documents of such underwater sarcophagus were issued in 1995 and it would be the decision of industry to use them for oil pollution protection.

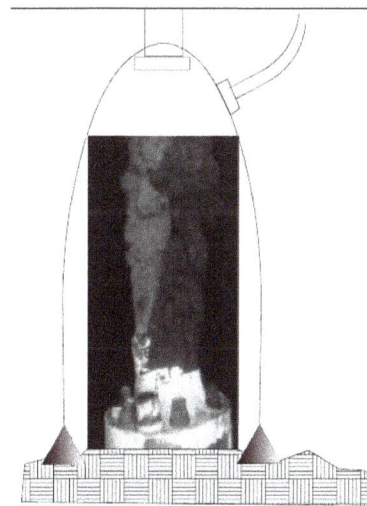

Figure 16. Underwater oil pollution protection device

The important component of proposed concept is the submarine which could work at depths 300-400 meters. Submarines would be the vessels converted from navy and adjusted as shown in Figure 4, or the multifunctional underwater tanker-rescuer could be built specifically for the discussed concept application. The information regarding two projects of these submarines was presented in [2]. The projects considered construction of underwater tankers with freight capacity of 9000 ton and 30000 ton. The schematic image of a tanker 9000 ton [2] is shown in Figure 17. The limited cargo capacity (9000 ton) was dictated by restriction of selected maximal submarine length of 170 meters. The designed underwater tanker consists of main master / navigator command cabin 1; rooms for living 2; engine room 3; pump room 4; cylindrical cargo tanks 5; main ballast tanks 6; ballast substitution tanks 7. The projected operational speed of a submarine was 14 knots, and the exploitation depth was 100 meters. The anticipated budget cost of construction was 19 million dollars. It had been supposed that the underwater tanker would be built within 45 months. Regarding the underwater tanker project of freight capacity 30000 ton the figures were as following: the length was 250 meters, the budget cost of building was 23 million dollars, and the predicted construction term was 50 months. These two projects demonstrated that the usage of underwater tankers would be the effective from financial point of view.

The attempt of creation of an unmanned autonomous automatic and self controlled submarine was provided in [29]. The proposed submarine would be programmed to dive to preset depths, move along preset trajectories, and return to the base. In addition to the autonomous properties, a remote control option is provided for the emergency situations or in order to perform special tasks.

The submarine is equipped with several sensors that can measure depth, orientation, attitude, location and speed. It is also equipped with an underwater video camera that can send wireless video pictures from underwater to the located above water surface monitor.

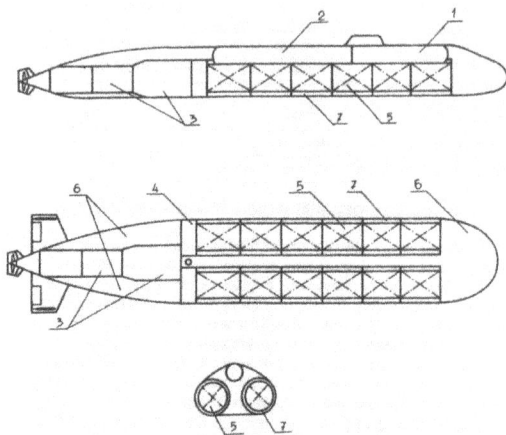

Figure 17. Underwater tanker. Cargo 9000 ton [2]

Various objectives of the unmanned autonomous submarine are to perform several tasks under water replacing crew, minimizing the cost of an underwater operations such as exploration, rescue, photography, and inspection of submerged structures (e.g. ship hulls, oil rigs, dams, etc.); monitoring various objects under water and transmitting live video and pictures to the operator on board of a commanding boat above water; being used as a carrier and base for underwater robotics, among other undersea functions

and tasks. Figure 18 (taken from [29]) provides the some structural details of invented submarine in accordance to [29].

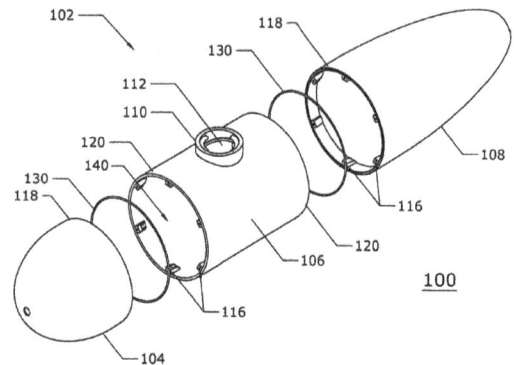

Figure 18. Submarine design in accordance to [29]

A submarine has several sections of hull 102: a forward hull section 104, a middle hull section 106, and an aft ward hull section 108. The submarine vessel can be assembled from two, three, or more hull sections with appropriate sealing devices 120. An opening 110 is formed in the upper side of one of the body sections, for instance the middle hull section 106, with a removable cover 112. The opening 110 is provided for access to the cabin 140 during assembly and servicing of the submarine 100. The removable cover 112 is provided to seal and protect the interior of cabin 140 of the submarine from the external water environment. Also additionally, the opening 110 facilitates assembly of the hull sections 108 using internal clamps 116. It is clear that the proposed submarine structurally based on approach which differs from one presented for submarine shown in Figure 17.

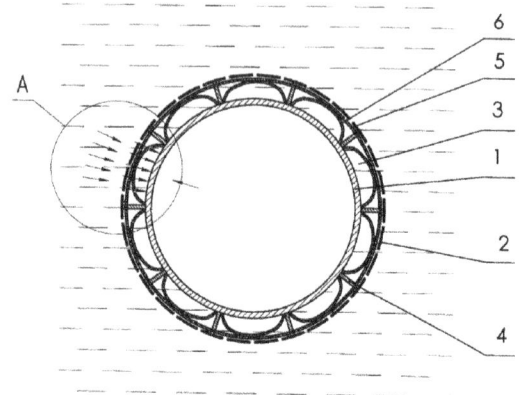

Figure 19. Freighter submarine, general section accordingly to [30]

The proper advanced protection of submarine metallic body with guaranteed zero leakage is suggested in [30]. Figure 19 (taken from [30]) shows the structural layout of the installation sheet lining of two layers of elastic material; Figure 20 shows the position of structural elements after sealing holes (when installing sheet lining of two layers of elastic material). The hull of the submarine cargo vessel contains a strong solid shell 1, the outer casing 2, forming intermediate space 3 that communicates with the external environment. The intermediate space 3 is divided into definite sections by the structural elements 4. The cover plate 5 (made of elastic material) is installed inside of intermediate space 3. The cover plate 5 consists of two or more layers 6 of a specified material lay overlapping each other. One edge of the layer 6 is fixed on the structural elements 4, and adjacent layers are fixed in opposite manner. The proposed idea works in a submerged position of a submarine as follows (see Figure 20). Let's assume that a hole as a result of accident has been formed in a solid prime shell 1. The hole is initiating the pressure drop in the volume between the sheets 5 and prime shell 1. Under the influence of pressure drop the loose part of the layer 6 of the lining 5 is moved to the surface of the shell 1, and in the same time moving along its surface. Thus, there is a displacement of bore in the plate 5 relatively to the hole in the solid shell 1. Finally the cover plate covers the hole in the solid shell 1 and the sealing of the holes would happen due to the elasticity of the material of the lining 5. The proposed novel tool would reduce the time of water penetration into submarine.

It must be highlighted that the crew starts necessary rescue action immediately for the hull integrity restoration so the invented tool should help keep the submarine safety.

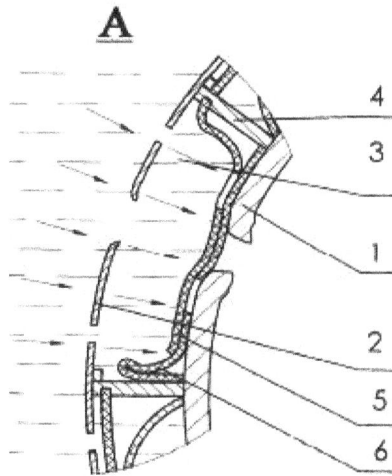

Figure 20. Freighter submarine, position of elements after sealing holes, accordingly to [30]

It is right time to say that an enormous amount of technical literature, research, development and engineering projects dealing with protection from underwater oil spills now exist, which has to be adequately organized and summarized. Of coarse, no one paper can include everything about the technique purposely struggling with underwater oil pollution. The emphases here are on effective and perspective engineering solution which is furnished in presented concept of the Three Component Salvage Unit.

A. For the oil tanker transporting industry it would be the first line of defence for oil spill prevention. The different situation is for offshore petroleum industry.

B. There is the proved concept of escape, evacuation and rescue measures developed and implemented in offshore petroleum industry (as an example, see [31], [32]), and the

concept (like in [31], [32]) is the first line of defence. Hence, the proposed herein Three Component Salvage Unit concept would be the second line of defence for oil spill prevention for offshore petroleum industry.

6. Conclusion

The pollution caused by underwater oil spillage has a significant impact on the environment. The majority of underwater sources of oil spills are connected to oil tankers and offshore petroleum platforms. The presented concept of Three Component Salvage Unit could successfully solve the problem of underwater oil pollution. The main elements of proposed concept are: a. the emergency – rescuer ship fully equipped with various techniques, ready to fix the damages in oil tankers and offshore platforms, b. the submarines which could execute two tasks – rescue and oil transporting, c. supportive unmanned devices which may be remotely operated or autonomous underwater vehicles. The formulation of mostly effective means, tools and devices is developed from analysis of the best industrial samples. The designed composition of each component is based on the perspective novel engineering solutions. The presented concept provides the guaranteed reduction or avoidance of underwater oil pollution.

Acknowledgements

First of all the author pays a special tribute to Professor David Tsagareli (deceased) who initiated the creation of presented concept and delivered the tremendous effort into implementation and promotion of this project.

The author is very much thankful to Dr. F.G. Davletyarov who discovered new horizons and opportunities of the concept application.

Special thanks to Professor K.K. Glukharev who inhaled the life in this concept.

The author would like to thank Professor A. F. Gal and M. Eng. O.N. Abramov for the input of their design knowledge and engineering experience.

Finally the author wishes to pay a special tribute to departed Dr. S.S. Pavlov who had energized and led his team to implement in industry a majority of elements and devices of presented concept.

References

[1] A.F. Gal, K.K. Glukharev, G.P. Nerubenko, D.V. Tsagareli, O.N. Abramov. Method of transporting of petroleum products and a device for its realization. Russian Patent 2037451. Oct 14, 1991

[2] V.P. Vinogradov, F.G. Davletyarov, K.K. Glukharev, A.V. Kuteynikiv, G.P. Nerubenko, V.V. Chernousov, D.V. Tsagareli. The underwater transportation of petroleum products to far north regions. Moscow, CNIITEN, 1993. 76 p.

[3] Samuel K. Skinner. William K. Reilly. The EXXON VALDEZ Oil Spill. A Report to the President. May 1989. The

National Response Team

[4] Office of Exxon Valdez Oil Spill (EVOS) Damage Assessment and Restoration. http://alaskafisheries.noaa.gov/oil/

[5] Thomas Höfer. Tanker Safety and Coastal Environment: Prestige, Erika, and what else? ESPR – Environ Sci & Pollut Res 10 (1) 1 – 5 (2003). Ecomed publishers, D-86899 Landsberg, Germany and Ft. Worth/TX, USA • Tokyo, Japan • Mumbai, India • Seoul, Korea.

[6] Curry L. Hagerty, Jonathan L. Ramseur. Deepwater Horizon Oil Spill: Selected Issues for Congress. Congressional Research Service. 7-5700. Report R41262. July 30, 2010. 48 p.

[7] Jonathan L. Ramseur, Curry L. Hagerty. Deepwater Horizon Oil Spill: Recent Activities and Ongoing Developments. Congressional Research Service. 7-5700. Report R42942. May 12, 2014. 16 p.

[8] Oil Spill Intelligence Report. Southern Cross University, Australia. Aspen Publishers. 2000.

[9] ITOPF. Oil Tankers Spill Statistics 2013. London, United Kingdom, 2014.

[10] Cheryl McMahon Anderson, Melinda Mayes, and Robert LaBelle. Update of Occurrence Rates for Offshore Oil Spills. OCS Report BOEM 2012-069 BSEE 2012-069. June 2012, 76 pages.

[11] Kerr, Richard A. (13 August 2010). A Lot of Oil on the Loose, Not So Much to Be Found. Science 329. (5993): 734–5.Bibcode:2010Sci...329..734K. doi:10.1126/science.329.5993.734. PMID 20705818.

[12] Dagmar Schmidt Etkin. Analysis of oil spill trends in the United States and worldwide. Proceedings, 2001 International oil spill conference. pp. 1291- 1300.

[13] Stanislav Patin. Environmental Impact of the Offshore Oil and Gas Industry. Ecomonitor Pub; 1 edition (December 1, 1999), 448 p.

[14] G.P. Nerubenko. Design of robot's drives. Textbook. Publisher: Shipbuilding Institute. Nikolaev.1985. 41 p.

[15] S.S. Pavlov, G.P. Nerubenko, B.N. Toropov. Advanced technology for modern ships. Shipbuilding SUDOSTROENIE. Leningrad.1988. 56 p.

[16] V.M. Rulevskiy, A.G. Yudintsev. The systems of electrical feeding of modern TV-controlled unmanned underwater apparatus. Proceedings of 5th Conference "Technological Problems of Worldwide Ocean Development" Section 1. Vladivostok, Russia, 2013. pp. 108 – 112.

[17] TETIS PRO. TV-controlled autonomous unmanned underwater apparatus OBSOR – 600. Catalogue, Moscow. 2014.

[18] Tadahiro Hyakudome. Design of Autonomous Underwater Vehicle. International Journal of Advanced Robotic Systems, Vol. 8, No. 1 (2011) ISSN 1729-8806, pp 131-139.

[19] Stokey, R. P. et al. (2005). "Development of the REMUS 600 autonomous underwater vehicle," OCEANS 2005. Proc. of MTS/IEEE, Vol.2 pp. 1301- 1304.

[20] Rules for Classification and Construction. I Ship Technology. 5 Underwater Technology. Germanischer Lloyd Aktiengesellschaft.. Hamburg 1 November 2009.

[21] Robert D. Goldbach. Apparatus and method for accommodating leaked oil within a double-hulled tanker after suffering grounding damage. US Patent 5520131. 1996

[22] Michael B. Geiger. Underwater self-propelled surface adhering robotically operated vehicle. US Patent 5947051, 1999.

[23] Richard Lawrence Ken Woodland. Autonomous marine vehicle. US Patent 6269763, 2001

[24] Jae-Hee Kim et al. Laser-guided underwater wall climbing robot for reactor pressure vessel inspection. US Patent 5809099. 1998.

[25] L.A. Naumov, I.N. Borovik, A.I. Borovik. An autonomous unmanned underwater apparatus MMT-2012. Proceedings of 5th Conference "Technological Problems of Worldwide Ocean Development" Section 1. Vladivostok Russia, 2013. pp. 46-48.

[26] Marine Robot Autonomy. Editor- Mae L. Seto. Department of Mechanical Engineering, Dalhousie University, Halifax. Springer New York 2013. 373 p.

[27] B.B. Klyachkin. Floating underwater vessel for collecting and storage of petroleum and petroleum products. Russian Patent. №:2021959. 1994.

[28] Eugene R. Barnett, Deep sea oil salvage means. US Patent 4,531,860. 1985

[29] Khaled R. Asfar et al. Unmanned autonomous submarine. US Patent 7290496. 2007.

[30] V.B. Shepelenko. Freighter submarine. Russian Patent RU 2488516. 2013

[31] Canadian Association of Petroleum Producers. Spill prevention and response in Atlantic Canada. Brochure. 2006-0012, 2006. 4 pages.

[32] Atlantic Canada Offshore Petroleum Industry Escape, Evacuation and Rescue. Guide. Canadian Association of Petroleum Producers (CAPP). June 2010. 2010-0017. 34 pages.

16

An appraisal of groundwater quality in selected areas in Warri Metropolis

Ogeleka Doris Fovwe[1, *], Okpako Solomon[1], Okieimen Felix Ebhodaghe[2]

[1]Department of Chemistry, Federal University of Petroleum Resources, Effurun, Delta State
[2]University of Benin, Geo-Environmental and Climate Change Adaptation Research Centre, Benin City

Email address:

dorysafam@yahoo.com (Ogeleka D. F.)

Abstract: Twenty four (24) groundwater samples were collected from some selected areas in Warri Metropolis, Delta State in the Niger Delta area of Nigeria. The areas covered include: Udu, Ajaminogha, Edjeba, Bendel Estate, Enerhen, Igbudu, Jakpa and Eboh. The quality of these private borehole water samples was assessed using methods recommended for water quality by America Public Health Association. The water samples were analysed for physico-chemical and microbial parameters so as to ascertain their potability. The results revealed that pH was relatively low in all the waters with values varying from 4.15 to 4.86 pH. All the groundwater samples could be termed 'freshwater' as revealed in the total dissolved solids (TDS) and salinity levels, which ranged from (61.33 mg/L to 277.0 mg/L) and (12.55 mg/L to 82.59 mg/L) respectively. The mean turbidity values varied from 0.11 to 1.17 NTU. The water samples were fairly soft as revealed in the results for total hardness (12 – 16 mg CaCO$_3$/L). Total iron, copper and zinc had concentrations levels within the local (Nigerian) and International [World Health Organization (WHO)] standards. Concentrations levels of these metals in the sampled locations were in trace quantity. However, lead, cadmium and chromium were not detected in any of the samples analyzed. It has been known that intake of water with low pH values could cause some severe health implications including gastrointestinal disorders and ulcers. This assessment was necessitated due to the fact that individuals resort to digging their privately owned boreholes, since the Government public utility water services have failed and do not supply potable or domestic waters to residences in these areas. In this vein, since residents in these areas use the untreated water from such borehole extensively for drinking and major domestic purposes, there may be likely health and environmental implications.

Keywords: Water Quality, Potability, Standards, Contamination, Safe Drinking, Parameters

1. Introduction

Water is an essential and useful natural resource that is needed by all forms of life for survivor, and without water, there would be no life. Safe water means water that would not harm the person drinking or using it. To be safe, the water must have sufficiently low or no concentrations of harmful contaminants to avoid sickening people who use it. Water may be naturally potable, as is the case with pristine springs, or it may need to be treated in order to be safe. In either instance, the safety of water is assessed with analysis which looks for potentially harmful contaminants while ensuring that certain regulatory standards are complied with [1].

Water contains lots of dissolved chemicals, and since ground water moves through rocks and subsurface soil, it has a lot of opportunity to dissolve substances as it moves. Water is the most abundant substance on earth and an integral part of living tissue (about 25% of living organisms is made up of solid matter while the remaining 75% is water). It is also the most valuable resource of man. Water is important to all living things as it covers seventy five (75%) of the earth's surface. The oceans contain 97.5% of the earth's water that is seawater containing about 3.5% salt (NaCl), which is too salty for humans to drink, the land 2.4%, and the atmosphere holds less than 0.001%. Only one (1%) of the earth's water is available for drinking; 2% is frozen [2].

Safe drinking water is defined as water with microbial, chemical and physical characteristics that meet World Health Organization (WHO) guidelines of national standards on drinking water quality. Water, which is not safe to drink, can carry diseases causing organisms, heavy metals and organic contaminants [3-5]. It is believed that many of the diseases associated with water contamination are caused by pathogens. Examples of pathogens are bacteria, viruses, and parasites. Pathogens are considered to be "communicable" because they have the ability to spread from one person to another by way of contaminated water and/or other vectors. The most common diseases spread through water are diarrheal diseases; examples include cholera, typhoid, paratyphoid, salmonella, giardiasis, and cryptosporidiosis. Many pathogens arise from animal and human feaces and from insufficient water supply. People who consume such water could become ill, and there is the risk of death. Unfortunately, even in areas where the water is known to be unsafe, people may still drink it out of desperation or since there are no other alternatives. The lack of potable water is often accompanied by lapses especially in sanitation, such as open sewers and limited garbage collection [6].

Sources of natural water include rainwater, spring water, well water, river water, lake water and seawater. The presence of any contaminant would make water impure and fall short of its characteristics quality. If the water becomes polluted, it may lose its value economically and aesthetically and can become a threat to human health and to the survival of aquatic organisms living in it and the wildlife that depends on it. Water pollution comes from two main sources: natural processes (rainwater runoffs, tornadoes, tsunamis, agricultural, land and mining operations) and anthropogenic activities (wastewater, sewage effluent, petroleum contamination).

However, the following activities may cause harmful chemicals / substances to enter surface water and likely percolate groundwater supplies.

1 leakage from waste disposal, treatment, or storage underground sites / tanks.
2 discharges from factories, industrial sites or sewage treatment facilities.
3 leaching from aerial or land application of pesticides and fertilizers on farms or fields.
4 accidental chemical / oil spills.
5 improper disposal of household wastes such as cleaning fluids, paint and motor oil.
6 seepage from septic tanks or suck away pits.

In Nigeria, the Government public utility water supply no longer functions and as such individuals resort to digging their own privately owned boreholes. Most groundwater borehole owners do not disinfect or treat the water from their supplies. The water from these sources is used untreated [7,8].

The quality of drinking water can be determined by its appearance, taste, and odour. Water quality parameters for which standards are established and determined are classified into two main groups: physico-chemical and bacteriological properties. Physico-chemical properties include: Organoleptic (odour, colour, taste, smell), physico-chemical (pH, temperature, total dissolved solids (TDS), total suspended solids (TSS), total solids (TS), salinity as chloride, bicarbonate, alkalinity, sulphate, nitrate, conductivity, turbidity, total hardness and so on), inorganics (light and heavy metals - calcium, magnesium, iron, lead, chromium, cadmium, copper and so on), organics (pesticides, oil &

grease, total hydrocarbon content (THC), polyaromatic hydrocarbon (PAH), polychlorinated biphenyl (PCB), total petroleum hydrocarbon (TPH) and so on), gross organics [biochemical oxygen demand (BOD), dissolved oxygen (DO), chemical oxygen demand (COD), total organic content (TOC), permanganate value (PV)]. Bacteriological properties include: total coliform, *E. coli*, *Clostruidum perfringes* and so on.

The most common standards used to assess water quality related to drinking water, safety of human contact and the health of ecosystems include: World Health Organisation (WHO), Department of Petroleum Resources (DPR), Federal Ministry of Environment (FME), Nigerian Industrial Standards (NIS), United State Environmental Protection Agency (USEPA). However, standard established by WHO is the most commonly referred to.

The aim of this study was to evaluate the physico-chemical and biological characteristics of different borehole water in some selected areas in Warri Metropolis in the Niger Delta area of Nigeria. This is with a view to ascertaining their potability since the water is not treated and are used extensively for drinking and major domestic purposes.

2. Materials and Methods

2.1. Sampling Location

Warri is located in Delta State, which is in the Niger Delta ecological zone of Nigeria. The Niger Delta area is a prominent coastal region of the southern part of Nigeria facing the Atlantic Ocean, covering an area of about 70,000 square kilometres and comprising of nine states namely: Abia, Akwa Ibom, Bayelsa, Cross River, Delta, Edo, Imo, Ondo and Rivers states. Warri is in one of the twenty five Local Government Areas (LGAs) of Delta State [9,10]. The city is subdivided into Warri South, Warri North and Warri South-East. Anthropogenic activities in this region include but not limited to gas flaring, refinery, acid rain/precipitation, burning of wood, fossil fuels, commercial, and so on. To explicitly describe climatic information within a particular area, analysis of long term climate historical data were obtained from the Nigeria Meteorological (NIMET) Centre, Oshodi Lagos with respect to coastal areas comprising Warri, Sapele, Bayelsa and Rivers, between 2000 and 2010. It was noted that the southern part of the country, experienced rainfall between 3000 – 4500 mm. There is hardly any month without rain in the Niger Delta Coastline. Generally, in the southern part of the country, monthly averages are usually above 300 mm from June to October but less than 50 mm from December to March, during which time only about 4 % of the annual rain is recorded. The map of the sampling locations is shown in figure 1 while the sampling points and geographical coordinates are presented in table 1.

Table 1. Sampling points and geographical coordinates

Sampling Location	Latitude	Longitude
Udu	N050 491 30.31	E0050 791 64.111
Ajaminogha	N050 521 75.81	E0050 731 21.311
Edjeba	N050 541 05.21	E0050 731 83.111
Bendel Estate	N050 541 07.81	E0050 761 42.411
Enerhen	N050 521 73.31	E0050 771 23.711
Igbudu	N050 511 82.91	E0050 761 46.911
Jakpa	N050 561 60.31	E0050 751 47.311
Eboh	N050 521 38.11	E0050 751 37.511

2.2. Water Sampling

Water samples were collected in triplicates for physico-chemical and heavy metal analysis from twenty four (24) boreholes in Warri Metropolis, Delta State in the Niger Delta area of Nigeria. The areas sampled include: Udu, Ajaminogha, Edjeba, Bendel Estate, Enerhen, Igbudu, Jakpa and Eboh. The sampling containers were washed with acid and rinsed with detergent and plenty of water to ensure there was no contamination from the sampling containers. Appropriate sampling protocol was duly followed as the first rush of water sample was not taken. Samples were collected after three minutes of constant running of the water. The sample containers were rinsed three times with the water to be sampled before collection. The water samples were stored in 1 L polyethylene bottles and preserved by cooling at 4°C while samples for metals was preserved with 1-2 mL of 1:1 nitric acid (AR) [11]. The nitric acid was added to the water samples for metal analysis because it leads to a drop in pH therefore the loosely bonded ions can be released for determination.

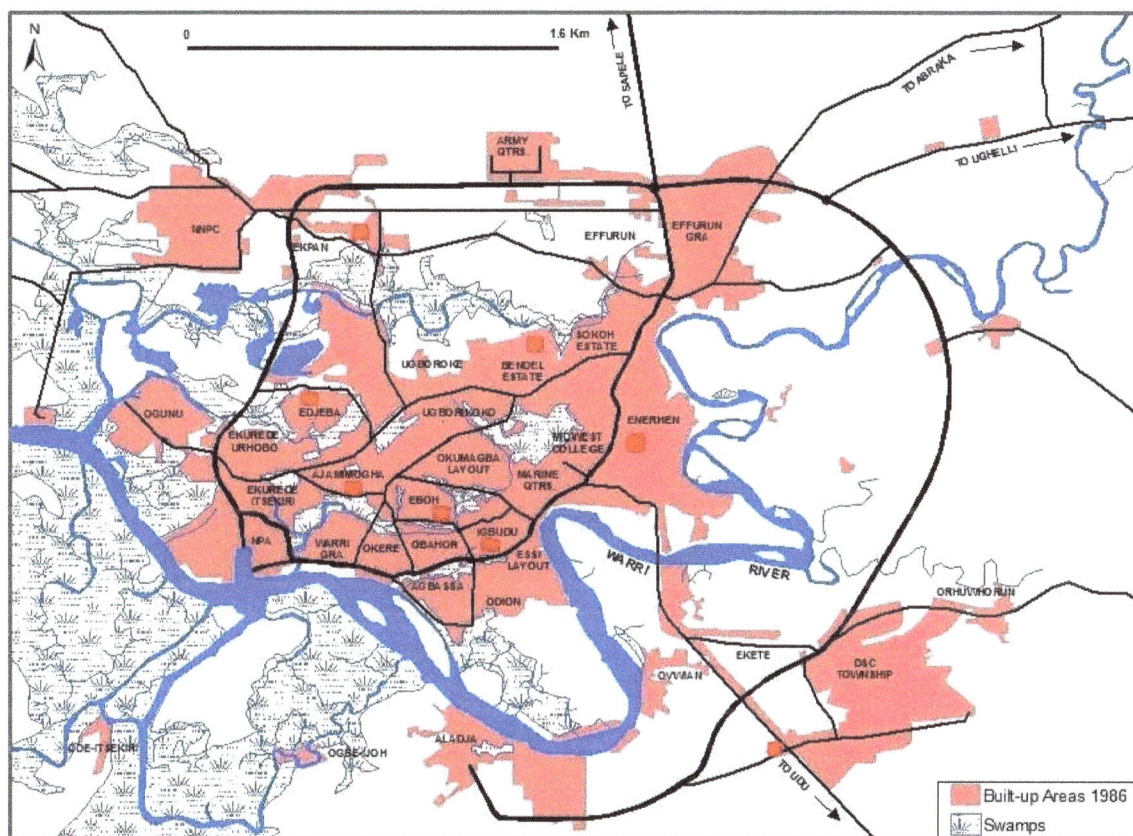

Map drawn by Professor Francis Odemerho, Southern Illinois University, Edwardsville, USA

Copyright: Urhobo Historical Society 2008

Fig 1. Map of Warri Metropolis showing sampling locations (red)

Table 2. Some indicating parameters determined for this assessment

Parameters	Analytical Methods
pH	pH, (APHA 4500 H$^+$, [12]
Temperature, °C	Thermometer (APHA, 2550-B)
Total dissolved solids (TDS), mg/L	TDS (APHA 2540-C)
Dissolved oxygen (DO), mg/L	DO (Winkler), (APHA 4500-O-B)
Salinity (Cl$^-$), mg/L	Mohr's Argentometric method (APHA 4500 Cl-B)
Conductivity, μS/cm	Conductivity (APHA 2510 B)
Sulphate, mg/L	SO_4^{2-} (Turbidimetric), (APHA 4500-SO_4^{2-}-B)
Total suspended solids (TSS) , mg/L	TSS (APHA 2540-C)
Nitrate, NO$_3$-N, mg/L	APHA 4600-NO$_3$-B

2.3. Determination of Metals in Water Samples

Water samples were digested using concentrated nitric acid (AR). The samples were mixed and 50 mL was transferred to a beaker to which 5 mL concentrated nitric acid was added and brought to a boil on a hot plate to the lowest volume possible (15 to 20 mL). Filtration was done after digestion. The filtrate was then diluted to volume with distilled water in a 50 mL volumetric flask [11]. The concentration of heavy, alkali and alkaline metals were analysed using Atomic Absorption Spectrophotometer (AAS) by direct aspiration into a standardized computer interfaced Schimadzu AAS-6701F.

2.4. Statistical Analysis

The mean and standard deviation of the various parameters were calculated using analysis of variance (ANOVA) in excel windows. Pictorial representation of some parameters was also done to indicate comparison with local and International standards. Variations in concentration levels between the different locations were also observed.

3. Results

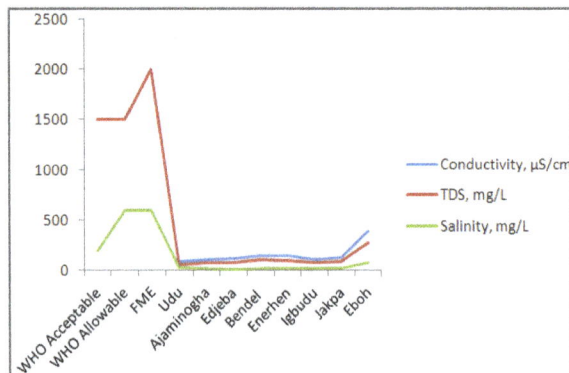

Fig 2. Mean concentration of salinity, TDS and conductivity levels in borehole water from Warri Metropolis

The results for twenty four (24) water samples analysed for potability in some areas in Warri, Delta State are presented Table 3. The water samples analyzed showed a short fall in pH when compared with local (Nigerian) and International (WHO) standards. Mean pH values were in the range of 4.15 (Edjeba) to 4.86 (Udu) pH units. Total dissolved solids (TDS) values were relatively low in most of the waters and were within acceptable limits (61.33 mg/L – 277.0 mg/L). All the waters sampled could be termed 'freshwater' as revealed in the salinity level, which ranged from 12.51 mg/L (Edjeba) to 82.59 mg/L (Eboh) (Figure 2). The mean turbidity values varied from 0.11 to 1.17 NTU. The waters were fairly soft as revealed in the results for total hardness (12 – 16 mg CaCO3/L). The heavy metals analyzed in the different samples had concentrations that complied with acceptable International and Nigerian standards (Figure 3). However, values observed for iron, copper and zinc were in trace quantity. Lead, cadmium and chromium were not detected in any of the samples analyzed. The water samples from all the stations were free of microbial contamination.

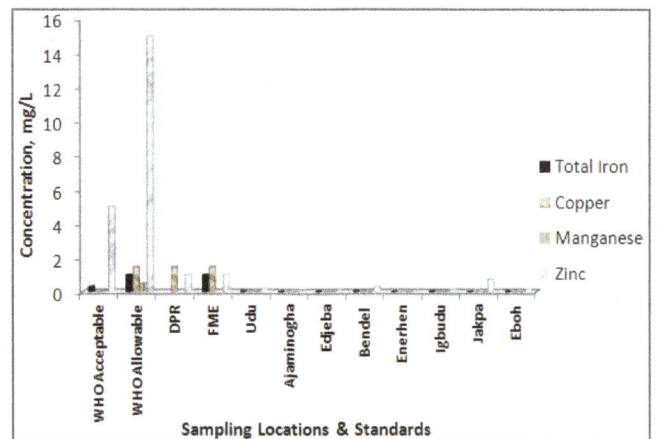

Fig 3. Mean concentration levels of heavy metals in borehole water from Warri Metropolis

Table 3. Values of Physico-chemical and biological parameters of groundwater samples in Warri Metropolis

Parameters	WHO Max Acceptable limit	WHO Max Allowable limit	DPR Limit	FME Limit	Udu	Ajaminogha
pH	6.5-8.5	6.5-9.2	6.5-8.5	6.5-8.5	4.86±0.03	4.43±0.04
Temperature, °C	N/A	N/A	N/A	N/A	28.9±0.08	29.0±0.06
Nitrate, NO_3-N, mg/L	10	N/A	N/A	N/A	0.04±0.01	0.01±0.001
Bicarbonate, mg CaCO3/L	500	500	N/A	N/A	4.88±0.01	<0.01
Sulphate, mg/L	200	400	N/A	N/A	1.05±0.001	0.94±0.001
Turbidity, NTU	5	5	N/A	10	0.56±0.04	0.14±0.01
TSS, mg/L	N/A	N/A	N/A	N/A	<1.00	<1.00
Hardness, mg CaCO3/L	500	500	N/A	N/A	16.0±0.3	16.0±0.3
Calcium, mg CaCO3/L	75	200	N/A	N/A	12.0±0.2	12.0±0.2
Magnesium, mg CaCO3/L	50	150	N/A	N/A	4.00±0.1	4.00±0.1
Total coliform, MPN/100 mL	Nil	Nil	N/A	N/A	Nil	Nil

Continued Table 3

Parameters	Edjeba	Bendel Estate	Enerhen	Igbudu	Jakpa	Eboh
pH	4.15±0.05	4.42±0.02	4.38±0.04	4.47±0.02	4.69±0.06	4.48±0.04
Temperature, °C	28.5±0.03	29.5±0.03	29.0±0.06	30.0±0.06	30.0±0.05	30.0±0.05
Nitrate, NO_3-N, mg/L	<0.01	0.07±0.02	0.33±0.05	0.09±0.02	0.15±0.04	0.21±0.11
Bicarbonate, mg $CaCO_3$/L	<0.01	<0.01	<0.01	<0.01	3.66±0.01	<0.01
Sulphate, mg/L	0.75±0.001	2.13±0.002	2.98±0.002	1.75±0.001	2.43±0.002	3.18±0.003
Turbidity, NTU	0.11±0.01	0.26±0.01	0.40±0.02	0.88±0.04	1.17±0.06	0.37±0.04
TSS, mg/L	<1.00	<1.00	<1.00	1.00±0.00	1.00±0.00	<1.00
Hardness, mg $CaCO_3$/L	12.0±0.2	12.0±0.2	12.0±0.2	12.0±0.2	16.0±0.3	12.0±0.2
Calcium, mg $CaCO_3$/L	8.00±0.1	8.00±0.1	8.00±0.1	8.00±0.1	12.0±0.2	8.00±0.1
Magnesium, mg $CaCO_3$/L	4.00±0.1	4.00±0.1	4.00±0.1	4.00±0.1	4.00±0.1	4.00±0.1
Total coliform, MPN/100 mL	Nil	Nil	Nil	Nil	Nil	Nil

Values are means of 3 replicates
N/A = Not available
TDS = Total dissolved solids

4. Discussion

Water is a very vital resource and many nations strive to protect the safety of their water and to increase access to potable water [13]. Some countries have laws governing water safety, with severe penalties for polluters. In developed countries, people may not put a great deal of thought into the source of their water since their tap waters are enriched with fluoride for health [2,14]. However, in developing countries and particularly in Nigeria, a large population does not have access to safe water and relevant legislations are more concerned with industrial effluent / wastewater regulations rather than household water supplies / sources.

Potable water is essential and if our body is not continuously supplied with water, the body becomes dehydrated and vital organs would deteriorate until they are no longer viable for human life. However, according to researchers, not all organisms and contaminants found in drinking water are harmful as long as they do not exceed safety values established by the World Health Organization (WHO) [15].

In developed countries, the Local Governments, public water systems, the states, and regulators work together towards the goal of ensuring that all public water supplies are safe. Public water systems have a greater responsibility to maintain sound treatment works and water distribution networks. They are responsible for ensuring that the water they supply does not contain contaminants at levels higher than the law allows. However, in Nigeria, most if not all of these legislations are not duly followed and as such monitoring the quality of potable water is not done at any tier of government, leaving the citizens to drink any water available without proper water quality assessment. Since the Government no longer provides public water services, individuals resort to digging their own boreholes and these are not regulated or treated.

The results from this study indicated that the drinking waters were potable and parameters measured were within acceptable regulatory limits except the relatively low pH values obtained in all the waters. It has been known that intake of water with low pH values could cause some severe health implications including gastrointestinal disorder and ulcer.

At low levels, contaminants generally are not harmful in our drinking water. A few of the naturally occurring substances may actually improve the taste of drinking water and may have nutritional values at low levels. Therefore, contaminants should be removed when absolutely necessary. However, at high concentration, heavy metals could lead to certain disruption in both humans and organisms. It is also important to remember that removing contaminants does not secure increased safety of human health [1].

Slight variation was noted in some parameters in the locations sampled and could be attributed to the intensity of different activities and population in these areas. Udu has less industrial activities, while businesses and a high population are concentrated in Jakpa and Enerhen areas. Edjeba axis has more of industrial activities while Ajaminogha, Bendel Estate and Eboh are residential zones though few business activities. Igbudu has a large market and lot of waste activities. Although turbidity, total iron, copper, zinc showed slightly significant levels, only pH had values below the specified range for both local and International standard.

The results obtained could be due to activities in the areas which include: improper sanitary conditions, sewage or septic tanks in close distance to the boreholes, poor waste management practices, acid rain/precipitation, mechanic workshops - disposal of spent oil and grease and spent batteries amongst others. Any water may be safe to drink if it complies with certain regulatory standards for both physico-chemical and microbial parameters. The guidelines / rules should be that values / concentrations should not exceed the required limits for it to be adjudged fit for human consumption.

5. Conclusion

Water is the essence of basic survival. Without it, life on Earth would cease to exist. In order to ensure that human life continues to exist, we must work together and do our part to improve the quality of drinking water. When the quality of drinking water is good, human health is also good. For the ground water samples covered in this assessment to be safe for drinking and other domestic use, they should be treated using the one or more of the following options: pH filters and lime to stabilize the pH while for low turbidity, the water should be passed through filter, which should be cleaned regularly. Similarly, in order to safe guard the health of humans; it is essential that the potable water be analyzed to ensure its integrity and / or standard before it is consumed. In addition, routine monitoring of drinking water should be carried out by individuals who are the consumers of such waters so as to detect any deviations from the set standards since the Government services are not functioning and cannot be relied upon.

References

[1] Gupta D.P., Sunita, J. and Saharan P., 2009. Physiochemical Analysis of groundwater of selected area of Kaithal City (Haryana) India. *Researcher*, 1(2): 1-5.

[2] Garg V.K., Chaudhary A., Deepshikha A. and Dahiya S., 1999. An appraisal of groundwater quality in some village of district Jind. *Indian J Environ Prot*. 19(4): 267-272.

[3] McGeer J.C., Brix K.V., Skeaff J.M., Deforest D.K. and Brigham S.I., 2003. Inverse relationship between bioconcentration factor and exposure concentration for metals: Implication for hazard assessment of metals in the aquatic environment. *Environ. Toxicol. Chem.*, 22: 1017-1037.

[4] World Health Organization (WHO)., 2006. Guidelines for drinking water quality. 3rd ed. WHO Press. Geneva, Switzerland. pp 398.

[5] Tawari-Fufeyin P. and Ekaye S.A., 2007. Fish species diversity as indicator of pollution in Ikpoba river, Benin City, Nigeria. *Rev. Fish Biol. Fisheries*., 17: 21-30.

[6] Riely, P.J. and Warren D.S., 1980. Money down the drain - A rational approach to sewage. *The Ecologist* 10:10.

[7] Federal Environmental Protection Agency (FEPA)., 1991. Guidelines and Standards for Environmental Pollution Control in Nigeria. University Press, Lagos. p. 238 - 245.

[8] Department of Petroleum Resources (DPR)., 2002. Environmental guidelines and standards for the petroleum industry in Nigeria (EGASPIN) Revised Edition. p. 277-288.

[9] Tamuno T., 2000. *The Niger Delta question*. Port Harcourt: Riverside Communications. *Tell* Magazine. (2005). 18 April, 16-18.

[10] Chinweze C. and Abiola-Oloke G., 2009. Women Issues, Poverty and Social Challenge of Climate Change in the Nigerian Niger Delta Context. Paper presented at the 7th International Conference on the Human Dimension of Global Environmental Change (IHDP Open Meeting), UN Campus, Bonn, Germany; 26-30 April.

[11] American Public Health Association APHA, America Water Works Association. AWWA, Water Environmental Federation WEF., 2005. Standard methods for the examination of water and wastewater. 21st ed.

[12] Folsom B.L., Lee C.R. and Bates D.J., 1981. Influence of disposal environment on availability and plant uptake of heavy metals in dredged material. Technol. Rep. EL-81-12.US Army Corps of Engineers Waterways Experiment Station, Vicksburg, MS.

[13] Routledge D. and Stewart D., 1998. Water: Essential for Existence. *Explore Magazine*. 8(5): 1-5.

[14] Shastri S.C., Bakra P.P. and Khan J.I., 1996. Industry environment and the law. R 13SA Publishers, Jaipur.

[15] Manivasakam N., 1996. Physical Chemical examination of water, sewage and industrial effluents 3rd ed, Pragati Prakashan, Meeret, India.

17

Quality of packaged drinking water marketed in Douala - Cameroon

Marlyne-Josephine Mananga[1], Marie Modestine Kana Sop[2], Etienne Nguidjol[3], Inocent Gouado[2, *]

[1]University of Yaoundé I, Faculty of Science, Department of Biochemistry, Yaoundé, Cameroon
[2]University of Douala, Faculty of Science, Department of Biochemistry, Box 24157 Douala, Cameroon
[3]Laboratory of physicochemical and microbiological analysis of water, HYDRAC Society, Box: 12806 Douala, Cameroon

E mail address:

gouadoi@yahoo.fr (I. Gouado)

Abstract: Quality and safety of water are very important for the health. The proliferation of conditioned drinking water in urban areas of Cameroon these years raises the problem of quality control. This research was then conducted in Douala town, to determine microbiological and physicochemical qualities of local and imported conditioned drinking water, in order to inform consumers and help them to make better choices. After the survey carried out with the population in Douala town, 32 drinking packaged water in polyethylene bags and bottles were collected and analyzed. The choice was made according to the amount of the persons using this water. The research of the total germs was done by the Plate Count Agar, the coliforms determined by the Eosin Methylen Blue and staphylococcus by the Chapman medium. The hardness, the concentration of calcium and magnesium were determined by complexometric method using tetra acetic diamine ethylen (EDTA). The pH, the turbidity and the conductivity were determined respectively by a pH meter, turbidimeter and a conductimeter. Chlorides and nitrates were measured by UV spectrometric method. According to the results obtained, consumers surveyed choose a type of brand water for economic reasons, 82 % of trademarks investigated had worst sanitary safety. The analytical results demonstrated that 76 % of water produced in Douala are soft (0°F<TH<15°F), weakly mineralized (total dissolved solid <500mg/L) with $[Ca^{2+}]$ <100mg/L and $[Mg^{2+}]$ < 50mg/L. However, 100 % of imported water analyzed were very hard (TH>43°F) and highly mineralized. Water packed in bags harbored numerous species of bacteria including *Pseudomonas cepacia, Acinetobacteria*, and nonpathogenic *Staphylococcus*; their densities were respectively, $23 \times 10^3 \pm 3605,55$ CFU.100 mL^{-1}, $22 \times 10^3 \pm 1732,05$ to $54 \times 10^3 \pm 4582,57$ CFU.100 mL^{-1} and $23 \times 10^3 \pm 2000$ CFU.100 mL^{-1}. Conditioned water produced in Douala (in bags) had the worse microbiological quality than the imported water. On the contrary they are softer than imported water.

Keywords: Drinking Water, Quality, Norms, Waterborne Diseases, Douala

1. Introduction

Water is second to oxygen as being essential for life. People can survive days, weeks, or even longer without food, but only about four days without water. Water makes up about 60% of the human body. Most of the living tissue of a human being is made up of water; it constitutes about 92% of blood plasma, about 80% of muscle tissue, about 60% of red blood cells and over half of most other tissues (Turgut *et al.*, 2005).

It covers some 70% of the earth's surface, with only 3% being from fresh water sources. With the world population growing and the increasing pollution of our natural resources,

we are facing a water crisis. The World Health Organization has estimated that over 1 billion people lack access to safe drinking water and about 4000 children die every day from water borne disease (Virkutyte and Sillanpa, 2006). The World Health Organization estimates that 500 million diarrhea cases reportedly taking place each year in children less than five years in Asia, Africa and Latin America (WHO, 1972-73). UNICEF estimated that about 60 % of child deaths worldwide are attributed to infectious or parasitic diseases related to water, the most frequent are gastro enteritis. About 2.3 billion of people suffer from infectious gastro enteritis each year (UNICEF, 2001). The extent of enteric diseases in different areas depends upon the extent to which water is exposed to contamination. The incidence of typhoid fever, bacillary dysentery, infectious hepatitis and other enteric infections are common and are transmitted through contaminated water. Cholera is still a wide spread water borne disease in some developing countries. There are numerous other diseases that are transmitted through polluted water. (Afia *et al.*, 2006; Delolme et al., 1992). According to WHO, drinking water should be clear, colorless, odorless, tasteless and free of pathogens or other toxic chemicals (OMS, 1994).

Douala, Cameroon's economic capital, although it has a distribution network of drinking water through a national company, has in recent years, an increase in production and consumption of drinking packaged water. These water are mainly composed of spring water, treated water and natural mineral water. Each type of water is subject to specials restrictions related to its operating system, its packaging or its composition (Touridomon et al., 2009). In the city of Douala, there are several brands of drinking packaged water in plastics bottles or in polyethylene bags, produced locally or imported.

Much work has already been conducted in several countries on the quality of drinking water in order to protect the consumers (Birley, 1991; Pathak et al. 1993; Boutin, 1993; Makoutode et al., 1999): Some tests have been done on the microbiological and, or physicochemical properties of drinking packaged water in Tunisia (Khosrof and Boudabous., 1992), in Ivory Coast (Blé et al., 2009), in Algeria (Djellouli et al., 2005) and in Burkina Faso (Touridomon et al., 2009). In Cameroon, studies were mainly focused on the quality of spring water, wells, rivers and the distribution network (Nola et al. 1998; Tatkeu 2001; Sena 2001). However, the review of the literature is lacking data necessary for water quality monitoring and information management.

This study was then conducted in the town of Douala in order to determine the microbiological quality and physicochemical composition of drinking water packaged in order to highlight potential risks of waterborne diseases that the populations could be exposed to.

2. Methods and Materials

2.1. Sampling

The Douala Town is located at latitude 4° North and longitude 9° East. The relief is made up of many sedimentary rocks. It is influenced by a classical equatorial climate and made up of ferralitic soils (Mamert, 2000).

From January to April 2009, a survey was conducted among 388 consumers, 20 districts and 89 outlets in the city of Douala. It aimed to identify drinking water brands consumed most and locate the various places of purchase. Thirty two (32) brands of packaged drinking water including 25 locally produced and imported, 07 were purchased in supermarkets, power supplies around schools, markets and

shops in the city of Douala. For each brand, three (03) samples were collected.

2.2. Methods

2.2.1. Microbiological Analysis

The Plate Count Agar medium (Humeau laboratory, ref. 550150300-54) was used for the enumeration of total germs in the water. The Eosin Methylen Blue (Scharlau laboratory, ref. 01-592) for enterobacteria. The Mannitol Salt Agar (Chapman medium) (Scharlau, ref. 01-116) was used for the detection of staphylococci. The API 20E galleries (Bioméreux laboratory, ref. 20100- 20160) were used for the identification of Enterobacteria.

2.2.2. Physicochemical Analysis

Chloride and nitrate ions were assayed by the spectrophotometric method using a spectrophotometer (Unicam 500). The hardness, the concentration of calcium and magnesium ion were determined by complexometric method with ethylen diamine tetra acetic acid as disodium (Na_2 EDTA).The bicarbonate content was performed from the values obtained from the full title alkalimetric. The temperature, pH, conductivity and turbidity were determined from an electronic temperature probe Testo 925, pH-meter Denver, conductivity by the conductimeter Hanna and turbidity were done by turbidimeter Hanna. A conversion by a corresponding factor of the conductivity allowed to assayed total dissolved solids (TDS).

2.2.3. Statistical Analysis

The software used was Statgraphics Centurion. The analysis of average and standard deviations allowed us to compare the physicochemical composition of different packaged drinking water. The multiple comparison test of Duncan has allowed us to compare the different drinking water standards.

3. Results and Discussion

3.1. Survey

Water packed in polyethylene bags are water consumed by a large portion of the population for economic reasons. However, the main reasons for the choice of imported water remain for their quality and availability.

3.2. Health Survey at Outlets

This study was conducted in 20 districts of the city of Douala. 82% of outlets inspected had inadequate sanitary conditions. The bags of water were spread on the ground, sometimes placed in freezers in contact with meat, fish or other foods. 52.81% of the outlets had very congested soil allowing the free movement of cockroaches, villains, lizards and mice over pallets of water. Some of these creatures came even sometimes and leave their droppings on top of these pallets behind them. In other outlets packaged water were exposed to the weather such as they were placed under the sunshine or in direct contact with rainwater. This result leads to a deposit of a huge layer of dust over the packaging of drinking water. 13.48% of the outlets were located near the garbage. Only 17.98% of the shops showed hygiene compliance. The floors and walls maintained the commercial space being ventilated with air conditioners, surfaces covered with tile.

3.3. Health Survey of Consumer

Eighty-seven per cent (87 %) of consumers revealed to take no hygienic precautions before ingestion of drinking packaged water. After our investigation, we can say that despite the fact that waterborne diseases reported by people could come from the quality of water consumed, but also that failure to follow basic hygienic rules by traders and consumers would also provide a huge source of contamination and spread of waterborne diseases.

3.4. Microbiological Analysis

In Cameroon, the drinking water standards apply to packaged drinking water are the guidelines values of WHO. Packaged water analyzed show that the water of local production do not meet the standards. Their total count rate is higher than two thousand (2000) total germs in 100 mL. According to WHO, water of good microbiological quality should contain less than two thousand (2000) total germs /100 mL.

According to table 1, 16 % of drinking packaged water of locally production are polluted. This rate is higher than that obtained by Feumba et al., (2009) on the water consumed in the district of Bonangang (13.33%) and Tatkeu (2001) on Water of network distribution of Cameroon (0%). This difference is due particularly to the influence of processing methods on the microbiological quality of water consumed and on the other hand, the state's distribution network at the time of sampling. However, some authors such as Blé et al., (2009) notice that most of these packaged water collected in polyethylene bag on the market in Abidjan have concentrations of coliforms outsized respectively values of 78/100 mL and 32/100 mL. Similarly, according to Khosrof et Boudabous (1992) mineral water of Tunisian show a pollution levels by 32%, 9.2%, 5.5% and 12% respectively due to the total germs, thermo tolerant coli forms, fecal streptococci and mold.

The results obtained in our study are consistent with those of Salim and Malik., (2008); Nola et al. (2000) and Sena (2001), who worked respectively on the bacteriological and physicochemical parameters of the mouth of the Oued Beni Messous, wells and spring water of Yaoundé and the bacteriological quality of water from springs and rivers of Dschang, Baleveng and Bafou (Cameroon). Considering these results and pollution levels observed in our study, we can say that the packaged water sold in the city of Douala are more suitable for consumption compared to Tunisian mineral waters, the waters of the Great Abidjan, marine waters, or still water wells and rivers of the Department of Menoua

(West Cameroon). It is noteworthy that 87.5% of samples tested were consistent over the WHO guideline value. The results reveal (table 2) few cases of non-compliance with the standards guidelines value of WHO (WHO, 1972-73). These water at concentrations higher or lower a number of germs that are indicators of pollution: *Pseudomonas cepacia* ($23x10^3 \pm 3605.55$ CFU/100 mL), *Acinetobacter spp* ($22x10^3$ to $54x10^3 \pm 1732.05 \pm 4582.57$ CFU/100 mL) and non pathogenic staphylococci ($23x10^3 \pm 2000$ CFU/100 mL). We observed that concentrations are substantially above those required by the WHO guidelines. All organisms identified are opportunists' pathogens germs. These germs do not have direct effects on health, but under certain conditions (temperature fluctuation). They can cause many health problems such as secondary infection or acute gastroenteritis (Maul et al., 1989). Several factors could explain the contamination of either count:

We have contamination of groundwater; the water is generally more vulnerable than its peak close to the soil surface (Guiraud and Galzy., 1984). The origin of bacteria in various types of groundwater is still debated, but those would come from mid-surface transport during the infiltration of surface water to groundwater (Nola et al., 2001).

Another factor of contamination could be a deterioration of facilities or the presence of cracks in the pipes that could cause infiltration of polluted water thereby polluting the catchment (Maul et al., 1989). The absence or inadequate treatment, inadequate investment and risk knowledge, poor maintenance of the production units of bottled water are all blocking factors and risk factors for groundwater pollution.

Exposure of polyethylene bags of water to solar radiation during the sale, contributes to bacterial growth because intense heat transforms the inactive chloride ions (Blé et al. 2009).

3.5. Physicochemical Analysis

The results are compiled in Tables 3, 4 and 5. It appears from these results that the majority of water brands analyzed had the pH between 6.5 and 8.5 which are values accepted by the WHO standards. The pH was 7.18 ± 0.56 (Table 5) for imported water, 6.56 ± 0.86 for the local production of water packaged in bags and $7.43 \pm 0,50$ for packaged water bottles. These changes in pH from one sample to another are due to the CO_2 content of the water. For example, we noted a pH of 6 at Badoit bottled water which was carbonated water. Indeed according to Touridomon al., (2009), a high concentration of carbon dioxide (CO_2) would contribute positively to lower the pH. The season climate, the geological land traversed and different treatments applied to safe drinking water influence the physicochemical composition of drinking water. The imported water are extremely hard, while the water of locally production are sweet (water packaged in bags) and moderately hard (water packaged in bottles). This is the case of trademarks Tanguy, Semme, Supermont, Natura, and Volcanic, Hesco water (Table 3). The analyzed imported water would be excellent for pregnant and lactating women, infants, the old persons (Benani–Nodot and Hardy, 2000). In the same idea, the results of the analysis of cations (calcium and magnesium) also show that these imported brands were rich in calcium ($[Ca^{2+}] > 150$ mg / L) magnesium ($[Mg^{2+}] < 50$ mg / L) while the water of local production is low in calcium and magnesium. The low hardness of these local water is due to the chemical composition of the ground traversed.

Test results show that the anions of water imported brands were rich in bicarbonate (HCO_3^-), especially the brand that has a Badoit content of 1076.04 mg / L. The levels vary from bicarbonates 391.97 ± 306.50 for imported water, 15.20 ± 11.23 for water bags and 155.88 ± 47, 35 for bottled water produced locally. The test results showed that NO_3^- nitrate content varies from one brand of water to another. However, all observed values of nitrate meet the standards of WHO. We are seeing very high values in samples of produced water with local water brand Wally has a nitrate content of 24.56 and 17.03 for the brand Sweet Water (Table 3). This could be explained by the fact that these water brands have their point of capture in areas close to industrial discharges. The nitrate content of water varies according to season (Makoutode et al., 1999). It has been demonstrated that if a pregnant women consumes regular water rich in nitrates, it could increase the risk of methemoglobinemy in newborn baby (Rodier and al., 1996). Compared to the water of local production, the overal mineral content determined from the conductivity shows that all imported water were highly mineralized. The mean levels ranging from 0.72 ± 0.62 for imported water 0.09 ± 0.09 for water bags and 0,21 ± 0.07 for bottled water. The turbidity is

0 for imported water, 0.15 ± 0.15 for water bags. However, Table 5 shows that there is no significant difference between the imported water and water bottle wrapped in local production. The material used for packaging water locally produced bagged could be responsible for turbidity observed at these packaged water. Indeed for packaged water in bags at high temperatures or in poor storage conditions. The bags could release microparticles, which escape direct packaging and come into contact with drinking water and promoting the development of its turbidity (Blé et al., 2009).

4. Conclusion

This study allowed us to determine the physicochemical and microbiological quality of packaged drinking water sold in the city of Douala. It appears from this study that the people surveyed choose a type of brand water for economic reasons. Retail outlets and consumers had inadequate sanitary conditions. The microbiological parameter deserves strong vigilance because it represents an immediate risk to the health of consumers. It is therefore necessary to warn people of health risks. The physicochemical parameters reveals that locally produced water are soft, low mineral content: calcium and magnesium. However, the imported water was extremely hard, and highly mineralized in calcium and magnesium. More detailed studies of these packaged drinking water will have an exact idea of the quality of packaged water sold in the city of Douala.

Table 1. Rates of water pollution according to WHO standards

Origin of water samples	Imported	Local production
Quantity of polluted brands	0	4
Quantity of analyzed brands	7	25
Pollution rate according to WHO	0 %	16 %

Table 2. Mean density of total bacteria (expressed as UFC/mL of water) and identification

Samples	Golden	Ndiba	Sky Water	Cool
Mean ± SD	$23.10^3 \pm 3605,55$	$54.10^3 \pm 4582,57$	$22.10^3 \pm 1732,05$	$23.10^3 \pm 2000$
Isolated species	*Pseudomonas cepacia*	*Acinetobacter spp*	*Acinetobacter spp*	No pathogenic *Staphylococci*

Table 3. Physicochemical parameters of drinking packaged water (local brands)

Parameters / Brands	Tur (mg/L of SiO_2)	pH	HTT (F)	Ca^{2+} (mg/L)	Mg^{2+} (mg/L)	HCO_3^- (mg/L)	Cl- (mg/L)	NO_3^- (mg/L)	NO_2 (mg/L)	Cond (μs /cm)	TDS (g/L)
Aquaba	0.00	6.89	0.4	1.6	0.96	12.2	2.87	4.66	0.08	224	0.15
Aquavita	0	6.11	0.35	1.4	0.13	17.08	0.51	0.81	0.03	57	0.04
Audes	0.40	7.00	5.40	21.6	12.96	24.4	5.21	2.39	0.12	368	0.247
Bethesda	0.41	4.27	0.617	2.45	0.23	0	2.33	8.57	0.07	57	0.04
Cool	0.08	6.80	0.9	3.6	2.16	7.32	0.15	2.25	0.00	140	0.09
Delices	0.02	7.05	5.4	21.6	12.96	12.48	13.36	1.43	0	54	0.03
Gecca	0.25	7.01	6.84	27.37	10.87	10.25	1.11	0.79	0.16	22	0.02
Glanneuse	0.01	7.01	4.6	18.4	11.04	26.108	4.68	0	0.01	135	0.09
Golden	0.17	7.02	0.5	2.00	0.59	1.61	0	0.87	0	20	0.01
Hesco Water	0	7.5	22.5	90	54	119.56	3.19	1.64	0	254	0.17
Namiwa	0	5.38	3.8	15.2	9.12	14.64	11.01	0.43	0	32	0.02
Natura	0	8.00	5.5	22	13.2	145.66	3.33	7.36	0	269	0.18
Ndiba	0.09	7.00	3.56	14.24	8.54	39.04	0.49	0.86	0	108	0.07
Sawawa	0.08	7.02	0.5	2	0.59	7.08	0.33	0.64	0.03	17	0.01

Parameters Brands	Tur (mg/L of SiO₂)	pH	HTT (F)	Ca²⁺ (mg/L)	Mg²⁺ (mg/L)	HCO₃⁻ (mg/L)	Cl- (mg/L)	NO₃⁻ (mg/L)	NO₂ (mg/L)	Cond (µs /cm)	TDS (g/L)
Semme	0	8.6	14.8	59.4	35.61	246.44	18.61	2.90	0.05	517	0.346
Sky Water	0.13	6.8	0.5	2.00	0.59	36.6	10.90	3.08	0	98	0.07
Source d'Orly	0.36	4.63	4.35	17.4	10.44	4.88	6.99	3.09	0.04	511	0.34
Source de vie	0.39	7.53	3.6	14.4	8.64	2.44	0.25	1.87	0.01	75	0.05
Source du Sahel	0	7.03	4.7	18.8	11.28	43.42	7.35	1.81	0.04	73	0.05
Supermont	0	7.00	21.8	87.2	52.32	122	3.50	7.73	0	228	0.153
Sweet Water	0.24	6.7	2	8.00	4.8	12.93	6.06	17.02	0.00	344	0.23
Tangui	0	7.00	9	36	21.6	136.92	2.62	0.11	0	304	0.204
Tati Water	0.09	7.00	2.7	10.8	6.48	6.48	25.13	5.30	0.37	0.03	0.06
Volcanic	0.08	7.03	9	36	21.6	164.7	2 .11	0.98	0.21	330	0.22
Wally	0.11	6.5	2.80	11.2	6.72	10.49	62.32	24.56	0	211	0.14

TDS: Total Dissolved Solids; Tur: Turbidity;   Cond: Conductivity;   HTT: Hydro Titrimetric Title

Table 4. Physicochemical parameters of drinking packaged water (imported brands)

Parameters Brands	Tur (mg/L of SiO₂)	pH	HTT (F)	Ca²⁺ (mg/L)	Mg²⁺ (mg/L)	HCO₃⁻ (mg/L)	Cl- (mg/L)	NO₃⁻ (mg/L)	NO₂ (mg/L)	Cond (µs /cm)	TDS (g/L)
Badoit	0	6	68	272	163.2	1076.04	19.77	4.10	0.02	1828	1.22
Beaupré	0.17	7.6	43.3	173.2	103.92	250.1	8.03	0	0	410	0.27
Contrex	0	7.6	110.3	441.2	264.72	322.08	10.29	1.77	0	2357	1.7
Evian	0	74	20	80	48	28548	868	300	0	527	0.35
Hépar	0	7.2	88.9	355.6	213.36	319.64	18.02	3.88	0.01	1668	1.12
Mont Blanc	0	7.5	99	39.6	23.76	319.64	2.30	0	0.01	212	0.14
Roxanne	0	7.00	15.4	61.6	36.96	170.8	11.3	1.94	0.08	359	0.24

Table 5. Duncan test of multiple comparisons of different physicochemical parameters according to each type of drinking packaged water.

Parameters	Imported water	locals water	
		bags	bottles
	average	average	average
Turbidity (mg/L of SiO₂)	0 ± 0ᵃ	0.15 ± 0.15ᵇ	0.012 ±0.01ᵃ
pH	7.18 ± 0.56ᵃᵇ	6.57 ± 0.87ᵃ	7.43 ± 0.50ᵇ
HTT (F)	50.83 ± 39.20ᵃ	2.82 ± 2.08ᵇ	13.77 ± 7.15ᶜ
Ca²⁺ (mg/L)	203.31 ± 156.79ᵃ	11.27 ± 8.33ᵇ	55.1 ± 28.61ᶜ
Mg²⁺ (mg/L)	121.99 ± 94.07ᵃ	6.27 ± 4.80ᵇ	33.06 ± 17.16ᶜ
HCO₃⁻ (mg/L)	391.97 ± 306.50ᵃ	15.20 ± 11.23ᵇ	155.88 ±47.35ᶜ
Cl⁻ (mg/L)	11.20 ± 6.01ᵃ	13.92 ± 7.43ᵃ	6.41 ± 5.56ᵇ
NO₃⁻ (mg/L)	2.10 ± 1.68ᵃ	6.37 ± 3.97ᵇ	3.45 ± 3.30ᶜ
NO₂ (mg/L)	0.03 ± 0.02ᵃ	0.05 ± 0.03ᵇ	0.08 ± 0.04ᶜ
Conductivity (µs/cm)	1077.29 ± 917.97ᵃ	139 ±104.43ᵇ	317±104.43ᶜ
Total dissolved solids (g/L)	0.72 ± 0.62ᵃ	0.09 ± 0.09ᵇ	0.21 ± 0.07ᶜ

The values which have in index exponent the different letters are significantly different (p<0.05) according to comparison multiple Duncan Test.

Acknowledgements

This study was realized with the partial support of Hydrocarbons-Analysis-Control society (HYDRAC).

References

[1]   AFIA ZIA, HAMIDULLAH SHAH, SALEEMULLAH, MOHAMMAD NOMAN AND M. SHAFI SIDDIQI, (2006). Evaluation of heavy metals and pathogenic microorganisms in drinking water of Peshawar Valley. Asian Environ. Technol., 10: 08-09.

[2]   AFNOR (Association Française de Normalisation) (1990). Eaux méthodes d'essais. 4ᵉᵐᵉ édition Paris 735.

[3]   BENANI-NODOT A., HARDY J. (2000). Eau de boisson et santé. *Les grandes sources de Wattwiller*. Médecin, Jeunesse et Sports.

[4]   BIRLEY M. (1991). Lignes et directives pour prévoir les implications pour les maladies transmises par des vecteurs du développement des ressources en eau. *Sci.eau*. 5 (1): 33-35

[5]   BLE L., MAHAMAN B., BIEMI J., MIESSAN G., LEKADJOU S. (2009). Etude de la potabilité des eaux de boisson conditionnées en Côte d'Ivoire : Cas des eaux de la région du Grand Abidjan. *European Journal of Scientific Research* 28 (4) : 552-558

[6] BOUTIN C. (1993). L'eau des nappes phréatiques superficielles, une richesse naturelle vitale mais vulnérable : l'exemple des zones rurales du Maroc. *Sciences de l'eau*, 6 : 357- 365.

[7] DEGREMONT (1989). Mémento technique de l'eau tome1 et tome 2.

[8] DELOLME, H., BOUTIN, J., ANDRE L-J, (1992). Eau douce et Pathologie. Médécine d'Afrique Noire: 39 (3).

[9] DJELLOULI HM., TALEB., HARRACHE-CHETTOUH D., DJAROUD S. (2005). Qualité physico-chimique des eaux de boisson du Sud-Algérien : Etude de l'excès en sels minéraux. *Cahiers d'études et de recherches francophones / Santé*. 15 (2) : 1-7

[10] FEUMBA R., KANA M., NGONO A., EBELLE R. (2009). Influence des méthodes de traitement sur la qualité microbiologique des eaux consommées au quartier BONAGANG (Douala-Cameroun). Sciences, Technologie et Développement 12 (1) 8-10.

[11] GIRAUD J. GALZY P. (1984). L'analyse microbiologique dans les industries alimentaires. *Collection génie alimentaire*, 92-119.

[12] KHOSROF S., BOUDABOUS A. (1992). Qualité microbiologique des principales eaux minérales tunisiennes. *Microb.Hyg. Ali*. 4 (11) : 4-16.

[13] MAKOUTODE M., ASSANI A., OUENDO E., AGUEH V., DIALLO P. (1999). Qualité et mode de gestion de l'eau de puits en milieu rural au Bénin : Cas de la sous- préfecture de Grand-Popo. *Médecine d'Afrique Noire* ; 46: 528-534.

[14] MAMERT, F.,(2000). Plan municipal de gestion de l'environnement de Douala, communauté urbaine de Douala

[15] MAUL A., VAGOST D., BLOCK J.C., (1989). Stratégies d'échantillonnage pour analyse microbiologique sur réseaux de distribution d'eau. Lavoisier Tec. Doc. (2-10) p.

[16] NOLA M., NJINE T., DJUIKOM E. FOKO S. (2000). Bacteria indicator dynamics in wells as influenced by well depth and well water column thickness, in Yaounde (Cameroon). *AJST*. 1 (2): 82-91.

[17] NOLA M., NJINE T., DJUIKOM E., FOKO S. (2001). Distribution des *Pseudomonas aeruginosa* et *aeromonas hydrophila* dans les eaux de la nappe phréatique superficielle en zone équatoriale au Cameroun et relations avec quelques paramètres chimiques du milieu. *Rev. Sci. Eau*. 14, 35-53.

[18] NOLA M., NJINE T., MONKIEDJE A., FOKO S. V., DJUIKOM E., TAILLEIEZ R. (1998). Qualité microbiologique des eaux de sources et des puits de Yaoundé. *Cahier santé*. 8 (5): 330-336.

[19] OMS (1994). *Directives de qualité pour l'eau de boisson*. 2^[eme]édition. OMS Genève 1-31.

[20] PATHAK S., BHATTACHERJEE J., RAY P., (1993). "Seasonal variation in survival and antibiotic resistance among various bacterial pollution in tropical rivers. *J Gen. Appl. Microbiol*. 39: 46-56.

[21] RODIER J., BAZIN C., BROUTIN J., CHAMBON P., CHAMPSAUR H., RODI L. (1996). *L'analyse de l'eau, eaux naturelles, eaux résiduaires, eau de mer*. 8^[eme]édition Dunod Technique 915-1199.

[22] SALIM H., MALIK AIT K. (2008). *Contribution à l'étude des paramètres physico-chimiques et bactériologiques de l'embouchure de l'Oued « Béni- Messous »*. Mémoire en vue de l'obtention du diplôme des Etudes Universitaires Appliquées en sciences de la mer.

[23] SENA J. (2001). *Evaluation de la qualité bactériologique des eaux de quelques sources et rivières des localités de Dschang, Bafou et Baleveng dans le Département de la Menoua*. Mémoire de maîtrise de Biochimie. Université de Douala. 34p.

[24] TATKEU J. (2001). *Evaluation de la qualité microbiologique de l'eau de consommation utilisée par une société de restauration de la ville de Douala: la Doual'air*. Mémoire de maîtrise de Biochimie. Université de Douala. 30p.

[25] TOURIDOMON S. BANAO I., GOUADO I., TAPSOBA T . (2009). Composition physico-chimique des eaux de boisson conditionnées commercialisées à Ouagadougou (Burkina Faso). *Cahier Santé* 1:1-10.

[26] TURGUT, S., B. KAPTANOG, G. TURGUT, G. EMMUNGIL AND O. GENC, (2005). Effects of cadmium and zinc on plasma levels of growth hormone, insulin-like growth factor I and insulin-like growth factor-binding protein Biol. Trace Element Res., 108: 197-204.

[27] UNICEF (2001). Planning for health and socioeconomic benefits from water and environmental programs. Water environmental sanitation. 25: 32-41.

[28] VIRKUTYTE, J. AND M. SILLANPA, (2006). Chemical evaluation of potable water in Eastern Qinghai Province, China: Human health aspects. Finland. Environ. Int., 32: 80-86.

[29] WHO, 1972-73. Guidelines for drinking water quality, Int. Health Criteria and Other Supporting Information.1st Edn., World Health Organization Geneva, 3: 101-146.

18

Exchanges of methane between lakes and the atmosphere in Hokkaido, Subarctic Climate Region, Japan

Masafumi Sasaki, Noboru Endoh

Department of Mechanical Engineering, Kitami Institute of Technology, Kitami, Japan

Email address:

sasaki-m@mail.kitami-it.ac.jp (M. Sasaki)

Abstract: Dissolved methane concentrations (*DM*) at twelve major lakes in Hokkaido, the northernmost island in Japan were observed in the open water season during 2006 ~ 2009 to estimate diffusive flux from lake surfaces to the atmosphere. An inverse relationship between lake size and *DM* was obtained in lakes in Hokkaido as was found for the European boreal lakes. All lake images larger than 0.001 km^2 were obtained by image processing of map data and area and number distributions were analyzed in order to calculate mass fluxes of diffusive methane. Total area of all (1,269) lakes in Hokkaido is 809 km^2. Regional diffusive flux of methane from lakes to the atmosphere in Hokkaido was estimated to be 0.581 Gg CH$_4$ yr^{-1}. Average diffusive flux density (per lake area) was about 0.718 g CH$_4$ m^{-2} yr^{-1}. This is a similar value to that in European boreal lakes on no-permafrost inland areas. Extremely high flux was found in Lake Abashiri-ko, one of highly eutrophic, meromictic lakes.

Keywords: Methane Flux, Lakes, Eutrophication, Global Warming

1. Introduction

Methane is well known as one of the green house gases (GHG). Lakes are significant source of methane to the atmosphere. In the previous work [1] [2], it was reported that the anomalous dissolved methane concentrations (*DM*) were frequently found even in Antarctica. In a subarctic climate zone, Hokkaido, the northernmost island in Japan, because of higher ambient and water temperatures and thicker organic sediments on the lake floors than those in Antarctica, it is expected that more active methanogenesis must be observed. If *DM* of a lake at the surface is supersaturated to the equilibrium for atmosphere, the lake acts as a methane source in the same manner of oceans [3].

In this study, field observations of *DM* were carried out mainly in twelve major lakes in Hokkaido in the open water season during 2006-2009. Diffusive mass fluxes of methane from all lakes in Hokkaido to the atmosphere are estimated from the results of the *DM* measurement and analysis of lake area and number distributions.

2. Methods

2.1. Site Descriptions

The locations of twelve observed lakes (circles) containing big-3 freshwater lakes in Hokkaido are illustrated in figure 1. The big-3 (70~80 km^2) freshwater lakes are Lake Kussyaro-ko, L. Shikotsu-ko and L. Toya-ko. The big-3 and L. Akan-ko are caldera lakes which have hot-sprig wells. Sampling stations are located on the shore of the lakes and off shore approached by a powerboat.

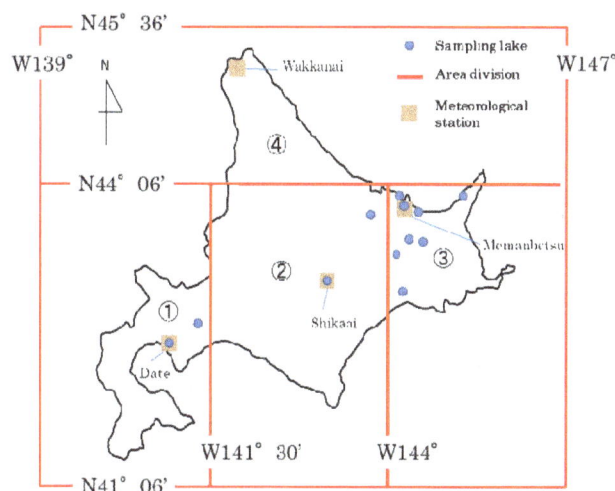

Fig. 1. Observed Lakes in Hokkaido, Japan are shown by circles. Representative meteorological stations 1~4 are shown in solid square.

2.2. Sampling Procedures

Surface water was sampled with a stainless steel pail and Van Dorn water sampler was used for sampling along water column. Temperatures and salinities of the water column were measured by a CTD. The waters were sealed in 500 ml glass vials after overflow procedure. Air was sampled in pre-vacuumed Tedler bags at the height of 1.5 - 2 meters from the ground.

2.3. Measurement of Dissolved Methane Concentrations (DM)

A headspace technique was applied to measure DM in water. A part of sample water in three glass vial (27 ml each) was replaced by pure nitrogen and 10 ml of sample water was remained in the vial. After each vial sample was heated and kept 60 °C for 20 minutes, headspace gas of 1.625 ml was injected to a gas chromatograph installed FID (GC-FID: Shimadzu GC-8A).

The measurements of DM were usually finished within 12 hours after sampling. When the samples had to be preserved for a few days or more, however, $HgCl_2$ was added as a preservative to the samples.

2.4. Measurements of Gas Concentrations in the Air

The sample air from a Tedler bag was directly injected to a 2 ml sample loop in the gas sampler which was installed in the same GC-FID mentioned above at the measurements of the methane concentrations in the air. Carbon dioxide concentrations were also measured through a methanizer located before FID. CO_2 escapes preferentially out of Tedler bag (3 % and 15 % of maximum reduction after one and three months respectively). The measurements of CH_4 and CO_2 concentrations were usually finished within 24 hours after sampling.

2.5. Estimation of Diffusion Flux of Methane from Lakes to the Atmosphere

A two-film model in air-water gas transfer was first proposed by *Liss and Slater* [4]. A great number of studies on air-sea exchange have been introduced and summarized by *Nightingale and Liss* [5]. Mass flux m (g CH_4 s^{-1}) is expressed as Eq. (1).

$$m = M_{CH4} \, k_{CH4} \, (DM - DM_0) \, A \quad (1)$$

where, M is molecular weight, k is transfer coefficient, DM is dissolved methane concentration (DM_0 is equilibrium DM of atmosphere) and A is area of water surface.

The transfer coefficient of CO_2 in freshwater in the Schmidt number Sc of 600 at 20 °C, k_{600} was calculated using the empirical formulation Eq. (2) for air-lake exchange proposed by *Cole and Caraco* [6] .

$$k_{600} = 2.07 + 0.215 \, U_{10}^{1.7}. \, (cm \, hr^{-1}) \quad (2)$$

The transfer coefficient of methane k_{CH4} can be obtained by the Schmidt number for methane, Sc_{CH4}. as Eq. (3)

$$k_{CH4} = k_{600} \, (Sc_{CH4} / 600)^{-n} \, . \quad (3)$$

$$n = 0.5 \, (U_{10} \geqq 3.6 \, m/s)$$

$$n = 2/3 \, (U_{10} < 3.6 \, m/s) \, by \, Deacon \, [7]$$

$$Sc_{CH4} = \nu_{freshwater} / D_{DM}. \quad (4)$$

Hourly mean wind speeds observed at four representative meteorological observatories were applied as U_{10} (equivalent wind speed at 10 m in height) to Eq. (2) during the open water season. These locations are shown in Fig. 1 (squares).

3. Results

3.1. DM Measurements

The results of measurements of DM in surface water are shown in Table 1. There are three big freshwater lakes (70~80 km^2), Lake Kussyaro-ko, L. Shikotsu-ko and L. Toya-ko in Hokkaido.

Table 1. *The results of measurements of DM in surface water in 2006~2009.*

Lake name	Latitude N	Longitude E	Area km^2	DM nmol L^{-1}
Kussharo	43°33.533'	144°20.350'	79.4	90
Shikotsu	42°46.300'	141°24.400'	78.4	34
Toya	42°34.993'	140°51.999'	70.7	27
Notoro	44°02.267'	144°10.388'	58.4	73
Abashiri	43°58.000'	144°10.000'	32.3	587
Mashu	43°34.694'	144°31.799'	19.2	41
Akan	43°27.000'	144°06.000'	13.0	114
Shikaribetsu	43°16.348'	143°07.133'	3.43	285
Mokoto	43°57.467'	144°19.933'	1.12	449
Rausu	44°01.826'	145°05.186'	0.43	501
Harutori	43°58.712'	144°24.361'	0.37	711
Tomisato	43°51.317'	143°45.200'	0.21	707

3.1.1. Lake Kussyaro-ko

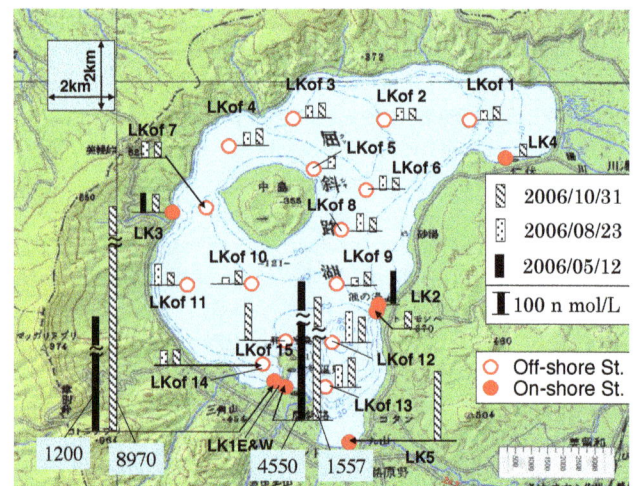

Fig. 2. *Surface DM distributions and seasonal changes in Lake Kussyaro-ko, 2006*

DM distributions at the surface and seasonal changes in Lake Kussyaro-ko in 2006 are shown in figure. 2. The surface DMs in northern area were lower than those in southern area, though the surface water is always super-saturated at every off / on shore station. Seasonal change of *DM* was not so clear in spite of change in water temperature through May to October.

Since L. Kussyaro-ko is an oligotrophic lake, lower methanogenesis activities at almost all stations are expected. However, there found extremely high *DM* (more than one μmol L^{-1}) along the shore of Wakoto Peninsula (St. LK1E and LK1W). Higher temperature than those at other stations suggested the existence of large amount of hot springs on the lake floor. One of the reasons of the *DM* anomaly may be higher activity of the methanogenesis because of such higher water temperature. Remarkable anomaly of methane concentrations in the air were seldom observed even around Wakoto Peninsula.

3.1.2. Lake Shikotsu-ko and Toya-ko

Both L. Shikotsu-ko and L.Toya-ko are typical oligotrophic lakes. *DM* distributions at the surface of L.Shikotu-ko on September 11 and that of L. Toya-ko on September 12 in 2007 are shown in figure 3 and figure 4, respectively. Both maximum *DM* are less than 50 nmol L^{-1} and *DM* are less than 30 nmol L^{-1} at almost all stations. It is considered that these *DM* distributions are typically shown in the oligotrophic lakes in Hokkaido, subarctic climate region.

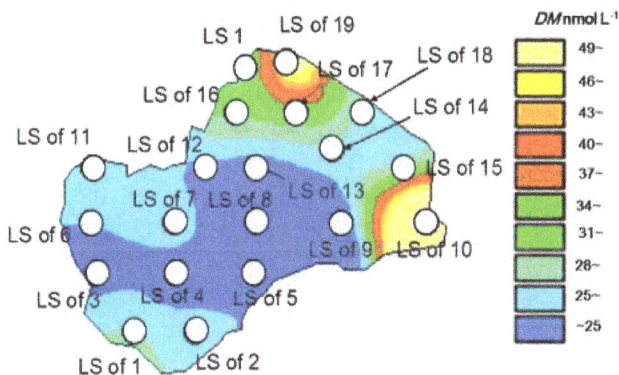

Fig. 3. *Surface DM distributions in Lake Shikotsu-ko, on September 11, 2007*

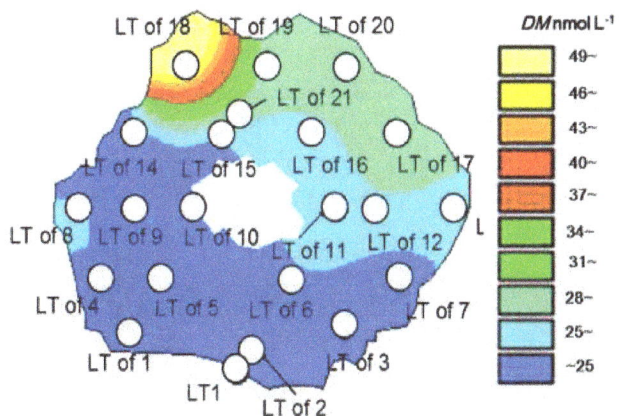

Fig. 4. *Surface DM distributions in Lake Toya-ko, on September 12, 2007.*

3.1.3. Lake Akan-ko

DM distributions at the surface and seasonal changes in 2006 in Lake Akan-ko, one of eutrophic lakes is shown in figure 5. Since seasonal changes in the surface *DM* can be

clearly observed, it seems that the *DM* strongly depends on water temperature.

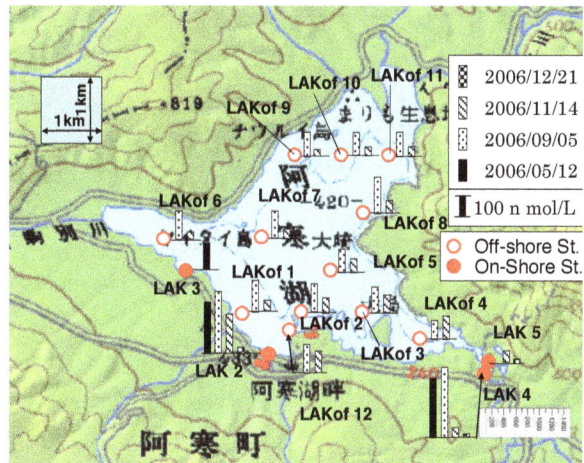

Fig. 5. *Surface DM distributions and seasonal changes in Lake Akan-ko, 2006*

3.1.4. Lake Abashiri-ko

Lake Abashiri-ko is highly eutrophic, meromictic lake. There is a clear saline layer at 5~7 meters deep. It has been eutrophicated by agriculture and pasturage along Abashiri inflow river. *DM* distributions at the surface in Lake Abashiri-ko are shown in figure 6. Super-saturated *DM* of 200 times or more than the equilibrium were found at all stations in summer. In early ice-covered season, *DM* under ice floe generally increased extremely (2~20 μmol L^{-1}) on-shore stations. Further study should be carried out to understand the relationship between these anomalous *DM* and the eutrophication or stratification.

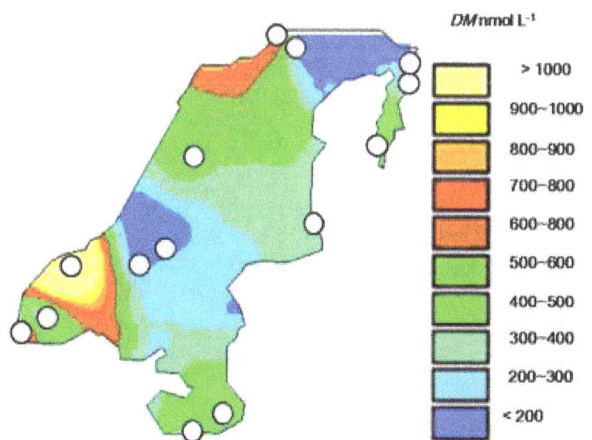

Fig. 6. *Surface DM distributions in Lake Abashiri-ko on August 9, 2007*

3.2. Relationship between DM and Lake Size

In order to determine average *DM* (=DM_{av}) in Table 1, a lake area was divided into pieces where each station located at near center as shown in figure 7 (Lake Kussyaro-ko as an example). The 'on shore' area was defined as the area of 0 ~ 10 m deep (painted out in figure 7) where the sun effectively lightened lake floor. Areal weighted mean DM can be defined as Eq. (5).

Fig. 7. Area divisions of Lake Kussyaro-ko (Circles show observation stations.)

$$DM_{av} = \sum_{K=1}^{Nst} (DM_K A_K / A) \qquad (5)$$

K: station number 1~$N_{station}$

Relationships between surface *DM* and lake area *A* are shown in Fig. 8. The dotted line in the figure shows Eq. (6) proposed by *Bastviken et al* [8]. As shown in Fig. 8, Eq. (6) does not express the *DM-A* relationship in Hokkaido, which has a steeper negative inclination as Eq. (7). One of the reasons of lower *DM* than those of Eq. (6) in beggar lake area zone is very low *DM* because of the big oligotrophic lakes as mentioned above. Another one of the reasons of higher *DM* than those of Eq. (7) in smaller lake area zone may be caused by higher water temperatures in Hokkaido than those in boreal lakes in Europe.

$$DM = 261.8\, A^{-0.227}. \qquad (6)$$

$$DM = 381.6\, A^{-0.498} \qquad (7)$$

(DM nmol L^{-1}, A km^2)

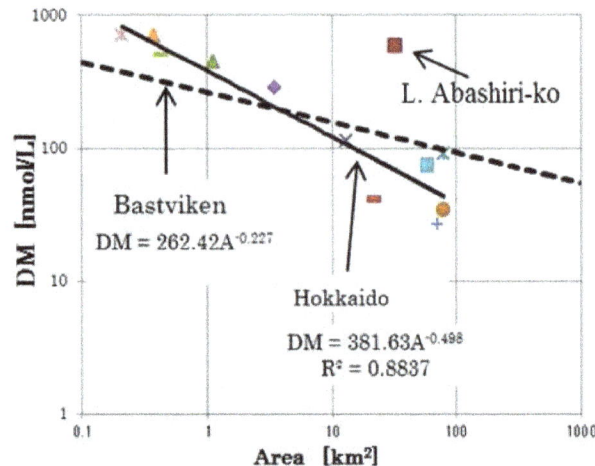

Fig. 8. Relationships between surface DM [nmol L^{-1}] and lake area A [km^2]

DM in L. Abashiri-ko becomes far from this *DM-A* trend (Eq. (7)) in Hokkaido as shown in Fig. 8. For example, *DM* in

L. Mokoto-ko, which is also highly eutrophic, meromictic lake like L.Abashiri-ko, corresponds to the relationship Eq. (7). A unique production and migration processes of methane might be expected in this lake. Further observations of DM should be concentrated to L. Abashiri-ko in the future.

4. Estimations and Discussions of Methane Flux from Lakes to the Atmosphere in Hokkaido

4.1. Lake Size and Number Analysis (Individual Lake Area and Regional Histogram of Lake Area)

Map data (http://maps.google.co.jp/) were referred for lake area analysis. The map images were gradated into two images (black & white) using 'Adobe Photoshop'. Lakes were expressed as black images. The ratio of black pixels to total pixels was obtained from the histogram of each image (also made by Adobe Photoshop). The product of the black pixel ratio and the analyzing area shows the lake area. Images of rivers and seas were removed by hand before analysis.

All lake images larger than 0.001 km^2 were analyzed and area and number (A-N) distributions were obtained in order to calculate mass fluxes of diffusive methane using a commercial particle image analyzer 'A-zokun' made by Asahi Kasei Engineering   Co. (http://www.asahi-kasei.co.jp/aec/business/sensing/product/azokun.html). Only scale conversion is required to apply this micron scale particles analyzer to kiro-meter scale lakes. The result of *A-N* analysis is shown in Table 2. Total area of all (1,269) lakes in Hokkaido is 809 km^2. Average lake to land area ratio is 1.03%, which is about 1/3 of global average (2.8% estimated by *Downing et al.* [9]). Class i (= 1~6) in Table 2 shows a lake size range between 0.001 ~ 1000 km^2. Representative value of *DM* for every class i was calculated using Eq. (7) for every average lake area.

Table 2. Area and number (A-N) distributions of lakes from a map analysis in Hokkaido

Class i	Lake area range		Number of lakes	Average area km^2	Total lake area km^2	Ratio of area to land
	Min km^2	Max km^2				
1	0.001	0.01	505	0.004	2.061	0.00003
2	0.01	0.1	544	0.033	18.121	0.00023
3	0.1	1	163	0.328	53.393	0.00068
4	1	10	46	3.256	149.754	0.00191
5	10	100	10	44.671	446.705	0.00570
6	100	1000	1	139.151	139.151	0.00177
Total			1269	0.638	809	0.01032

4.2. Estimations of Annual Methane Flux from Lakes in Hokkaido

Hourly mass flux Δm_{ij} at lake size class i was estimated by Eq. (8) (see Eq. 1~4) for a certain k_{CH4} calculated with an hourly mean wind speed U_{10} and monthly water temperature t_s at each representative meteorological station (j=1~4) as shown in Fig. 1.

$$\Delta m_{ij} = M_{CH4}\, k_{CH4ij}\,(DM_{ij} - DM_{0j})\, A_{ij} \times 3600. \quad (g\ CH_4\ hr^{-1})\ (8)$$

Annual mass flux m_{ij} (g CH_4 yr^{-1}) can be estimated by summation of hourly mass fluxes Δm_{ij} during the whole open water period. It is assumed that there was no air-lakes exchange during the ice-covered season. Total annual mass flux in Hokkaido is expressed as the total sum of m_{ij}.

$$m = \sum_{j=1}^{4} \sum_{i=1}^{6} m_{ij} \quad (g\ CH_4\ yr^{-1}) \quad (9)$$

Seasonal changes in mass flux of methane from lakes to the atmosphere in Hokkaido are shown in figure 9. Higher flux trend in summer is mainly caused by higher water temperature (= higher k_{CH4}) in this analysis.

Total annual diffusive methane flux in Hokkaido was estimated to be 0.581 Gg CH_4 yr^{-1}. Average flux density (per lake area) is 0.718 g CH_4 m^{-2} yr^{-1}. This is similar value to those in European boreal lakes on no-permafrost inland areas (0.923 for Swedish lakes by Bastiviken [8] and 0.784 for Finnish lakes by Juutinen [10]). And this average flux density is about five times greater than that from lakes in Syowa Oasis, in East Antarctica [2],

The annual methane flux in L. Abashiri-ko was estimated to be 0.066 Gg CH_4 yr^{-1}. The value corresponds to 11% of total annual flux in whole Hokkaido, while the lake area of Abashiri-ko occupies only 4%. The methane flux density in L. Abashiri-ko was 2.029. Such extremely high flux density suggests a unique mechanism of methane cycle. Further observations and analysis must be required for Lake Abashiri-ko.

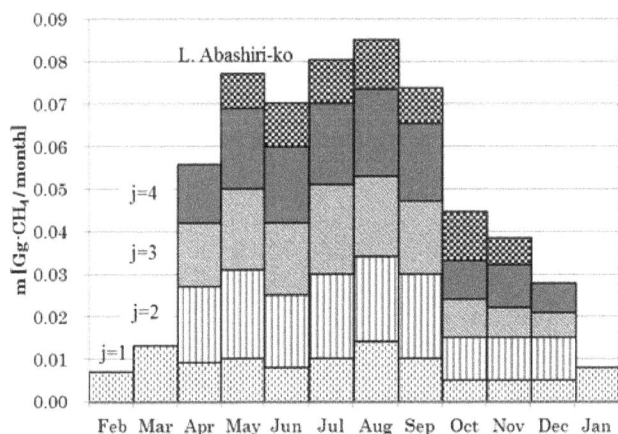

Fig. 9. Seasonal changes in mass flux from lakes in Hokkaido.

5. Conclusions

Observations of dissolved methane concentrations (*DM*) at twelve major lakes in Hokkaido, were carried out in the open water season during 2006 ~ 2009. Using the results and area histogram analysis of whole lakes larger than 0.001 km^2, diffusive methane flux from lake surfaces to the atmosphere were estimated. The study was concluded as follows,

1) An inverse relationship between lake size and DM was

obtained in lakes in Hokkaido as was found for the European boreal lakes. The relation slope was steeper than that for the European lakes.

2) Total area of all (1,269) lakes larger than 0.001 km^2 in Hokkaido is 809 km^2. Regional diffusive flux of methane from lakes to the atmosphere in Hokkaido was estimated to be 0.581 Gg CH_4 yr^{-1}. Average diffusive flux density (per lake area) is 0.718 g CH_4 m^{-2} yr^{-1}, which is also comparable value for the European boreal lakes.

3) The annual methane flux in L. Abashiri-ko was estimated to be 11% of total annual flux in whole Hokkaido, while the lake area occupies only 4%. The methane flux density in L. Abashiri-ko was more than five times greater than that expected from the *DM-A* trend in Hokkaido. Such extremely high flux density suggests a unique mechanism of methane cycle. Further observations and analysis must be required for Lake Abashiri-ko to understand the carbon cycles.

Acknowledgements

We thank Shota Hirano, Yohei Baba, Toru Wakamatsu, Takuya Jin and Takahiro Kato for valuable assistances in the field works and the analysis. A part of this work was supported by the Japan Society for the Promotion of Science (#18651011).

References

[1] Sasaki, M., S. Imura, S. Kudoh, T. Yamanouchi, S. Morimoto and G. Hashida (2009), Methane efflux from bubbles suspended in ice-covered lakes in Syowa Oasis, East Antarctica., *J. Geophys. Res.*, 114, D18114, doi: 10.1029/2009JD011849.

[2] Sasaki, M., N. Endoh, S. Imura, S. Kudoh, T. Yamanouchi, S. Morimoto and G. Hashida (2010), Air-lake exchange of methane during the open water season in Syowa Oasis, East Antarctica., *J. Geophys. Res.*, 115, D16313, doi: 10.1029/2010JD013822.

[3] Liss P.S. and L. Merlivat (1986), Air-sea gas exchange rates: Introduction and synthesis, in The Role of Air-Sea Gas Exchange in Geochemical Cycling, edited by P. Buat-Menard , pp.113-127, D. Reidel, Norwell, Mass.

[4] Liss P.S. and Slater P. (1974), Flux of gases across the air-sea interface, *Nature* 247, pp.181-184.

[5] Nightingale P.D. and Liss P.S. (2003), Gases in seawater, in The Oceans and Marine Geochemistry, ed H. Elderfield, Vol 6 of A Treatise on Geochemistry (eds Holland and Turekian) Ch 3, 49 – 82.

[6] Cole J. J. and N. F. Caraco, 1998, Atmospheric exchange of carbon dioxide in a low-wind oligotrophic lake measured by the addition of SF_6, *Limnology and Oceanography*, 43(4), 647-656

[7] Deacon, E.L. (1977), Gas transfer to and across an air-water interface, *Tellus*, 29, 363-374.

[8]   Bastviken, D., J. Cole, M. Pace and L. Tranvik (2004), Methane emissions from lakes: Dependence of lake characteristics, two regional assessments, and a global estimate, *Global Biogeochemical Cycles*, vol. 18, GB4009, doi: 10.1029/2004GB002238.

[9]   Downing, J. A., Y. T. Prairie, J. J. Cole, C. M. Duarte, L. J. Tranvik, R. G. Striegl, W. H. McDowell, P. Kortelainen, N. F. Caraco, J. M. Melack and J. J. Middelburg (2006), The global abundance and size distribution of lakes, ponds, and impoundments, *Limnology and Oceanography*, 51(5), 2388-2397

[10]   Juutinen, S., M. Rantakari, P. Kortelainen, J. T. Huttunenn, T. Larmala, J. Alm, J. Silvola and P. J. Martikainen (2009), Methane dynamics in different boreal lake types, *Biogeosciences*, 6, 209-223.

19

The Impact of Karkheh Dam Construction on Reducing the Extent of Wetlands of Hoor-Alazim

Samira Fuladavand[1], Gholam Abbas Sayyad[2]

[1]Faculty of Agriculture, Shahid Chamran University of Ahvaz, Ahvaz, Iran
[2]Department of Soil Science, Faculty of Agriculture, Shahid Chamran University of Ahvaz, Ahvaz, Iran

Email address:
Fuladvandsamira@yahoo.com (S. Fuladavand)

Abstract: Hoor-Alazim wetland in the southwest part of Iran at Khozestan province is located in Iran and Iraq borderline. The wetland extent depends on the entering river sediments and its supply. Hoor-Alazim is affected by the torrential sediments of rivers such as Tigris and Euphrates in Iraq and Karkheh in Iran. By 1998, Iran has started to exploit from the largest dam on Karkheh River, this river is one of the largest watery sources in Hoor-Alazim wetland. To supply with water of the dam caused to decline the quantity of entering water to the wetland and following it, the wetland extend has been reduced remarkably. Although various studies have been performed about the Hoor-Alazim wetland and it's expansion, no study has been yet conducted about the quantity of the entering water to the wetland. Therefore, this study aimed to examine the quantity (discharge and rainfall) entering water to the wetland during a certain statistical period in 3 hydrometric stations of Karkheh watershed basin (Hufel, Neisan, Hamidieh). To achieve the purposes of this research, researchers used the recorded data and information from the Khuzestan Water and Power Authority during the last 40 years using SPSS, GIS, and RS. Also, these influenced the wetland expansion and also finally the dust production in Ahvaz city was examined. Results from this study showed that the main factor which reduced wetland expansion is the reduction of entering water to it caused by the Karkheh River drainage. Despite the rain rate in the recent years which hadn't changed significantly, the amount of the entering water to the wetland was reduced. According to the results, the wetland expansion has changed from 900 km^2 by 1991 to about 300 km^2 by 2008.

Keywords: Dam Construction, Karkheh Dam, Wetlands of Hoor-Alazim

1. Introduction

The water shortage in dry areas has negatively affected welfare, economic and politic practices. Population growth and increased public demands for new agricultural lands have led to the policy of increasing agriculture and performing projects related to the water resources like dam construction and water transportation in order to increase agriculture products [14, 15]. One of the specified examples of this kind of natural ecosystem destruction is Tigris and Euphrates aquifers. It can be said that it is one of the largest destructors. The underlying area has witnessed more than 60 types of engineering projects during 3 last decades (including dam construction, or skewed channels to remove seasonal floods and using irrigate water of the lands). It resulted to the reduction of the entering water into the basin and its destruction which has reduced significantly its area [9, 17].

Hoor-Alazim pond is located at the west – east part of Khuzestan province in Iran – Iraq boundary [10]. Karkheh River which originated from Lorestan Mountains, after considerable distance in Azadegan plane and an area called Hamidieh the Karkheh noor branch separated of it and divides to two Hoofl and Nisan branches at Soosangerd city, each of these branches divides into several branches at Hoor. The area of the pond is influenced by flood regimens of Karkheh rivers in Iran and Tigris and Euphrates in Iraq [1, 4]. About $\frac{2}{3}$ of the area is in Iraq and $\frac{1}{3}$ of it is in Iran, and it called Hoor AlHoveizeh in Iraq [11]. This pond is in 47^0 longitudes and 31^0 altitudes [20]. In 1377, Iran began to construct the largest dam on the Karkheh River. This is one of the biggest water resources of Hoor-Alazim lagoon that has been designed to irrigate 320,000 hectare lands of Khuzestan plane [16]. Jone et al. [2005] reported that the area of Mesopotamia marshy lands which have been more than 8000 km^2 in 1966 have decreased to about 750 km^2 in 2002. Jamei et al. [2006] using remote techniques measured that the area of Hoor-Alazim ponds has decreased from 1991 to 2002. According to them, the ponds area has decreased from 900 km^2 in 1991 to 400 km^2 in 2003 [2]. This study was performed to investigate the quantitative transformations of the entering waters into the Hoor-Alazim pond during a statistical period, and also to study the entering flow into the pond before and after the exploitation of Karkheh dam. Although there have been some studies about quantitative changes of the entering water values to the pond, so far there hasn't been a study considering the impact of this trend on the area of Hoor-Alazim pond.

2. Material and Method

2.1. The Studied Area

The studied area is a part of Karkheh River basin located at the end of it. Karkheh is the third full-water river of the country that is one of the most important rivers of the aquiferous basins of the Persian Gulf. The Karkheh lops runoff a wide area of Ilam, Kermanshah, Lorestan, Hamedan and limited areas of Kurdistan and Khuzestan, then enter its into Khuzestan plain passing Bakhtar Dezful Shosh, Soosangerd and Boostan cities, and enters into Hoor-Alazim pond along with some branches. The basin area is about 50727 km^2. This river doesn't enter into the Persian gulf directly rather it first enter into Hoor-Alazim then communicates the Persian gulf via Arvand River [2, 18].

2.2. Hydrometric Station

In this study the data from 3 hydrometric stations at the end of the Karkheh watershed basin were used. Table (1) shows the geographical coordinate systems of the selected hydrometric stations of this study. The number of used years to study the entering water into the pond is 52 ones during 1987 – 2009. In addition to the mentioned hydrometric station in tablet the recorded rainfall data in Hamidieh rainfall-metric station were used.

In this study the flow position in Hydrometric stations was studied in two months and yearly time scales. In general, the below steps were followed:

1) Providing water entering set in monthly and yearly scales using Excell software.
2) Independent – sample T test of the rainfall before and after the construction of Karkheh dam, using SPSS software.
3) Independent – sample T test of the Karkheh water entering before and after Karkheh construction, using SPSS software.

Table 1. Selective correlations of Hydrometric stations

Neisan	Hufel	Hamidieh	Station identity data
48°-11'-40"	49°-49'-30"	50°-70'-26"	Length
31°-34'-34"	31°-03'-20"	31°-46'-19"	Wide
13.97	12.61	24.5	Height (m)
1987	1987	1955	Establishment year

2.3. Pond Extraction Using Remote Measurement

To assess the watery and plant covering surfaces of the pond the data from three sets of Landsat satellite 7-images with following characteristics were used:

1) Landsat TM measurement block 38 – 164, in May 1991
2) Landsat ETM measurement block 38 – 164, in May 2002
3) Landsat TM measurement block 38 – 164, in August 2009.

It is worth noting that the used software in this section includes:

ENVI 4 software to process images, radiometric and geometric correctness and classification. Arc GIS software to measure area, and extracted layers of the environment.

3. Discussion and Results

Regarding the research objective, at first the flow regimen at the end of the Karkheh basin has been studied. (1), (2), (3) diagrams show the yearly water entering set of changes in the hydrometric station. The river water entering has been decreased over time. This reduction has been significantly from 1998 – 1999 years (during Karkheh exploitation).

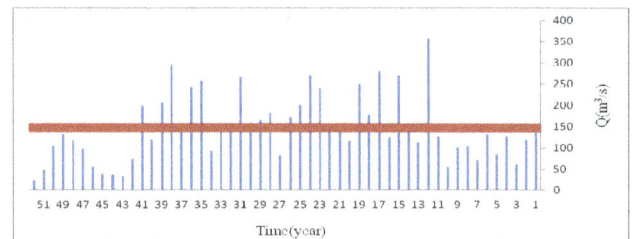

Diagram (1). The mean annual water entering series of changes in Hamidieh Hydrometric station.

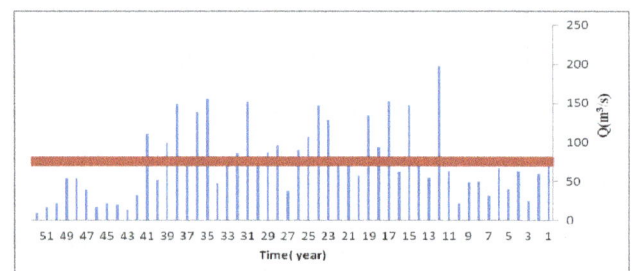

Diagram (2). The mean annual water entering series of changes in Hufel Hydrometric station

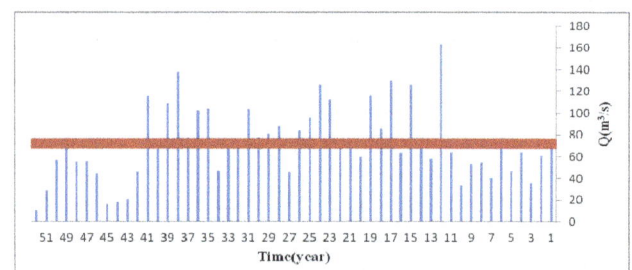

Diagram (3). The mean annual water entering series of changes in Neysan Hydrometric station

A summary of water entering and mean rainfall comparison test is shown in table (2). In this table, the comparing means include time series of rainfall and water entering for two statistic periods before (1987 - 1998), and after Karkheh dam exploitation (1998 – 2009).

The means of the rainfall time series for the period before and after Karkheh exploitation were 179 and 249 mm, respectively. The difference was at the statistical level of 1%. The water entering means before and after dam exploitation were 168 and 69 $\frac{m^3}{s}$, respectively.

The partial difference in the water entering mean is more than rainfall mean differences. This indicates that water entering mean differences are affected by rainfall changes and Karkheh savings. As table (3) shows to clarify the weight of difference between rainfall and entering flow means, the simultaneous data of rainfall in Hamidieh meteorologic station and flow in its hygrometric station after Karkheh exploitation (1998 – 2009) were in non-dimensional order. The non- dimensional ratio of these time series resulted from dividing each year to their means is as the following ratio: ($\frac{P}{\bar{P}}$) and ($\frac{Q}{\bar{Q}}$). Results from simultaneous tests of these ratios in two separated groups show that the effect of the entering flow to the pond is higher than the rainfall. In other words, the weight of reservoir saving index is higher than the rainfall index in entering flow to the pond. Results of this study show that entering flow to the pond after dam exploitation has decreased sharply. The statistical analysis showed that the water entering value in selected hydrometric station, and also the rainfall rate during the study has decreased, but the reduction in water entering has been higher than rainfall.

Table 2. Results of statistical analysis in Hamideieh hydrometric stations.

Variable	Satistical period	Number of years	Mean	Std. deviation	F	Sig
*Rainfall	1966 - 1998	32	249	85	25	0.007
	1998 – 2009	11	179	59.6		
*Entering flow	1987 - 1998	41	167.63	72	31	0.00
	1998 – 2009	11	69.28	38		

*rainfall in terms of (mm) and water entering in terms of m³.

Table 3. Results of statistical analysis of the simultaneous rainfall and water entering after dam exploitation in Hamidieh hydrometric and Metrologic stations.

Variable	Satistical period	Number of the year	Mean	Std. deviation	F	Sig
*Index Rainfall	1998 – 2009	11	0.77	0.25	20	0.012
*Index Entering flow	1998 – 2009	11	0.47	0.25		

*rainfall in terms of (mm) and watering in terms of m³.

Figure (1). Lansat TM measuring image, during May 1991

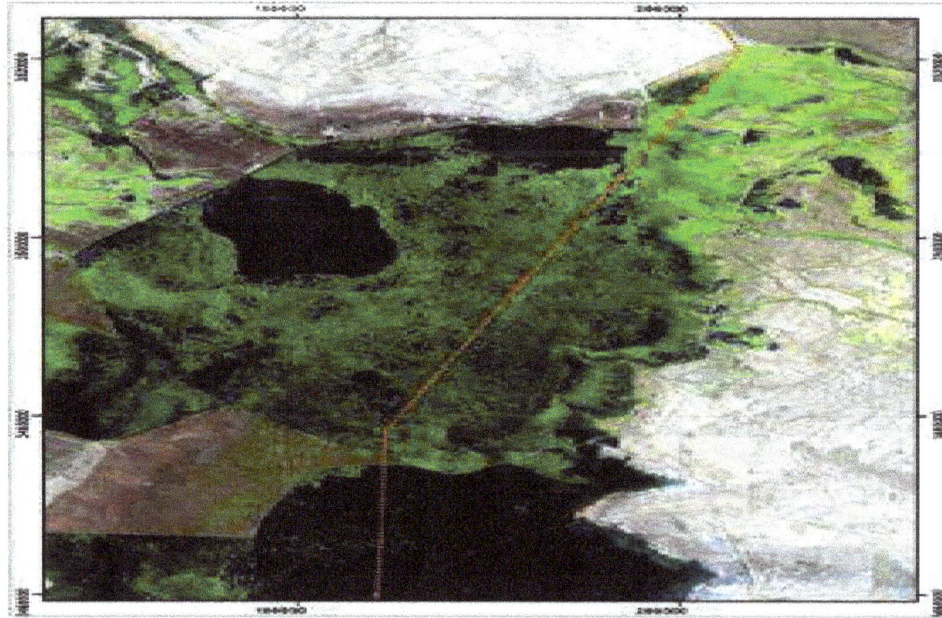

Figure (2). Landsat ETM+ measuring image, in May 2002

The pond surface changes. Figures (1) and (2) illustrated the Landsat TM measuring images during May 1991, and Landsat ETM during May 2002. It is indicated the pond water surface, and plant covering changes during these years. Based on the remote measurements, the vegetal plant mass observed in green and water observed in black. Also, the pond dry sections are in purple. As the figures shown, the area of the pond has decreased extremely from 1991 to 2002. Regarding the mentioned issues in previous sections, it can be said that the main reason of the area reduction is the decreased water entering into the pond, and also the decrease in its quality. Landsat TM measuring images during May 1991, and Landsat ETM+ measuring during May 2002 have been displayed that indicate the changes in water and vegetation level of the pond during these years. Figure (3) shows the measuring image during August 2009 that displays the change in water level and vegetation, in this figure the specificity of the images is higher because of its newness. Based on the above mentioned topics the main cause of the pond area reduction is the reduction of the entering water.

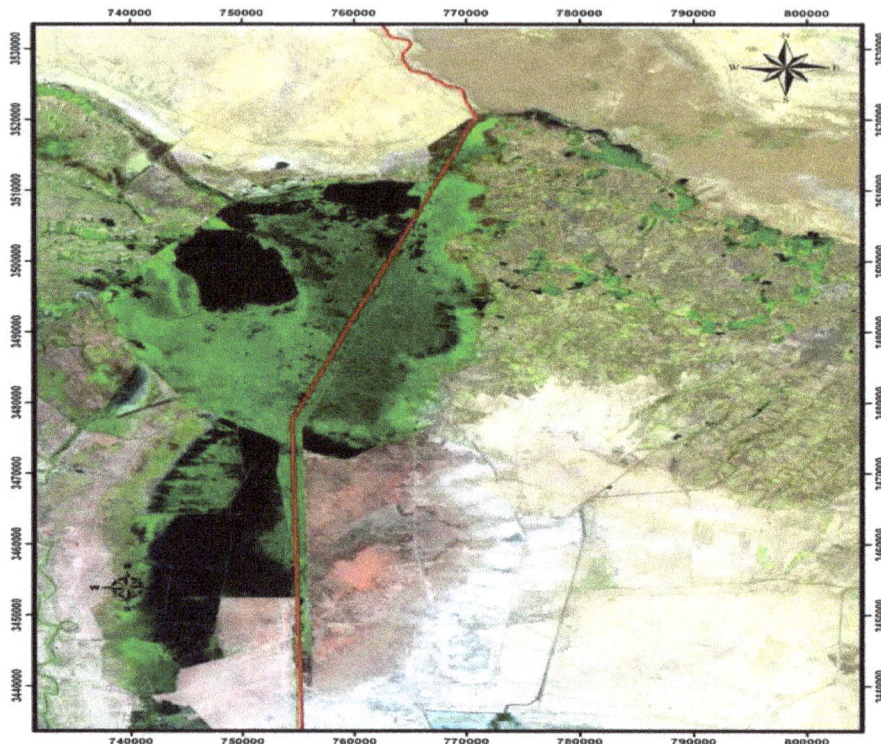

Figure (3). Landsat ETM+ measuring image, in August 2009

Table (4). *Hoor-Alazim vegetative and watery bulk areas (Km²)*

Time of sat. transition	Vegetative bulk	Watery bulk	Sum
1991	583.5	317.5	901
2002	180.5	217.2	397.2
2009	269.12	21.81	290.93

In Table (4) Hoor-Alazim pond vegetation and watery mass based on the data from remote measurement techniques are presented. Results show that the area in '81 year has reduced from 1991. The pond area has reached from 900 km² in 1991 to 300 km² in 2009.

4. Conclusions

Hoor-Alazim is served as a local and natural native environment and it is one of the most important and various ponds in the world, because of some incorrect human policies such as dam construction, and also natural unfavorable elements and feeding river pollutions, the pond is seemingly eroding and gradually becomes smaller and more unstable. Also the studies by Jones et al. (2005) showed that dam construction on the river paths have caused a reduction in area by 85% [9]. Ghobadi et al. (2012) also indicated that 72% reduction in area induced by increase in agricultural lands, demands for water and man interference in high area of the pond [21]. Results showed that the main reason of reduction in the area is the reduction in the water entering into the pond because of the Karkheh tankage. As the rainfall rate has not significantly changed during last years, the water entering into the pond has decreased considerably.

References

[1]   Augustine M.F., Warrender C.E, 1998, Wetland classification usingoptical and radar data and neural network classification, Int. J. of Remote Sensing. Vol. 19, pp. 1545-1560.

[2]   Chen, X., 2002, Using remote sensing and GIS to analyze land cover changeand its impacts on regional sustainable development, Int. J. RemoteSensing, Vol. 23, pp. 107-124.

[3]   Clark, P., and Magee, S, 2002, The Iraqi Marshland (A Human and Environmental Study), Amar International Charitable Foundation London, UK, pp. 20-27.

[4]   Coppin, P., Jonckheere, I., Nackaerts, K., Muys, B., and Lambin, E., 2004, Digital change detection methods in ecosystem monitoring; a review, International Journal of Remote Sensing. Vol. 25, pp. 1565-1596.

[5]   Dugan, P.J (ed), 1993, Wetland Conservation: a Review of Current Issue and Required Action, Vol. 1, first ed., IUCN/Gland, Switzerland, pp. 65-70.

[6]   Friedl, G., Wuest, A., 2001, Southeastern Anatolia Project in Turkey – GAP. Seminar for Doctoral Students at the ETH Zurich, pp. 21-24.

[7]   Friedl, G., and Wuest, A, 2001, Southeastern Anatolia Project in Turkey–GAP, Seminar for Doctoral Students at the ETH Zurich,   pp.   21-24.

http://www.eawag.ch/research_e/apec/damsworldwide_e.html.

[8]   Ghobadi, Y., Pradhan, B., Kabiri, K., Pirasteh, S., Shafri, H. Z.M., and Sayyad, G. A., Use of Multi-Temporal Remote Sensing Data and GIS for Wet land Change Moni tor ing and Degradation, IEEE Colloquim on Humanities, Science & Engineering Research (CHUSER 2012), December 3-4, 2012 Kota Kinabalu, Sabah, Malasia (2012).

[9]   JAMAB, 1999, Comprehensive assessment of national water resources: Karkheh River Basin, JAMAB Consulting Engineers in Association with Ministry of Energy, Iran (In Persian).

[10] JAMAB, 2006, Water balance report of Karkheh River Basin area: preliminary analysis, JAMAB Consulting Engineers in Association with Ministry of Energy, Iran.

[11] Jamei, M., 2003, Introducing the Khuzestan Wetlands and estuaries, first ed., Water Research Council Publications, Khuzestan Water and Power Authority, pp.39-44.

[12] Jamei, M., Hemadi, K., Hosein Zadeh, M., 2010, Assessment of water resources Hoor-Alazim wetland in order to use in the land-use plans with the techniques of remote sensing (RS), Article Summaries of the First National Conference on Water Resources Management in coastal areas, Sari Agricultural Sciences and Natural Resources University, Water Science Engineering Department, pp. 52-66

[13] Jones, C.K., Sultan, M., Al-Dousari, A., Salih, S.A., Becker, R. and Milewski, A., 2005, Who did what to the Mesopotamian Marshlands? Inferences from Temporal Satellite Data. The Proceedings of Annual Meeting of Geological Society of America Programs, 37(7), pp. 231-6.

[14] Jones, C., Sultan, M., Yan, E., Mileweki, A., Hossein, M., Al-Dousari, A., Al-Kaisy, S., and Becker, R., 2008, Hydrological impacts of engineering projects on the Tigris-Euphrates system and its marshlands, Hydrology, 353, pp. 59-75.

[15] Masih, M. D. Ahman. S. Uhlenbrook, H. Turral, and P. Karimi, 2009, Analyzing stream flow variability and water allocation for sustainable management of water resources in the semi-arid Karkheh river basin, Iran, Physic and Chemistry of the Earth, Vol. 34, pp. 329-340.

[16] Meghdad, A., 1998, Water Quantity in Border Marshlands between Iran and Iraq, Vol. 1, first ed., Baghdad University, Baghdad, pp. 104-108.

[17] Mokhtari, S., Soltani, H., Yavari, A., 2009, Self Organizing at Hoor-Alazim/ Hoor Alhoviezeh wetland with emphasis on ecology Landscape, Journal of Natural Geography Researches, No. 70, pp. 93-105.

[18] Ramesht, M., H., 1987, Natural Geography Hoor-Alazim, first ed, Art Publications, Tehran, pp. 15-24.

[19] Shabankari, M., Halbian, A.H., 2007, Studying the process of degradation, instability and ecological changes of Hoor-Alazim wetland. Humans and Environment, No. 5, pp. 31-44.

[20] UNEP, 2001, The Mesopotamian Marshlands: Demise of an Ecosystem Early Warning and Assessment Report, UNEP/DEWA/TR.01-3 Rev.1. Division of Early Warning and Assessment, United Nations Environmental Programe, Nairobi, Kenya.

[21] Zarasvandi, A., Carranza, E.J.M., Moore, F., Rastmanesh, F, 2011, Spatio-temporal occurrences and mineralogical–geochemical characteristics of airborne dusts in Khuzestan Province (Southwestern Iran), Geochemical Exploration, 4884, pp. 2-14.

20

Stochastic simulation of shallow aquifer heterogeneity and it's using in contaminant transport modeling in Tianjin plains

Lingling Liu, Lixin Yi[*], Xiaoqing Cheng

College of Environmental Science and Engineering, Nankai University, Tianjin, China

Email address:

liulingling@mail.nankai.edu.cn (Lingling Liu), yilixin@nankai.edu.cn (Lixin Yi), ccxq_0106@163.com (Xiaoqing Cheng)

Abstract: Shallow aquifers of Tianjin Plain formed by alluvium, marine and lacustrine sedimentary sequences, and resulting complex structure impose challenges to modeling groundwater flow and contaminant transport in it. To solve the problem and prove its feasibility, this study utilizes TProGS (Transition Probability Geostatistical Software) to describe hydrogeological structure of engineering sites, and then simulates contaminant transport by integrated using MT3D (Modular Three-Dimensional Transport Model) with traditional layered assignment approach and transition probability geostatistical approach respectively. The results show that aquifer structure on local scale is effectively described by TProGS and there is a smaller plume distribution in modeling with transition geostatistical approach than that with traditional layered assignment approach, it's also more in line with the groundwater flow direction. It illustrates the advantages of stochastic simulation in detailed conceptualization of hydrogeological structure. Furthermore, it demonstrates that integrated utilizing stochastic simulations and MT3D is more practicable than traditional approach in engineering practice for both probabilistic estimation of hydraulic conductivities and probabilistic assessment of contaminant plume capture at a heterogeneous field site.

Keywords: Stochastic Simulation, Heterogeneity, TProGS, MT3D, Tianjin Plains

1 Introduction

Constraints in accurate and realistic groundwater flow modeling are caused by the difficulty of characterizing the geological structure [1]. And the subsurface heterogeneity heavily influences the distribution of contaminants in the groundwater system [1]. However, the aquifer properties can vary as much as two orders of magnitude within a ten-foot radius and the scale of heterogeneity is often smaller than the data availability (e.g. Borehole spacing) [2]. Therefore, the selection of appropriate approach for accurate conceptualization of geological structure and description of aquifer heterogeneity is the critical work for modeling contaminant transport.

One approach for dealing with model heterogeneity is stochastic simulations based on multiple equally plausible candidate realizations of the site heterogeneity [3]. Stochastic simulations are particularly well-suited to local scale models since the resulting complex heterogeneity is more representative of actual stratigraphic deposition [3]. Among multiple methods of generating stochastic simulations of

hydraulic parameters, the transition probability approach of indicator geostatistics is the relatively robust and practical one for representing heterogeneity in 3-D aquifer stratigraphy. The powerful tool called TProGS (Transition Probability Geostatistical Software) developed by Carle operates on the basis of transition probabilities [3] and it has been successfully applied to simulate highly heterogeneous subsurface systems by constraining the simulation to borehole data [4].

Compared to the traditional indicator methods, the distinct strength of TProGS is the simple and direct incorporation of explicit facies manifestations like means lengths, volumetric proportions and juxtapositional tendencies into model [1]. In other words, the transition probability approach provides the conceptual framework to incorporate the geologic interpretation into the development of 3-D Markov chains models of spatial variability [5,6,7,8,9,10]. Furthermore, TProGS also provides the consideration for asymmetric tendencies such as fining-upwards, which is common in geological processes but not be captured by simply using a variogram model. Generally speaking, the transition probability approach couples geologic knowledge and mathematical manipulations to overcome the shortcomings of the traditional indicator geostatistics methods.

The objectives of this paper are to prove that if it is more appropriate to conceptualize aquifer structure in detail by utilizing TProGS and then to illustrate the advantages and significances for the Environmental Impact Assessment of engineering project and groundwater pollution assessment with integrated utilization of stochastic simulations andMT3D.

2. Study Sites

Two sites we selected located at Tianjin urban, which is situated in northeast of North China Plain (see Fig.1). At present, with the rapid expansion of the city and population growing, extensive groundwater exploitation activities are taking place, and many new factories have been set up in research area, this process ultimately caused an unprecedented pollution risk to groundwater resources, especially the shallow aquifer that plays an important role in local ecological environment has been polluted seriously already. Accurate assessing the impacts of new factories to groundwater, and correct predicting contaminant transport in groundwater flow are big challenges for local government and scientists. The two sites we choose in this study are located in difference places, one is located in urban centre, at where the shallow groundwater have disturbed and contaminated. The other is located in core region of Binhai New Develop Area, this place is currently developing, many factories and buildings promise to erect in near future. In the traditional groundwater flow simulation, all these two places were conceptualized as the same aquifer structure except for geometry.

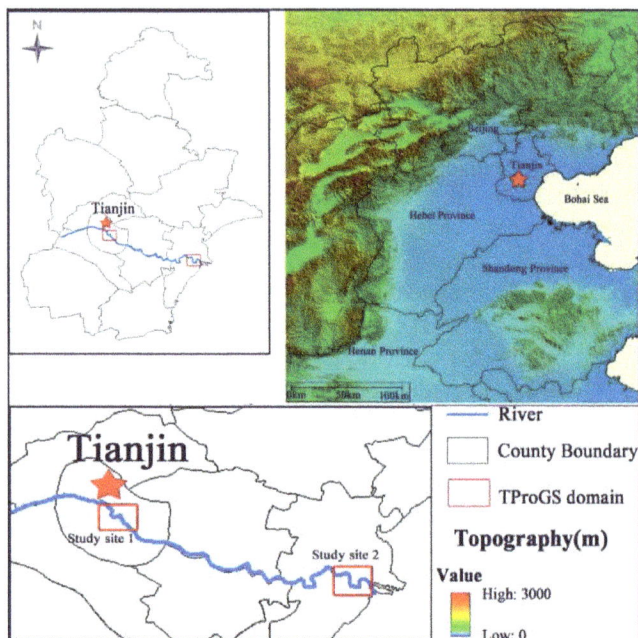

Fig 1. Location of two study sites in Tianjin plains

The problem of identify hydrogeologic structure on engineering site scale has not been well solved in local

environmental impact assessment. As stated in study objective, this study attempt to identify the differences of the hydrogeological structure between two sites not far from each other, and evaluate its effect on groundwater flow and contaminant transport prediction.

The hydrostratigraphy of the study site is characterized by the shallow aquifer system and the deeper fresh aquifer system, these two groundwater system hydraulically isolated with a thick impermeable clay layer. This study focuses on the shallow aquifer system which has a more sensitivity to anthropogenic activity. The shallow aquifer system composed of Holocene and late Pleistocene sedimentary sequences resulting from the deposition of continental alluvial, lacustrine, palustrine and marine sediments, which features by interbedded silt, clay, fine sand and sandstone. Two continuous transgressive sandy deposits are the main high permeability aquifers. Major marine transgressions and regressions and the resulting shifts of facies are the chief determinants of the sequences of high- and low-permeability sediments. Complicated sedimentary sequence causes difficulties to apply layered assignment parameters method for aquifer properties in this area, but it is an ideal place for stochastic simulation application.

3. Theoretical Backgrounds

TProGS (Transition Probability Geostatistical Software), included in the software Groundwater Modeling System (GMS), is a software package based upon the transition probability geostatistics approach to simulate spatial variability through the implementation of Markov Chains. It allows the simulation of multiple realizations through two mutually dependent steps: sequential indicator simulation (SIS)[11] and simulated quenching [6,12]. SIS algorithm establishes the initial configuration using transition probability-based co-kriging estimation [13]. And then the simulation quenching optimization algorithm will iteratively improve conditional simulation in terms of matching simulated and modeled transition probabilities, in other words, it will reshuffle the initial configuration[12].The following steps are necessary to accomplish the TProGS implementation[3]:GAMEAS processes borehole data and calculates bivariate geologic characteristics (e.g., material proportions, transition probability, etc.). MCMOD develops one- and three- dimensional Markov chain models of spatial variability. TSIM generates three dimensional, cross correlated geological realizations.

3.1. Transition Probability

The transition probability is defined as

$$t_{j,k}(h_\emptyset) = \Pr\{k(x+h)|j(x)\}$$

where $h_\emptyset$represents the lag distance in direction $\emptyset$; j, k are two categories, maybe j=k; x is a location in space. As shown from above equation, the transition probability can be described as "the probability of category k occurs at the location x+h based on the condition that category j occurs at location x".

3.2. Markov Chain

Markov chain assumes that the future condition only depends on the current condition, not the past and the result of the n time only was impacted by the n-1 result during the transition process of some factors in a system. 3-D Markov chain model simply but efficiently offers a mathematical method for the simulation of geostatistics spatial variability. Mathematically, the Markov transition probability matrix in primary direction Φ was defined with the exponential matrix formal:

$$T(h_\emptyset) = \exp(R_\emptyset h_\emptyset) = \begin{bmatrix} t_{11,\emptyset} & \cdots & t_{1k,\emptyset} \\ \vdots & \ddots & \vdots \\ t_{k1,\emptyset} & \cdots & t_{kk,\emptyset} \end{bmatrix}$$

where k is the number of categories; $h_\emptyset$ represents the lag distance in direction Φ; $R\Phi$ denotes the transition rate matrix in direction Φ, it is defined as:

$$R_\emptyset = \begin{bmatrix} r_{11,\emptyset} & \cdots & r_{1k,\emptyset} \\ \vdots & \ddots & \vdots \\ r_{k1,\emptyset} & \cdots & r_{kk,\emptyset} \end{bmatrix}$$

where $r_{jk,\emptyset}$ denotes the transition rate from category j to category k per unit length and $j \neq k$. It is obtained via the formula:

$$r_{jk,\emptyset} = \frac{\partial t_{jk}(h \to 0)}{\partial h_\emptyset}$$

And when category j=k, that is $r_{jj,\emptyset}$, it denotes the self-transition rate of category j. It is defined through:

$$r_{jj,\emptyset} = -\frac{1}{\overline{L_{j,\emptyset}}}$$

where $\overline{L_{j,\emptyset}}$ represents the mean length of the category j in the direction $\emptyset$, it is conceptually defined as

$$\overline{L_{j,\emptyset}} = \frac{\text{total length of j in direction } \emptyset}{\text{number of embedded occurrences of j}}$$

However, $t_{jk,\emptyset}$, the element in transition probability matrix $T(h_\emptyset)$, cannot be directly calculated by $t_{jk,\emptyset}(h_\emptyset) = \exp(r_{jk,\emptyset}h_\emptyset)$. The eigenvalue analysis of the transition probability matrix, as the following equation shows, is essential.

$$\exp(Rh) = T(h) = \sum_{i=1}^{k} \exp(\lambda_i h)Z_i$$

λ_i, Z_i denote the eigenvalue and eigenvector of the transition rate matrix respectively.

So for the current system composed of four categories, the transition rate elements are calculated by:

$$t_{jk}(h) = p_k + z_{jk,2}\exp(\lambda_2 h) + z_{jk,3}\exp(\lambda_3 h) + z_{jk,4}\exp(\lambda_4 h)$$

where p_k is limiting probability of category k.

Once the transition probability curve for the vertical direction was accomplished, the combination of Walther's Law and geologic knowledge is indispensable for the development of the transition probability curves in lateral (x and y) directions. It is because that rarely is the quantity of data adequate to develop an accurate model in lateral directions. And Walther's Law states that vertical successions of deposited facies represent the lateral succession of environments of deposition. However, one issue will arise when applying vertical transition trends to lateral directions is how to work out asymmetric vertical trends like fining upwards. Take a typical condition in fluvial deposition for instance, sand tends to deposit on gravel. The transition rate of gravel to sand is greater than sand to gravel because of the fining upward trend. Therefore, in order to determine a transition rate for lateral conditions, the transition rate between the two categories should be equivalent or symmetric as defined via:

$$r_{jk} = \frac{p_k}{p_j} r_{kj}$$

When finish the calculation of Markov chains in three primary directions, the MCMOD will be utilized to enable the generation of 3-D Markov chain model. So the transition rate in faculative direction is defined as:

$$|r_{jk,\emptyset}| = \sqrt{\left(\frac{h_x}{h_\emptyset}r_{jk,x}\right)^2 + \left(\frac{h_y}{h_\emptyset}r_{jk,y}\right)^2 + \left(\frac{h_z}{h_\emptyset}r_{jk,z}\right)^2}$$

$$\forall j, k \neq \beta$$

where β is the background material and hx,hy,hz is the lag distance of the x,y,z direction respectively, they are the components of the lag length in facultative direction:

$$h_\emptyset = \sqrt{h_x^2 + h_y^2 + h_z^2}$$

And the last required is that the discrete-lag distance Markov chains should be transferred into the continuous-lag distance Markov chains by:

$$R_\emptyset = \frac{\ln[T(\Delta h_\emptyset)]}{\Delta h_\emptyset}$$

3.3 Sequential Indicator Simulation and Simulated Quenching

The initialization step utilizes the sequential indicator simulation (SIS) algorithm described by Deutsch and Journel [13], except that a transition probability-based indicator cokriging estimate is used to approximate local conditional probabilities by

$$\Pr\{k \text{ occurs at } x_0 | i_j(x_\alpha); \alpha = 1, \cdots, N; j = 1, \cdots K\}$$

$$\approx \sum_{\alpha=1}^{N} \sum_{j=1}^{K} i_j(x_\alpha) w_{jk,\alpha}$$

where N is the number of data, K is the number of categories, $w_{jk,\alpha}$ represent a weightingcoefficient, and $i_j(x_\alpha)$ represents the value of an indicator variable

$$i_j(x_\alpha) = \begin{cases} 1, \text{if category j occurs at } x_\alpha \\ 0, \text{otherwise} \end{cases} \quad j = 1, \cdots, K$$

The transition probability-based cokriging system of equations [5,14] for computing the weighting coefficients is

$$\begin{bmatrix} T(x_1 - x_1) & \cdots & T(x_N - x_1) \\ \vdots & \ddots & \vdots \\ T(x_1 - x_N) & \cdots & T(x_N - x_N) \end{bmatrix} \begin{bmatrix} W_1 \\ \vdots \\ W_N \end{bmatrix} = \begin{bmatrix} T(x_0 - x_1) \\ \vdots \\ T(x_1 - x_N) \end{bmatrix}$$

where,

$$W_\alpha = \begin{bmatrix} w_{11,\alpha} & \cdots & w_{1K,\alpha} \\ \vdots & \ddots & \vdots \\ w_{K1,\alpha} & \cdots & w_{KK,\alpha} \end{bmatrix}$$

The quenching step attempts to solve the optimization problem of

$$\min \left\{ 0 = \sum_{i=1}^{M} \sum_{j=1}^{K} \sum_{k=1}^{K} \left[t_{jk}(h_l)_{MEAS} - t_{jk}(h_l)_{SIM} \right]^2 \right\}$$

where " O " denotes an objective function, the h_l denotes $l = 1, \cdots, M$ specified lag vectors, and"MEAS" and "SIM" distinguish measured and simulated (measured from the realization)transition probabilities, respectively [6,12,13,15].

4. Simulation Processing and Results

4.1. Collection and Collation of Borehole Data

There are totally 49 borehole logs with varying depths utilized to describe the spatial variability in study sites and the two sites with size of 52 km^2 and 78 km^2 respectively occupy 28 and 21 borehole numbers. In order to gratify for the software's limitation which is imposed to keep data processing at a reasonable level, the totally eight geological formations has been grouped into four categories based on the similar grain size and hydraulic conductivity: muddy clay, silt clay, silt and silt sand.

Among the above four categories, the most repeated material which is "silt clay" was designed as the background material that fills in the remaining areas not occupied by other units.Fig.2 presents the locations of the borehole logging and the material textural distribution over the defined grid frame of the first study site according to the previous categorization.

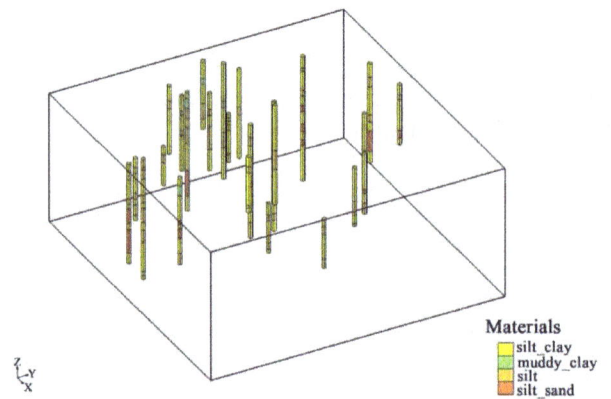

Fig 2. Borehole cross-section distributions in oblique view

In order to match the vertical grid spacing more closely to the thicknesses of the actual layers, the vertical thickness of the grid cells, determined by the user, must be as small as possible. So the two study sites were respectively conditioned on two 73×71×32/93×84×35 cell grids with the same discretization of 100m×100m×2m. And Table 1 shows their hydrogeological parameters obtained on the basis of geological exploration data.

Table 1. The hydrogeological parameters of the materials

Property	Material			
	Silt clay	Muddy clay	Silt	Silt sand
Horizontal hydraulic conductivity（m/d）	0.2	0.5	0.9	1.2
Vertical hydraulic conductivity（m/d）	0.06	0.2	0.3	0.36
Porosity	0.3			
Longitudinal dispersivity(m)	20			
Recharge concentration(mg/l)	20000			

4.2. Development of Geological Structure Model

The collated borehole data will be firstly passed to a utility within TProGS called GAMEAS to compute a set of transition probability curves with a given sampling interval. When it completed a successful run, the results will be read into the corresponding data fields in the "Vertical Markov Chain" dialog to develop a Markov chain model for vertical direction.

Amongst the five alternative methods of generating Markov Chains, the "Edit the embedded transition probabilities" was selected for that it is a more intuitive method of generating Markov chains and is conducive to sites with data because embedded occurrences can be easily tallied from borehole data[2].The computation result of transition probability matrixes in vertical direction of both study sites were respectively constituted as

$$t_{jk,z} = \begin{bmatrix} 3.675 & 1 & 0 & 0 \\ 0.048 & 11.246 & 0.762 & 1.190 \\ 0.021 & 0.979 & 4.512 & 0 \\ 0 & 0.917 & 0.083 & 5.750 \end{bmatrix}$$

$$t'_{jk,z} = \begin{bmatrix} 7.862 & 0.923 & 0.077 & 0 \\ 0.311 & 8.295 & 0.551 & 0.138 \\ 0.235 & 0.735 & 3.319 & 0.030 \\ 0 & 0.889 & 0.111 & 7.164 \end{bmatrix}$$

The diagonal elements represent self-transitions, that is, the transition probabilities from one category to itself. They coincide with corresponding vertical lens lengths (see Table 2)since the stacked beds of the same category are assumed not distinguishable from a single bed. And the off-diagonal elements per line satisfy

$$\sum_{k=1}^{K} t_{jk,z} = 1, \qquad j \neq k$$

Table 2. The material proportions and lens lengths of two study areas

Material	Proportion		Lens length	
	Site 1	Site 2	Site 1	Site 2
Muddy clay	0.015	0.238	3.675	7.862
Silt clay	0.699	0.555	11.246	8.295
Silt	0.218	0.127	4.512	3.319
Silt sand	0.068	0.080	5.750	7.617

The values in Table 2 show the proportions and lens lengths of four categories in both study sites. The material proportion is represented by the probability corresponding to the flat part of the diagonal curve shown in Fig. 3. And the lens length of material is expressed as the point where a tangent line from the early part of the curves on the diagonal intersects the horizontal (lag distance) axis on each curve. And it should be noted that lens length value of silt clay in site 1 is larger than site 2(11.246 > 8.295), but the value of muddy clay is smaller(3.675 < 7.862). This indicates that the silt clay is more continuous in vertical direction in site 1 but muddy clay in more continuous in site 2. The phenomenon is displayed in Fig. 4 obviously.

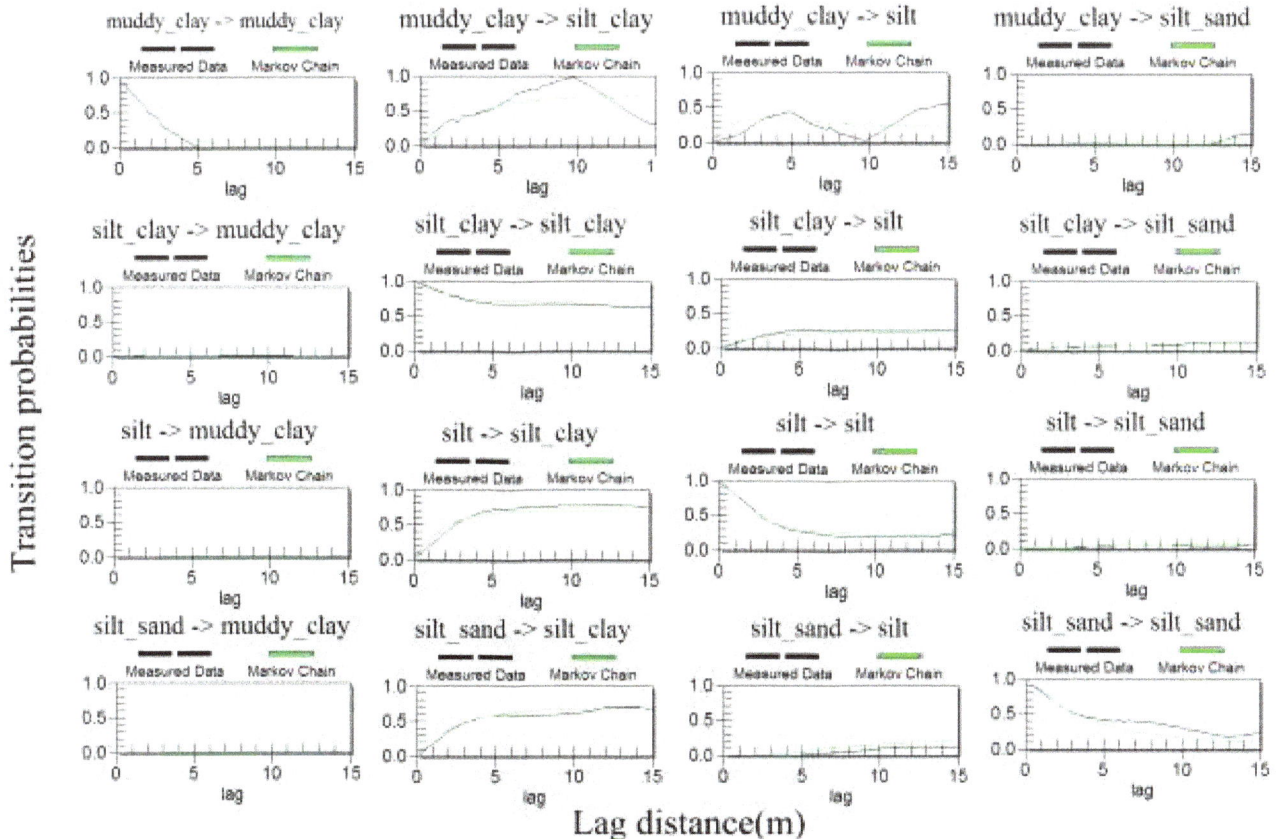

(a)

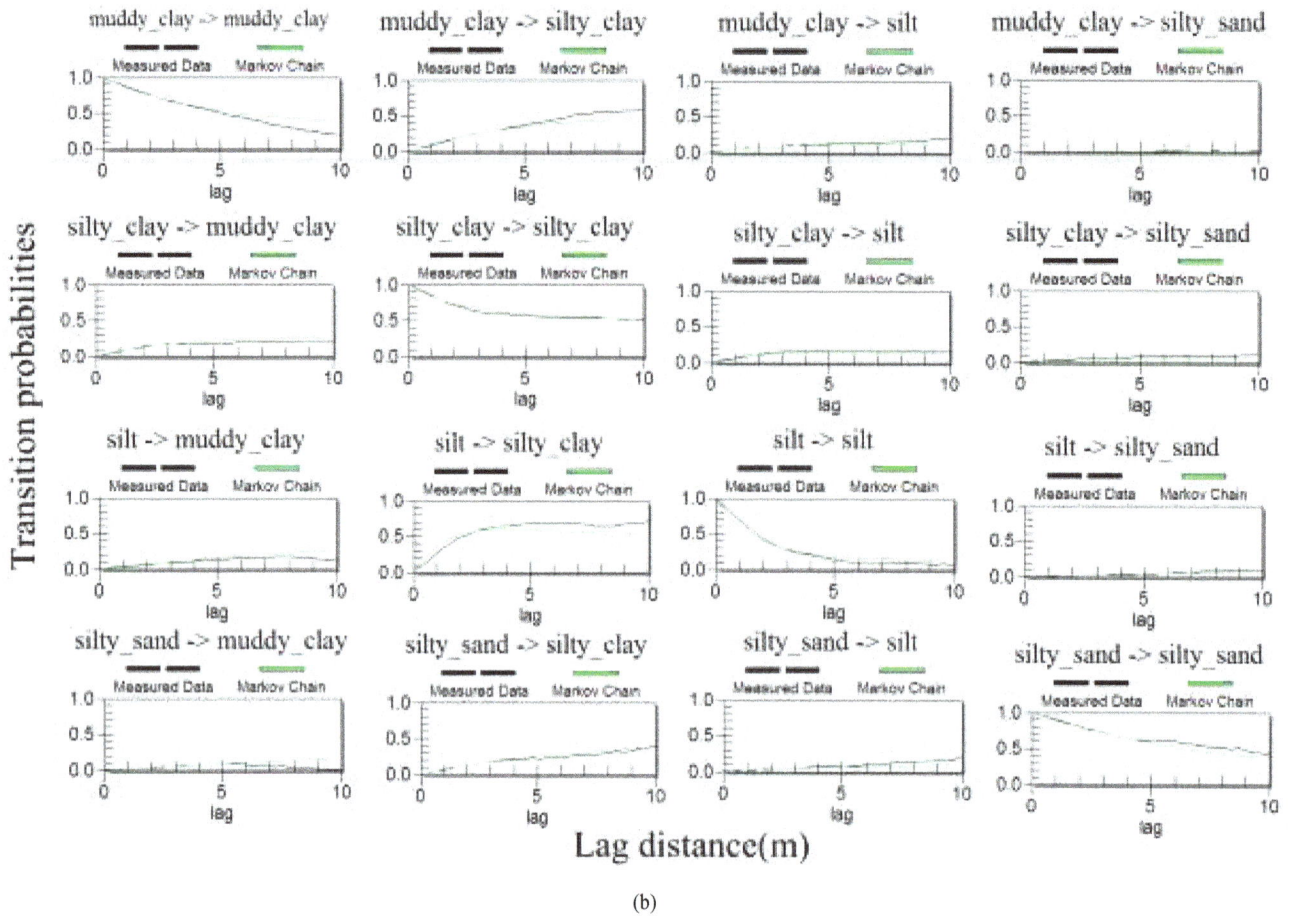

(b)

Fig 3. Measured transition probabilities curves (black lines) and Markov chain model (green lines) for a set of borehole data with four catagories: a represents the result of site 1; b represents result of site 2

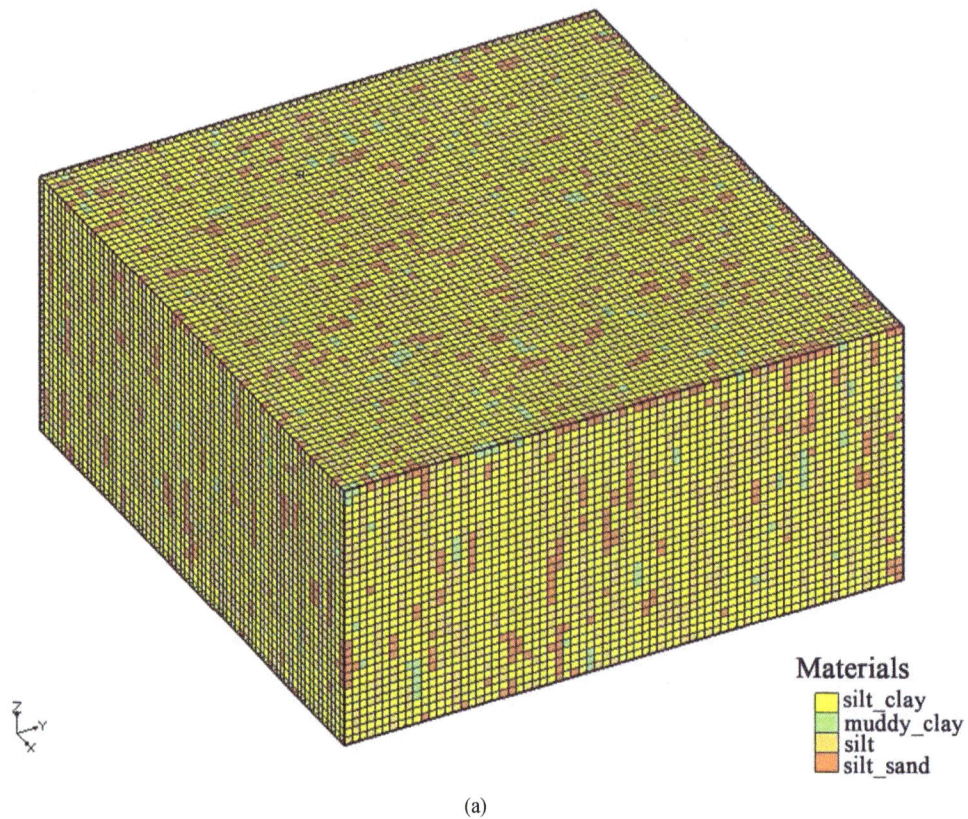

(a)

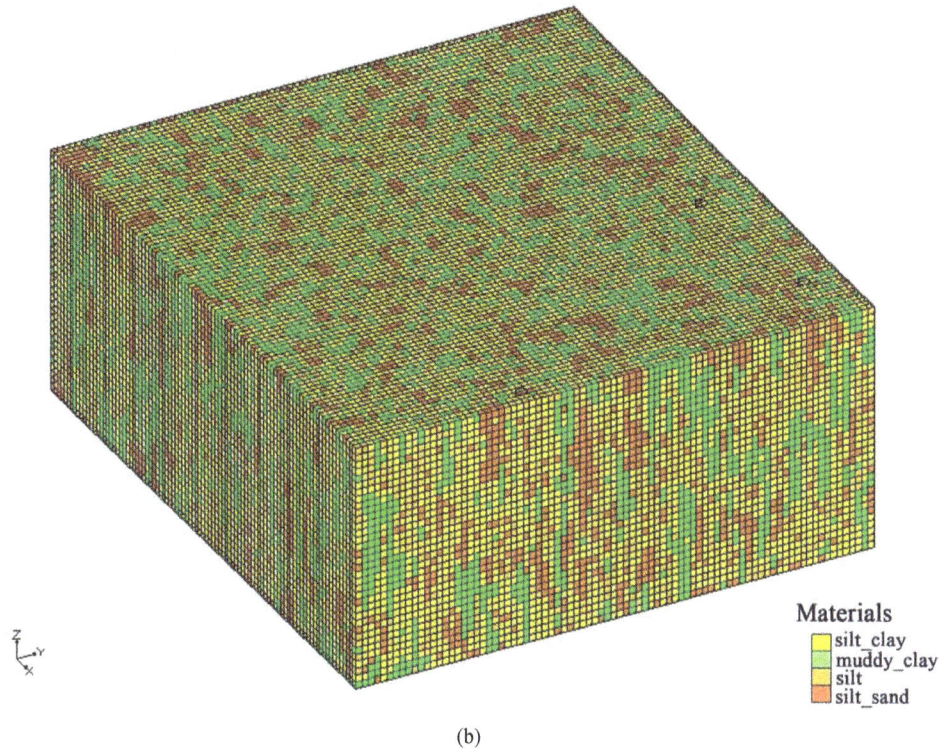

(b)

Fig 4. 3-D illustration of TProGS realization #1 filled with material colors: a represents the result of site 1; b represents result of site 2

With the utility of MCMOD, the measured vertical transition probability curves associated with above transition probabilities of both study sites are shown by the black lines in Fig.3. The diagonal self-transitional curves start at a probability of 1.0 decreasing with distance and the off-diagonal transition probability curves start at zero probability increasing with distance and they are eventually flatten out at some distance.

Finally the TSIM algorithm finished the generation of three-dimensional, cross correlated conditional simulations. Figure 4 shows one of the realizations of stochastically generated geologic units that honor known data.

4.3. Simulations of Contaminant Transport

MT3D, developed by Zheng [16], is a 3-dimensional finite-difference calculation program that was used to model contaminant transport in groundwater systems. It combines the advection, dispersion and chemical reactions of contaminants comprehensively and it is able to deal with the complex sources and sinks and boundary conditions. For that, it can accurately model the contaminant transport process in confined, unconfined and leaky aquifers. Thereby, in recent years, Zheng and Wang[17] and Clement[18] proposed the improvement based on the previous program. It has been widely accepted in research of hydrogeology and water environment modeling.

On the basis of previous experience of the impact assessment of groundwater environment, we presume that there is a refuse landfill with area of 500m×500m in site 2 and the leachate permeates into underground with a fixed concentration. We utilize two different conceptual methods to describe the geological structure of local site, in other words, different approaches for getting aquifer properties will be used in MT3D. Finally, contamination transport results of them will be compared intuitively to illustrate the advantages of the combination of TProGS with MT3D.

Fig.5 shows the side view of a MODFLOW grid with three curvilinear layers interpolated by elevations of true layers obtained through the borehole data. Fig.5 (a) is the result obtained through layered assignment method. On the basis of local lithology characteristic and experiences of previous numerical modeling, the model is conceptualized three layers and respectively assigned the aquifer properties (permeability coefficients, porosity, long dispersion, etc.). And Fig.5 (b) shows the result obtained through the combination of transition probability geostatistical approach with HUF (Hydrogeologic Unit Flow) package [19] in MODFLOW. The grid used in it is the same with upper one, but stochastic simulation program calculates the detail geological structure in a background grid which has a large number of layers.

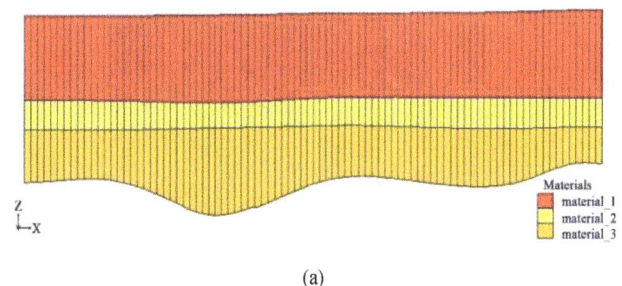

(a)

(b)

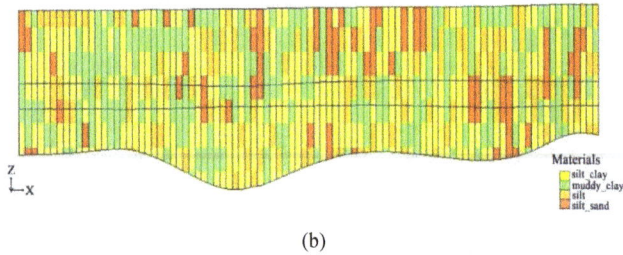

Fig 5. Side view of geological structure of site 2 with two different conceptual methods: a represents the result with layered assignment method; b represents result of combination of transition probability geostatistical approach with HUF package

And then the geological conceptualization results and aquifer properties are utilized in MT3D simulations with the same boundary and starting conditions. The values of hydrogeological parameters are obtained based on the values shown in Table 1. The results of contaminant transport modeling with two methods are respectively shown in Fig.6(a) and Fig.6(b). Obviously, the pollution plume distribution in second condition is smaller and the movement direction of contaminant in it is consistent with the flowing direction of groundwater.

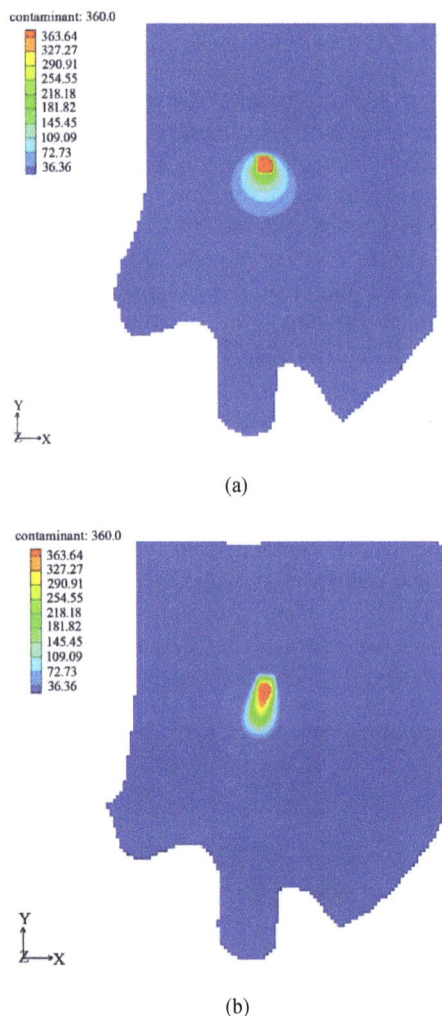

(a)

(b)

Fig 6. Plan view results of MT3D simulations with layered assignment approach (a) and transition probability geostatistic approach (b)

5. Conclusions

Hydrogeological structure differences on local scale between two engineering sites were identified by TProGS accurately. And compared with the pollution plume distribution of layered assignment approach, the affected area of contaminant with stochastic simulation method is smaller, and the movement direction of contaminant is consistent with the flowing direction of groundwater. Therefore, the modeling results with stochastic simulations have more realistic representation to the local situation. It illustrates that the application of stochastic simulations and its combination with MT3D is more practicable than traditional approach in engineering practice for both probabilistic estimation of hydraulic conductivities and probabilistic assessment of contaminant plume capture at a heterogeneous field site.

Acknowledgements

This research was funded by the National Natural Science Foundation of China (Grant No. 41276112).

References

[1]   Aarts E, Korst J (1989) Simulated annealing and Boltzmann machines: A stochastic approach to combinatorial optimization and neural computing. John Wiley & Sons, New York

[2]   Anderman ER, Hill MC (2000) The U.S. geological survey modular groundwater model-Documentation of the hydrogeological unit flow (HUF) package. U.S. Geological Survey Open-File Report 00–342

[3]   Arciprete D, Bersezio R, Felletti F, Giudici M, Comunian A, Renard P (2012) Comparison of three geostatistical methods for hydrofacies simulation: a test on alluvial sediments.Hydro 20:299–311

[4]   Carle SF (1996) A transition probability-based approach to geostatistical characterization of hydrostratigraphic architecture. Dissertation, University of California

[5]   Carle SF (1997) Implementation schemes for avoiding artifact discontinuities in simulated annealing. Math Geol29:231–244

[6]   Carle SF (1999) T-PROGS: Transition probability geostatistical software. Univ of Calif, Davis

[7]   Carle SF, Fogg GE (1996) Transition probability-based indicator geostatistics. Math Geol28:53–476

[8]   Carle SF, Weissmann GS, Fogg GE (1998) Conditional simulation of hydrofacies architecture:a transition probability approach/Markov approach. In: Hydrogeologic models of sedimentary aquifers. SEPM Spec Publ.pp147-170

[9]   Clement TP (1997) RT3D:A modular computer code for simulating reactive multi-species transport in 3-dimensional groundwater aquifers. Pacific Northwest National Laboratory, Washington

[10]   Deutsch CV, Cockerham PW (1994) Practical considerations in the application of simulated annealing to stochastic simulation. Math Geol 26:67–82

[11] Deutsch CV, Journel A (1992) GSLIB: Geostatistical Software Libary and Users Guide. Oxford University Press, New York

[12] Engdahl NB, Vogler ET, Weissmann GS (2010) Evaluation of aquifer heterogeneity effects on river flow loss using a transition probability framework. Water Resour Res. doi:10.1029/2009WR007903.

[13] Koch J, He X, Jensen KH, Refsgaard JC (2013) Challenges in conditioning a stochastic geological model of a heterogeneous glacial aquifer to a comprehensive soft dataset. Hydrol Earth Syst Sci Discuss 10:15219-15262

[14] Ritzi RW (2000) Behavior of indicator variograms and transition probabilities in relation to the variance in lengths of hydrofacies. Water Resour Res. 36:3375–3381

[15] Seifert D, Jensen, JL (1999) Using sequential indicator simulation as a tool in reservoir description:issues and uncertainties. Math Geol 31:527–550

[16] Weissmann GS，Fogg GE (1999) Multi-scale alluvial fan heterogeneity modeled with transition probability geostatistics in a sequence stratigraphic framework. Hydrol 226:48–65.

[17] Walker JR (2002) Application of transition probability geostatistics for indicator simulations involving the MODFLOW model. Dissertation, Brigham Young University

[18] Zheng CM (1990) MT3D:A modular three-dimensional transport model for simulation of advection, dispersion and chemical reactions of contaminants in groundwater systems. U.S.Environmental Protection Agency

[19] Zheng CM, Wang PP (1999) MT3DMS:A modular three-dimensional multi-species transport model for simulation of advection, dispersion, and chemical reactions of contaminants in groundwater systems: documentation and user's guide. U.S. Army Engineer Research and Development Center

21

Ocean Resources' Ascendancy over African States Friendship Relations: Can UNCLOS Help Resolve Current or Future Maritime Boundary Delimitation Dispute Issues

Djibril Moudachirou

Institute of International Law, School of Law, Wuhan University, Wuhan, China

Email address:

jibrilfofana@yahoo.com

Abstract: Despite African states massive support to the event of UNCLOS, their envy to possess ocean resources is gradually getting entangled in the implementation of UNCLOS provisions regarding the delimitation of maritime boundaries upon which they can enjoy sovereign rights or claim sovereignty. The complacency of these provisions is enraging African states on their attempts to appropriate maritime resources. These states do adopt divergent interpretations which entail conflicts that negotiations fail to resolve. Moreover, considering that inviolability principle, uti possidetis principle and even the principles of delimitation adopted by the Court to resolve maritime delimitation issues often result on biased outcomes, it is imperious to think about a concrete way to favor negotiation on a win-win basis. While this paper briefly highlights African states' efforts on the realization of UNCLOS, it does find out some attitudes that encourage and complicate friendly neighborhood relations. It therefore suggests stepping forward on cooperating through joint development agreements to explore and exploit maritime resources found on the disputed zones.

Keywords: African States, UNCLOS, Maritime Resources, Maritime Boundary Dispute Settlement

1. Introduction

Since immemorial time, the sea cause caused curiosity and passion through its mystery and its perfume of adventure. Nowadays, although this interest still exists, it is renewed and enriched by the revelation of countless wealth that the sea contains. Today, every state is concerned by the sea and the resources beneath it. Oceans are considered as a plinth for sustainable development for states. Either coastal or landlocked, states become aware of a growing ascendency that oceans can have on the future. Thus, states are carried away by sea's covetousness so that every coastal state wants to nationalize portions of sea including those beyond its jurisdiction. Hence 70% of the globe i. e. the sea is no more considered as an open access space. It is considered as a cake to be shared by coastal states. With this in mind, unilateral appropriation or delimitation of maritime zones has been the practice of coastal states before the event of UNCLOS. In addition, more than a cake, it is source of tensions and conflicts between states around world since every coastal state want to get a big share of cake.

In order to codify ocean's usage and stop abusive nationalization of parts of ocean the First United Nations Conference on the Law of the Sea (UNCLOS I) was convened. Through this Conference, every sovereign state is given opportunity to attend the discussions on the new rule that could govern ocean's usage. Few African states attended that Conference since the majority was under the colonization. However, African states massively attended and made great contributions to subsequent Conferences, i. e. UNCLOS II and UNCLOS III which resulted to the final outcome of the new rule governing the current sea usage: United Nation Convention on the Law of the Sea (UNCLOS). From its signature on 10 December 1982 through its entering into force on 16 November 1994, UNCLOS continues to attract states. Actually, UNCLOS counts 166 states parties including the majority of African states. That is to say that states understand that oceans are key to sustaining life on the planet. Moreover, the importance of oceans for sustainable development has been recognized in the outcome documents of various conferences and summits on sustainable development [1]. However, the best comprehension of that fact should necessarily pass through the best practice on implementing UNCLOS by states parties because the regime for oceans and seas established by UNCLOS deals with a wide range of issues on ocean affairs and recognizes that the problems of ocean space are closely interrelated and need to be considered as a whole. Although in general UNCLOS' provisions are more or less comprehensive, some of them are ambiguous and vague, left to the interpretations of states. For example, the lack of clarity or the imprecision of provisions relating to EEZ and the CS delimitation constitutes one of the key problems leading to disputes between states. This paper finds out and discusses some of these issues in highlighting African states attitude and practice which does not comply with UNCLOS' provision. The analysis of this practice affirms that African states are stingy in maritime boundary delimitation issues because they give more preference to the economic outcome they could gain from maritime resources exploitation. Therefore, political and military considerations are neglected during negotiations because each party wants to internalize the disputed zone rather than sharing it with its neighbor. Thus the idea of agreeing a practical arrangement to cooperate and manage the resources beneath the disputed zone are ignored or discussed in vain.

2. Attitude of African States on the Regulation of the Law of the Sea

By the time the First United Nations Conference on the Law of the Sea (UNCLOS I) was convened, a good number of African states was enduring their dreadful situation of colonization. Even before UNCLOS I, during the time that the issue of 'liberalization of the sea' from the idea of 'the sea appropriation' was dividing Europeans, Africa was in the ascendancy or the imprisonment of the Europeans. Thus, historically it can be noted that the law of the sea was dominated by a few European powers who applied it in their territories, as well as in the colonies they controlled. Once the struggle of 'sea liberalization' triumphed upon 'sea appropriation', the philosophy of open access to seas beyond narrow territorial waters became the new rule of ocean management. This new rule permitted Europeans to explore and scramble to marine resources wherever they could be found until the end of the 19th century. In order to codify this new rule of ocean usage, a conference was convened (UNCLOS I). Considered as an outsider of the long process of development of the international regime designed to regulate the use the seas, Africa was offered an opportunity to participate to that Conference though many African states were still under colonial domination. Unfortunately, among the few states which gained their independence [2] at that time some states [3] refrain from attending the Conference. As result of this situation, African support for the four conventions [4] adopted at that Conference is very low. For instance, 11/54 African states ratified the Convention on the territorial sea and the contiguous zone, 13/54 African states

ratified the Convention on the high seas, only 10/54 African states ratified the Convention on the continental shelf and 11/54 African states ratified the Convention on fishing and conservation of the living resources of the high seas. Even though this attitude can be understood by the lack of experience of the majority of African states in the laws governing the sea, the status quo observed through the attitude of African states has detonated the voices of some African diplomats and international law scholars for a reform of the international legal order. These calls added to those of former colonial territories on other continents have been responded during UNCLOS III negotiations since no African states voted against or abstained when UNCLOS was adopted on April 30, 1982. At UNCLOS III, the status quo previously observed had been replaced by the participation of several African intellectual heads and members who made great beneficial contributions to the international community as well as individual state. [5]

2.1. African States Attitude After UNCLOS

A decade after the adoption of UNCLOS it can be noticed that the situation has drastically changed in Africa. In fact, despite an economic crisis that Africa was experiencing in the early 1990s, Africa made the greatest contribution to the coming into effect of UNCLOS in terms of ratifications. By the end of 1992, 26 states (or 48%) of the then 54 African states had ratified UNCLOS. Many states have enacted their own maritime legislation and most coastal states have declared 200 mile jurisdictional zone either exclusive economic or exclusive fisheries in nature. Moreover, the need to apprehend maritime resources pushed African coastal states to extend their claims over water traditionally recognized as free seas. With this in mind, coastal states are claiming specific maritime areas enclosed by boundaries pertaining to some regulatory competence such as control over fishing or drilling hydrocarbons on the continental shelf. The majority of African coastal states make use of straight baselines. Most states claim a territorial sea of 12 nm in accordance with Article 3 of UNCLOS whereas the other states claim wider territorial seas although they are all parties to UNCLOS. For instance, from Atlantic Ocean to Indian Ocean there are states which claim 50 mile territorial sea and fishing zone [6] whereas other are claiming just 12 mile territorial sea and fishing zone due to their restricted coastlines with their neighbors [7]. Furthermore, while some coastal states are claiming more, i. e. 200 mile territorial sea and fishing zone [8] there are states that believe that 200 mile EEZ can be regarded as their legal largest possible EEZ right under Article 57 of UNCLOS [9]. In addition, almost 2/3 of the states claim a contiguous zone extending in most cases up to the maximum allowed by Article 33 (2) of UNCLOS, which is 24 nm from the baselines. Meanwhile, only 9 African states have fulfilled their obligation regarding the depositing of charts and lists of geographical coordinates with the UN Secretary- General. Altogether, 26 African coastal states are involved in claims submission to the Commission on the Limits of the Continental Shelf.

With the event of UNCLOS and its ratification it is also remarkable that many African states do hesitate to comply with UNCLOS' provisions and implement them. For example, Benin, Somalia, Congo and Liberia are still claiming a 200nm territorial sea, and Togo is claiming a 30nm territorial sea whereas UNCLOS limits territorial seas to 12nm. As far as it concerns the contiguous zone, except the Gambia and Sudan which are claiming 18nm instead of 24nm, many African states do not focus on this area of the sea. Perhaps, they already grabbed a large space of territorial sea and there is no need for claiming appropriate contiguous zones. African states' failure to delimit their contiguous zone could be seen as their irresponsibility regarding various enforcement jurisdictions such as immigration, health, sanitary and pollution controls on their waters. Indeed, contiguous zones' delimitations could help African states prevent or combat illegal environmental pollution, drug trafficking, illegal smuggling and piracy that can be qualify as contemporary significant problems for several African states. The regime of EEZ is generally accepted by states parties to UNCLOS throughout the world and African states widely respected the limit of 200nm. Although several states miss legislations on this maritime zone, they may have simply declared it or even delimited the EEZ boundary with their neighbor. Considering the CS, many African states are willing to enjoy the opportunities of extending their shelf beyond 200nm though relatively only few states have specifically incorporated the text of Article 76 of UNCLOS in their national legislation. In addition, some states would have to update their national legislations on CS delimitation in order to adapt them to the exigencies of UNCLOS rather than those of the 1958 Geneva Convention on the CS.

What can be the reasons justifying these attitudes from African states? Does UNCLOS create more problems than it resolves? The main reason of these attitudes could result from the ambiguities of UNCLOS since it leaves states without clear guidance in some issues. And in order to fill the gaps left by UNCLOS states interpret UNCLOS provisions according to their interests.

2.2. UNCLOS Ambiguities Regarding Maritime Zones' Delimitation

Regardless of the low rate of African states' participation to UNCLOS III, the lack of consensus on boundary delimitation issues led to evident ambiguities during the drafting of UNCLOS provisions on boundary delimitation. Although UNCLOS offers some guidance concerning the way to delimit maritime areas such as Territorial Sea (TS), Exclusive Economic Zone (EEZ) and Continental Shelf (CS) as well as the rights of coastal states upon these areas, it does remain silent on the preferred method of boundary delimitation. For example, it is true that Article 15 of UNCLOS tries to apprehend the issue of overlapping claims on territorial sea between coastal states by favoring equidistance method with the exception when there is an 'historic title or other special circumstances' on the area in dispute, it has been silent on what 'historic title' or 'special

circumstances' could refer to. Also, supposed that we know all the elements that these terms could encompass, could all these elements be taken in consideration while considering these terms? For instance, could economic factor be taken into account like geographical factors or coastal geography in any maritime boundary dispute? Then, the determination of relevant circumstance seems to be an issue in maritime boundary dispute settlement. Thus, though special circumstances play a major role in maritime boundaries delimitation and therefore fundamental in international law, UNCLOS left the burden to states asserting such circumstances to provide evidence of their claims. Interestingly, through judicial decisions and awards it could be contend that special circumstance like geographical configuration of the maritime areas is a given. And for this reason, the ICJ asserted that it is not an element opened to modification by the Court but a fact on the basis of which the Court must effect the delimitation [10].

In practice, the consideration of special circumstances could benefit more African states, particularly the coastal states situated in the western part in the gulf of guinea. Indeed, the western African coast is convex. This concavity poses problems of enclaves to many coastal states when they are confined between two other states. To this, the presence of large number of islands and largely scattered coastlines seem to complicate maritime dispute settlement between states. Thus, the application of equidistance principle without taking into account special circumstances could be detrimental to some African states. Cameroon has remarked that and highlighted that if the Court drew a strict equidistance line, it would be entitled to practically no EEZ or Continental Shelf. To bring the Court to understand its anxiety Cameroon invoked some reasons. For example, Cameroon alleged concavity of the gulf of guinea in general and of its own coastline in particular created its own virtual enclavement which should constitute a special circumstance. After the Court having dismissed this allegation Cameroon found out that the proximity of Bioko Island to its coast should be considered as a special circumstance requiring an adjustment of equidistance line. But once again the Court found this allegation irrelevant since the island in question lies out of the zone to be delimited.

In order to come to an equitable solution concerning the Continental Shelf and the EEZ delimitation, Articles 74 and 83 of UNCLOS call on states to effect this delimitation through 'agreements'. In the case they fail to do so, they should give preference to concrete 'provisional arrangements' under international law for a 'transitional period'. If these provisions are significant in the management of these zones, it is important to notice that they are generally silent on the method to be used for equitable delimitations. Thus, could the terms "an equitable solution" used in the formulation of Articles 74 and 83 presuppose a new "equitable principle" upon which CS and EEZ should be delimited? In other words, does "equitable solution" means "equitable principle"? And if yes, is this new principle different from the one used for territorial sea delimitations, i. e. equidistance principle? Could states call for the application of equidistance principle on disputes regarding the CS and EEZ delimitation too? All these questions demonstrate the imprecision and the vagueness of these provisions and give opportunity to states to interpret them according to their interests. And under such interpretations some states could request the application of equidistance/special circumstances method whereas other may claim equitable principle. Hence, the vagueness of these articles is source of disagreement between states in dispute during negotiations. Therefore, divergent point of views between states concerning the interpretation of these provisions generally open the opportunity to a third party, i. e. Court and Tribunals to provide their interpretations. In this way, ICJ seized such an occasion to clarify what the applicable criteria, principles and rules of delimitation are when there are disputes regarding maritime zones with coincidence jurisdictions. Thus, according to ICJ, similarly to the equidistance/special circumstances method developed in the delimitation of territorial sea, "an equitable solution" regarding the delimitation of the CS and the EEZ should be done under the equitable principles/relevant circumstances method. Despite the fact that international courts and tribunal try to redress the vagueness of UNCLOS provisions on EEZ and CS delimitations, the practice of African states does show that these states are not ready to understand the clarifications made by courts and tribunal on UNCLOS gaps. They are rarely satisfied with the decisions and awards of these courts and tribunals though they finally comply with. Like sovereignty on a piece of land, African states are really reluctant to make concession on the flexibility left to them by UNCLOS regarding the scope of provisions on EEZ and CS delimitations. They usually cling on conflicting interpretations and neglect the alternative way offer to them by UNCLOS, i.e. provisional arrangements such as joint development agreement or unitization on disputed zones though these zones contain valuable natural resources. Otherwise, when they think about these arrangements, the negotiations to conclude them are either deadlocked or unfeasible for one party may intend to get 'a lion share' in the detriment of its neighbor. What can be the reasons of such an attitude? Several reasons could explain this attitude. Amongst them economic factor is focused more than political and technologic factors.

3. Some Explicative Reasons

Presumably the overwhelming ratification of UNCLOS by African states could be seen as a symbol of awareness of the outcome they could get by being part of the new regime governing the sea. However, this awareness could not be seen as the expression of their will to promote the law of the sea of the international community. Rather, through political, technologic and particularly economic reasons, we could contend that these ratifications do appear like African states readiness to comply with the transformation of the law of sea.

3.1. Political Reasons

The direct impact of decolonization is that African states got their independence from colonial powers. Henceforth, they are new sovereign states, equal member of international community like their former colonial powers. These states can attend Conferences and discuss issues relating to the regulation of new international order. Of paramount importance, since many states have been freed from colonization, their number could be meaningful for votes. Moreover, conscious with the big gap of development between these states and their former colonial powers, the modification of previous equilibrium particularly within international organizations could be seen as the first priority. Thus, any issue relating to international economic order becomes the concern of African states. Therefore, due to the economic advantage of the sea, the participation and the contribution of these states to the transformation of the order governing the usage of ocean and seas is no more an issue of negotiation. In this context, it is useful to understand that African states would no more accept laws established without them on their behalf. For example, classical rules relating to the usage of ocean and seas such as limitless freedom of sea and 3 mile width of territorial sea have been contested in order to gain a future from the sea. Conjointly with other developing countries, African states constitute valuable and influential majority [11] to block American, European and soviets decades of rulings on sea usage during Geneva Conference in 1958. And this will has been expressed out by the Peruvian Prime Minister in August 1970 through the following wordings:

Undoubtedly, the balance is favoring us. Qualitatively, our case is better. No one can conscientiously deny that the means to make a living and the welfare of a people are greater than utilitarian purposes. And from the quantitative point of view, the number of developing states is greater by far than the number of industrialized countries. Things being this way, ail we need to do is remain united in the face of the intentions of those big powers that want to impose on us norms that are convenient for them [12].

Another political reason can be related to the exigencies of securing these new states. Once independent the search and the maintenance of national security is the key role of every state. Thus, the new challenge for African states while internalizing parts of seas is to consider the threats which could derive from sea usage. For example, if sea serves international community for transport of indispensable items for our survival, it is also worth mentioning that sea constitutes the lieu of harmful conducts (economic pillage, pollution, piracy and illegal fishing). Therefore, African states also understood that national security considerations do occupy a good place in every national politics. Moreover, in a time the law of the sea restricted territorial sea to 12nm with the idea of an open access sea rights and the right to extend the CS beyond 200nm, these new states seem to be aware of all forms of aggression on seas. And even though they ignore this, they cannot forget the invasion of their former colonial powers through ocean. However, if national security has to be understood through military aspects, the need to equip national army is therefore vital in order to defend portions of seas that new states internalized. Remarkably, despite the existence of the will to defend these portions, African states have to improve in this sense since they are facing cruel lack of equipment as well as qualified human resources to carry out this mission.

3.2. Economic Reasons

The existence of billions of tons of living and non-living valuable resources that ocean overflows with stimulates the envy from every states to get a portion of sea. African states are aware of this and express they envy to take part to the great competition of exploration and exploitation of ocean's richness. Although they are missing abilities to compete with developed countries as well as some developing countries in the globe, this awareness is really important since they could also benefit from the openness of the high sea at least through fishing. Moreover, article 2 of 1958 High Sea Convention delightfully provides the freedom of high sea [13]. This could help these states protest less about pillage of living resources by industrialized states.

Knowing that terrestrial resources are rare and in order to restrict their foreign dependences on energy supply, African states also understood that the struggle for oceans' internalization constitutes an important economic stake for their survival and their future since off-shore resources (oil and gas, manganese, iron, nickel) exploitation becomes an industrial reality nowadays. The discovery of these resources and their geographical distribution throughout the world seem to be unequal and thus poke up states' covetousness and threaten harmony between neighboring states. Moreover, this covetousness has cooled down with their formulation into conventional law (UNCLOS). For example, the envy of off-shore hydrocarbon exploitation has given birth to the institution of the Continental Shelf and the right of its extension beyond 200nm. Also, the institution of EEZ and the determination of the Area and its resources "common heritage of mankind" can be regarded as the manifestation and the chance given to states to exploit these resources wherever they may be on the oceans. However, despite the considerable progress in codification ocean management UNCLOS's ambiguities attest that ample scope remains to be clarified. Hence, states' practice demonstrates different interpretations of certain provisions of the law of the sea and, therefore, replacing comity by enmity between coastal states.

4. Do African Coastal States Refer to UNCLOS's Dispute Settlement Mechanisms

As discussed above, UNCLOS offers some guidance to coastal states in the delimitation of maritime boundaries. UNCLOS mentions that territorial sea delimitation dispute should be done in accordance with article 15, delimitation

dispute concerning the EEZ should follow the provision of article 74 and delimitation dispute concerning the Continental Shelf should be done in accordance with article 83. Due to the ambiguities revealed above, African coastal states are facing to delimitations problems. Thus, when delimitation disputes arise between them, the first question to address is how to deal about these provisions? And sometimes the answer to this question results to tensions between these states whereas UNCLOS provides them with tools to respond this question. These tools or mechanisms can be seen through Part XV of UNCLOS.

4.1. Settlement of Dispute Under UNCLOS

In order to preserve peace and friendly relationship among states UNCLOS reiterates international community traditional will to settle any dispute by peaceful means. Thus, UNCLOS requires its coastal states members to resolve their disputes by applying Part XV. And in this context article 279 states that:

States parties shall settle any dispute arising between them concerning the interpretation or application of this Convention by peaceful means in accordance with article 2, paragraph 3, of the Charter of the United Nations and to this end, shall seek a solution by the means indicated in article 33, paragraph 1, of the Charter.

In other words, UNCLOS enshrines UN Charter's goal of seeking and preserving peace when the latter states in article 2, paragraph 3 that "all members shall settle their international disputes by peaceful means in such a manner that international peace and security, and justice, are not endangered". What are these peaceful means?

Chapter VI of the UN Charter listed peaceful means like negotiation, enquiry, mediation, conciliation, arbitration and judicial settlement. In addition, states can resort to regional agencies or arrangements, or other peaceful means of their own choice [13]. The formulation of article 33(1) of UN Charter impliedly indicates that the list of peaceful means of dispute settlement is not exhaustive and thus, opens the avenue for interpretation in order to be understood. The only guaranty that it offers is that it absolutely requires states to refrain from any use of force in dispute settlements. Also the free choice of methods of dispute resolutions offered to states by this article does connote that states are not bound to pursue these methods in series. Nevertheless, these methods constitute without much doubt the most frequently used in dispute settlement. Moreover, the application of these methods in series though it is not an obligation upon states, does constitute a rule under UNCLOS. For example, article 283 (1) of UNCLOS states: "When a dispute arises between States Parties concerning the interpretation or application of this Convention, the parties to the dispute shall proceed expeditiously to an exchange of views regarding its settlement by negotiation or other peaceful means (…)".

Such a phrasing of UNCLOS though it does not impose a duty to states parties to resolve their disputes through exclusively negotiation, it does show a preference to resolve disputes through negotiation. Or better, it does impose a duty to states in disputes to commence resolving their disputes through negotiation. And in the case negotiations fail, to seek a solution to their dispute by other peaceful means such as inquiry, mediation, arbitration and judicial settlement. Thus, we can say that UNCLOS makes negotiation as a sine qua non condition to the application of any other method of dispute settlement. Interestingly, with particular focus to maritime boundary delimitation, this position has been supported by ICJ in the North Sea Continental Shelf Cases when it states:

The Parties are under an obligation to enter into negotiations with a view to arriving at an agreement, and not merely to go through a formal process of negotiation as a sort of prior condition for the automatic application of a certain method of delimitation in the absence of agreement; they are under an obligation so to conduct themselves that the negotiations are meaningful, which will not be the case when either of them insists upon its own position without contemplating any modification of it [14].

Under this finding of the Court, it could be noted that not only negotiation is a prerequisite to maritime boundary dispute settlement but negotiation should be carried by states in order to end with practical solutions to their dispute. This does imply that negotiation should be done in good faith. In addition this finding could be considered as requiring states to exhaust all the non-binding methods before submitting their disputes to a binding third party method such as arbitration, ICJ, ITLOS also available for them under Section 2 of Part XV of UNCLOS.

4.2. African States' Preference in Choosing Their Dispute Settlement Method

Without recalling past African maritime boundary disputes which have been resolved through arbitrations and ICJ, here we will rely on current manifestation of disputes between African states on one hand and invoke some potential disputes which will attract our attentions in the coming decades on the other hand. This could help us not only apprehend the venues through which African states prefer resolve their disputes but more importantly we could also deduce African states practice on seeking maritime natural resources.

4.2.1. Current African Maritime Boundary in Dispute

Here we will mention some ongoing disputes between African states on their maritime boundary delimitations and the venues of dispute settlement they have chosen to resolve their disputes.

a　The case between Somalia and Kenya.

Global interest for the richness of the Western Indian Ocean has risen over the past decade, affecting the hydrocarbon resources and fish sector in East Africa. Kenya in 2012 leased eight offshore blocks to oil exploration companies. Seven of these are located in a contested area in the Indian Ocean. Thus, by doing so, Somalia contends that Kenya has contravened Somali Law no. 37 which defines Mogadishu's continental shelf, its 200 nautical miles and

territorial seas and firmly protests Kenya's behavior. The Somali government even announced it had completed surveys of the disputed area and planned to start issuing offshore oil and gas exploration licenses by 2015.

Somalia wants that the boundary line in the territorial sea based on a median line as specified in Article 15, since there are no special circumstances that would justify departure from such a line and that, in the EEZ and continental shelf, the boundary on the basis of the three-step process the Court has consistently employed in its application of Articles 74 and 83. Meanwhile, Kenya wants a maritime boundary based on a straight line emanating from the Parties' land boundary terminus, and extending due east along the parallel of latitude on which the land boundary terminus sits, through the full extent of the territorial sea, EEZ and continental shelf, including the continental shelf beyond 200 nautical miles. This map explains the two divergent points of views.

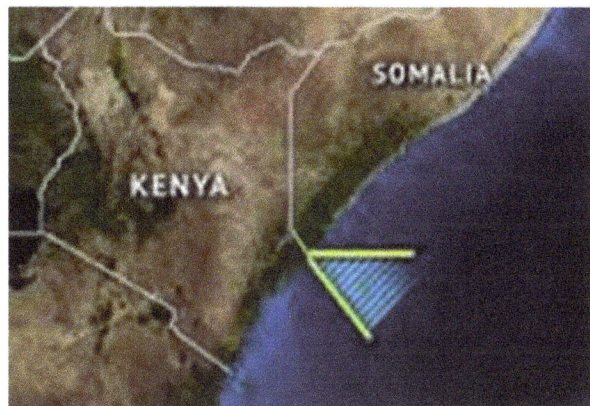

Figure 1. *Map showing Kenya and Somalia boundary claims.*

Source:　　http://www.odinafrica.org/news/139-african-maritime-border-disputes.html.

Therefore, following this divergent point of views on their maritime boundary limits, Somalia seized the ICJ on 28 August 2014 and asks the Court:

to determine, on the basis of international law, the complete course of the single maritime boundary dividing all the maritime areas appertaining to Somalia and to Kenya in the Indian Ocean, including the continental shelf beyond 200 [nautical miles]". The Applicant further asks the Court "to determine the precise geographical co-ordinates of the single maritime boundary in the Indian Ocean [15].

b　The case between Ghana and Ivory Coast.

When Ghana found oil in 2007 and began production in 2010 at the Jubilee Field – located in the Atlantic Ocean 60km off Ghana shore – Ivory Coast petitioned the U. N. to complete the demarcation of its maritime boundary with Ghana. However, tensions escalated in April 2013, when Ivory Coast announced it had struck oil in an area adjacent to Jubilee Field – Ghana's largest oil field – and sent a correspondence to the government of Ghana casting doubt on the existing median line that divides both countries' waters. The governments of both countries set up a commission in November to come up with technical solutions to end the dispute over the area – which reportedly contains reserves of 2 billion barrels of oil and 1.2 trillion cubic feet of natural gas, according to official statistics from both countries [16].

After more than five years of negotiations and many attempts of bilateral conciliation between Ghana and Ivory Coast, these two states have finally decided to submit their maritime boundary delimitation dispute to the arbitration of the International Tribunal for the Law of the Sea (ITLOS) under UNCLOS Annex VII by the end of 2014. Ivory Coast is requesting the Tribunal to order Ghana to suspend any explorations and exploitations activities on an offshore disputed zone which is situated on their international boundary demarcation line. In April 2015, Ghana has been ordered to suspend drilling in waters next to an existing oil and gas field development in the so-called 'ten fields' belonging to Ghana and which are expected to start producing oil by the middle of 2016. The tribunal ruled that any further work in the area would hurt Ivory Coast's interests ahead of a final judgment on where the maritime border lies. However, both countries were ordered to present further evidence and arguments in May in the next stage of the dispute [17], which could last until 2017.

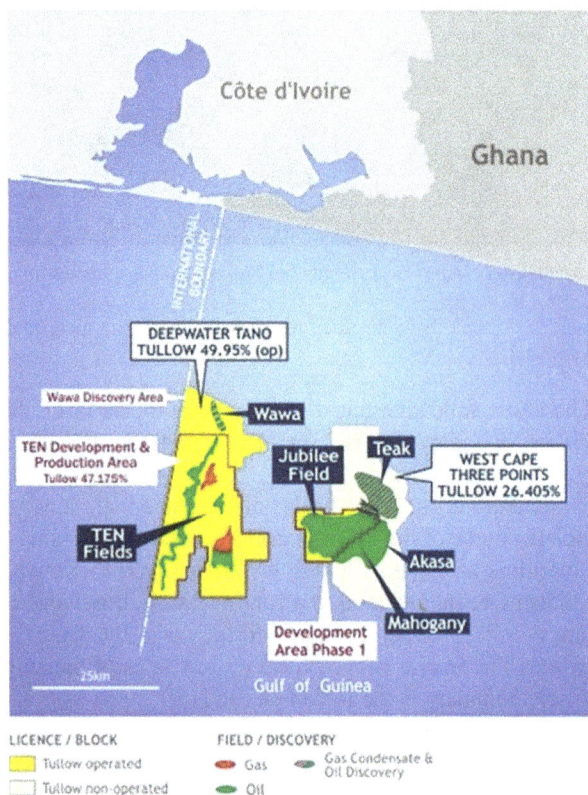

Figure 2. *Map showing Ghana and Cote d'Ivoire boundary dispute with regard to oil fields' exploitation.*

Source:
http://ressourcesafricaines.blog.lemonde.fr/files/2015/05/Carte_Tullow_Oil_Ghana1.jpg.

  c The case between Democratic Republic of Congo (DRC) and Angola.

The sliver of DRC coastline that reaches to the Atlantic is

wedged between Cabinda and Angola proper. Though no specific blocks have been named, the fields most likely to be affected by the dispute are those offshore from mainland Angola's coast and off the southern Cabinda coast, the Angolan enclave just north of the DRC's coastline. Included in this triangle are the Pitanguiera, Safueiro, Bananeira, Sapesapeiro, Essungo and Soyofields offshore Angola's mainland. To the north, the Kambal, Livuite, M'bili, N'tene, Sanzamo N'Kassa, N'dola and Minzu fields also might be affected [18]. These blocks are controversial, however, because the angle at which they are aligned seriously restricts DRC's access to the sea. About 65% of Angola's petroleum comes from the regions off Cabinda, and DRC has long been claiming that it deserves a portion of the income [19].

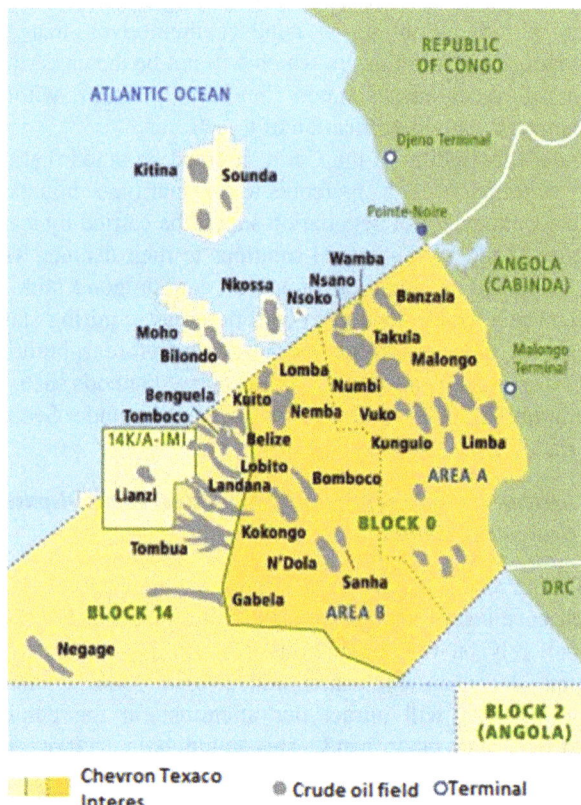

Figure 3. *Map showing Angola and DRC boundary dispute.*

Source: http://menasborders.blogspot.com/2010/12/border-focus-angola-and-drc.html

After vain temptations to conclude a boundary agreement, these two States began negotiations in May 2003 and signed their first Memorandum of Understanding (MoU) in August 2003. This agreement established joint technical committees mandated to prepare proposals to resolve maritime border disputes. In 2004, the two countries created, in principle, the common interest zone (CIZ) as a new special exploration area [20]. The Angolan government approved this initiative in September 2004 [21], but DRC only did so in November 2007 [22]. Although the DRC ratified the MoU, the decision was not unanimous. Senator Lunda Bululu opposed it because the area and coordinates of the CIZ were imprecise and the members of the Congolese Assembly did not have information on the extent of hydrocarbon reserves or the blocks where production was already underway. The MoU was all the more disadvantageous to the DRC because it did not provide compensation for the loss of a share of the royalties already received by Angola from blocks under production. Unfortunately, The CIZ has not resolved the dispute between Angola and the DRC for it is in a standby because the two States are sticking on their divergent points of views.

DRC wants the present maritime space taken from 40 km off the coast to 200 km, or an expanse of 4,000 square meters, an area that covers the oil zone, where Angola draws 500,000 barrels per day as DRC produces only 20,000 barrels/day. By doing so, DRC aims to receive its fees and take possession of half of oil deposits from two blocks exploited by several multinational companies for Angola. Then, to prevent this, Angola has approved the establishment of an agreement on the delimitation of its maritime borders with DRC on condition that the strict respect of the agreements signed between Portugal and Belgium are taken into consideration. To counter this assumption, DRC had referred to international arbitration by sending a request to the United Nations, for the extension of its continental shelf, within the delimitation of its maritime borders, in accordance with the sea rights. Moreover, it should be noted that both Gabon and DRC are rejecting Angola's proposed changes to the maritime borders, in which Angola seeks to increase its continental shelf length by more than 200mi. Gabon and DRC have written to United Nations secretary-general Ban Ki-moon rejecting part of Angola's 6 December 2013 submission to the UN Commission on the Limits of the Continental Shelf (UN-CLCS).

d The case between Gabon and Equatorial Guinea.

The dispute concerns Mbanie, Cocotiers and Congas, three small islands in CoriscoBay, just north of the Gabonese capital Libreville, near the border with the continental territory of Equatorial Guinea. The dispute has been simmering away quietly since 1972 and has prevented oil companies from carrying out a full exploration of the nearby offshore waters.

However, it came to a head in February 2003, when Gabonese Defense Minister Ali Bongo, actual Gabonese President, visited the Corisco Bay islands and reasserted his country's territorial claim to them. By the way, Nicholas Shaxson, a Berlin-based expert on oil and gas issues in Africa confirmed that "There are fields on both sides of the Corisco Bay area. These wells generally have reserves of several hundred thousand barrels of oil and there are very probably more wells of a similar size here,". In addition, When Ali Bongo visited Mbanie, a 30-hectare island inhabited by a handful of fishermen, and declared it part of Gabon, there was a swift reaction from Equatorial Guinea's Prime Minister Candido Muatetema Rivas said in a radio broadcast: "My government expresses its deep concern and its indignation regarding Gabon's illegal occupation of the small island of Mbanié."

Finally, on 23 January 2004, Equatorial Guinea and Gabon have agreed that a UN mediator should settle their dispute over a handful of small islands that hold the key to potentially oil-rich offshore waters [23]. Still, it should be noted that the establishment of a maritime boundary in hydrocarbon-rich Corisco bay is on stand-by between these two countries because the disputed islands in Corisco bay where the Rio Muni land boundary enters the sea complicate the problem.

4.2.2. Potential Maritime Disputes Between African States

Although ICJ helped resolve the dispute between Nigeria and Cameroon the implementation of ICJ's judgment could result in creation of other disputes between some neighboring states. The implementation of this judgment under equidistant line boundary will result in cutting through an oil field licensed by Nigeria. This is the first matter which should be resolved in order to lessen problems from happening throughout the Bakassi peninsula. Then Cameroon will have to address their boundary which has both a lateral aspect between the main land part of Equatorial Guinea, Rio Muni, and an opposite aspect between Cameroon and Equatorial Guinea's Bioko Island. This could also result on a dispute between these states. Furthermore, the equidistant-based boundary lines between Sao-Tome & Principe and both Equatorial Guinea and Gabon resulted to a slightly overlap zones between these three states due to Equatorial Guinea and Gabon dispute. This could also trigger a dispute if any resource is discovered in this zone.

Another muddling situation can be observed between the Republic of Congo and Gabon since there is no boundary agreement between these two states yet. Moreover, within the Republic of Congo and the Democratic Republic of Congo (DRC) lies the enclave of Cabinda which belongs to Angola, but partly claims by the DRC. This dispute creates many controversies and tensions that spoilt the relationship between Angola and the DRC. Negotiations between these states resulted to a fake memorandum of understanding on resources' exploitation which couldn't be implemented until today. We will discuss this dispute later in the course of this study. For the moment, let's say that there is no formal boundary agreement between these three states whereas potential maritime resources are lying on this area. However, there are unitization agreements that provide for equally shared revenues in this area between Angola and the Republic of Congo while the Democratic Republic of Congo has to fight toughly in order to gain such a deal. In addition, Uganda and DRC continues to dispute over the Rukwanzi Island in Lake Albert and other areas on the Semliki River with hydrocarbon potential.

Farther on the south part of Africa, from the Cape of Good Hope to the Horn of Africa, it is remarkable that although it does have a swift resolution of maritime boundary issues Namibia and South Africa, there are potential disputed islands in the river mouth and both the river itself and the immediate offshore area which are rich in diamonds and oil. Furthermore, it could be noted that South Africa and Mozambique has an incomplete international boundary agreement since these two states still have to determine the line of their maritime boundary. And the demarcation of this line could in some extent affect the interest of Madagascar due to its offshore presence in this area. In addition, Madagascar could also be disturbed by the presence of some French islands in this area, notably Ile Juan de Nova and Ile Europa and in some extent Iles Glorieuses which are also claimed by Madagascar. Thus, it will be difficult to guess any maritime boundaries agreement on the Mozambique Channel until the sovereignty issue is settled between France and Madagascar. Moving down to this area, it should be noted that Kenya still have problems to negotiate a maritime boundary with Somalia. Currently, the dispute between these states has been submitted to the ICJ.

The western part of Africa will register some of the most significant maritime boundary disputes due to the pre-eminent position of the West African coastline on the Gulf of Guinea. Indeed, the location of Benin-Niger-Nigeria tri-point is unresolved and a number of Gulf of Guinea maritime delimitations are yet to be achieved. As far as it concerns the turbulences on maritime boundary issues we can observe that Ghana and Ivory Coast have awaken the dormant situation on which the western part of Africa was by submitting their dispute to ITLOS. And perhaps, these two states did show the way to other states which may be supporting difficult maritime cohabitation with their neighbors on the Atlantic Ocean due to their maritime boundaries which have to be delimited.

A quick check of maritime delimitation cases submitted to ICJ by African states places Africa in the second position with a record of 7/22 cases (or 22% of all ICJ's cases from 1967 to 2014).

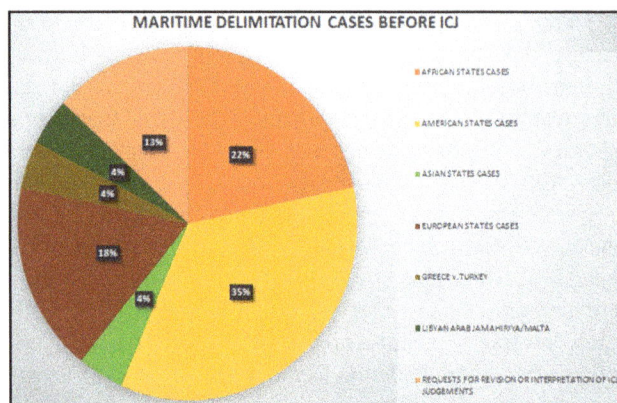

Figure 4. *Diagram of comparison of maritime delimitation cases before ICJ.*

Although America occupies the first position with 8/22 cases (or 35%), it could be permitted to contend that African states are keen on resolving their boundary disputes by judicial means. I wonder whether this avenue could be the best for resolving maritime boundary disputes since African waters are known to be rich in maritime resources. Generally, I think that African states and particularly those in the western part could win more by cooperating and friendly managing this area through practical arrangements rather than exhibiting their disputes to a third party. Moreover, this

cooperation could help them conjointly address insecurity and environmental issues on this area since it is proved that this task can not be efficiently undertaken by only one state.

To sum up, I can say that these potential disputes reaffirm the awareness of African states of the economic interest of the sea. Obviously, overlapping maritime claims will continue to increase in number in the following decades and this continent will also be facing the problems of disputes' resolution like the current event in Asian waters. Today, a state, be it a failed state or a civilized state, either a developed state or under developed state is not ready to concede to its neighbor a portion of empty land or a portion of land rich of natural resources. A third party dispute resolution often comply states to make such a concession although painfully. Therefore, why not develop the tradition of negotiation, mediation, conciliation in resolving a kind of dispute where disputed states have common economic interest? Moreover, natural resources are the 'gift of God', so when it comes to belong to different states, why not find out an alternative way to enjoy it friendly? In my view, Joint Development Agreements between states could help to come up to this end.

4.3. African States Practice on Apprehending Maritime Resources Straddling on Their Boundaries

The first practice or strategy to apprehend maritime resources in dispute by African states consists in denying the existence of boundary delimitation agreements between them on the disputed areas. And when it comes that they accept such an agreement, they challenge its binding force. The very imminent role that agreements play on dispute settlement between states brings international community to codify agreements on the basis of the law of treaty through the Vienna Convention on Law of the Treaties, 1969. Indeed, even in 1982 at the Montego Bay Conference, states did understand that agreements remain the best way of delimitation and dispute settlements either at international level or even at domestic level. That's why Article 83(4) UNCLOS, 1982 clearly assures that "where there is an agreement in force between the states in dispute, questions relating to the delimitation shall be determined in accordance with the provisions of that agreement." Perhaps, aware of this provision and in the case its application in a dispute may prevent a state from getting fully or in part the zone in dispute that African states challenge the existence of a boundary delimitation agreement or the binding force of such an agreement. However, this kind of practice didn't prevent the ICJ from granting the rich oil zone of Bakassi to Cameroon. Indeed, despite the challenge of Nigerian upon the existence or the binding force of territorial and maritime agreement between Nigeria and Cameroon, ICJ relied on the Anglo-German Demarcation Agreement of 11 March and 12 April 1913 [24], Yaoundé Declarations I & II and the Maroua Declaration [25] contracted by and on behalf of the two countries to delimit territorial and maritime boundaries between Nigeria and Cameroon. Moreover, in Guinea-Bissau and Senegal dispute, the Tribunal did take into consideration the validity of a colonial boundary delimitation agreement to delimit maritime boundaries between these two states by recognizing that:

The Agreement concluded by an exchange of letters on 26 April 1960, and relating to the maritime boundary, has the force of law in the relations between the Republic of Guinea-Bissau and the Republic of Senegal with regard solely to the areas mentioned in that Agreement, namely the territorial sea, the contiguous zone and the continental shelf' [26].

Therefore, we could notice that the existence of a delimitation agreement has mostly served as a point of departure of third parties' dispute settlement. So, either a colonial agreement or agreement concluded between sovereign states, terrestrial or maritime, bilateral or multilateral, the existence of a boundary delimitation agreement has influenced the ICJ and Tribunals decisions. Despite this fact, the DRC is still standing on the rejection of a colonial treaty in its maritime boundary delimitation's dispute with Angola.

Another practice which is growing now in African maritime dispute consists in challenging the method of boundary line. While some coastal states stand for straight baseline delimitation, other contends that maritime boundary should be effected on the basis of median line. This is the current case between Somalia and Kenya on their maritime boundary delimitation dispute on the Indian Ocean. While Kenya is defending that this delimitation should be based on straight line, Somalia affirms that a straight line delimitation will threaten its sovereign rights, and thus stand for a delimitation based on median line according to UNCLOS Article 15 as far as it concerns their territorial sea limits, respectively Articles 74 and 83 for the limits of their EEZ and their CS. In fact, by doing so, Somalia could grab the rich oil disputed zone on which Kenya already leased eight offshore blocks to oil exploration companies and of which seven are located in the contested zone in the Indian Ocean. In addition, we have to notice that the problem becomes difficult to solve where natural resources are found to straddle the delimited maritime boundaries. In this case, even though there is an agreement prescribing the boundary line, African states do raise some interpretational challenges as if at the time they were concluding this agreement there was some misunderstanding upon the method used. By the way, if agreement should be understood as an act whereby the states in dispute upon a boundary decide, in writing, to demarcate their boundary based on specific delimitation method negotiated, acceptable and agreeable to them, can we then admit tacit boundary delimitation agreements in an area full of natural resources? The Ongoing dispute concerning delimitation of the maritime boundary between Ghana and Ivory Coast in the Atlantic Ocean before ITLOS seems to be an illustrative case which could help us deduce another tactic of African state to conquest maritime resources. In that case, Ghana is contending that there is a continuously agreed equidistance-based boundary delimitation line between Ghana and Ivory Coast since almost four decades without any protest from Ivory Coast. Or, in fact there is no written or formal agreement proving the existence of such a line though Ghana states that an Ivorian boundary delimitation law (1977) contends that Ivory Coast's maritime boundaries should be delimited either on median line or equidistance line. Ivory Coast has rejected such evidence by stating that the 1977 law is approximate and that the median line or the equidistance line mentioned on this law cannot presume a final delimitation of Ivory Coast's maritime boundaries with its neighbors. This law is not indefinite. It is considered as a provisional law. Therefore, by this way, Ivory Coast is claiming maritime boundary delimitation based on equitable principle. This principle could help Ivory Coast get access to maritime resources on the disputed zone where Ghana grants licenses to oil companies and is expecting its first production by 2016. From this latter case, we could deduce the practice that consists of challenging equidistance line by equitable line. This trend is also adopted by Kenya and Somalia on their maritime boundary delimitation dispute actually pending before the ICJ. Thus, we could say that Somalia by asking a three-step delimitation procedure to the Court is claiming equitable-based maritime boundary delimitation while Kenya is defending equidistance-based line delimitation.

4.4. The Issue of Mineral Resources' Presence in Disputed Area in Boundary Delimitation

There is no need recalling the importance of mineral resources for the states. States need these resources not only as source of revenue for their nation but these resources could also help them release from energy dependence. Undoubtedly, the Court also understands this fact when it states in the Libya/Malta case, referring to mineral resources, that they "are the essential objective envisaged by States when they put forward claims to sea-bed areas containing them. [27]" However, with regard to this consideration, the issue which should be addressed by Court and Tribunals while proceeding on the merits upon maritime boundary delimitation is as followed:

Can the presence of mineral resources on disputed area influences a third-party's decision on maritime boundary delimitation? In other words, can the presence of maritime resources on a disputed zone be regarded or considered as a special circumstance while seeking an equitable result of boundary delimitation?

Despite the fact that the presence of mineral resources in a disputed zone constitutes a vital consideration in boundary delimitation dispute for the parties, the invocation of this fact seems not to convince the judges of ICJ and tribunals. This position is reasonable since modern international law does no more accept the rule of capture of mineral resource. Obviously, if the Court and Tribunals were to accept the relevance of the presence of mineral resources in disputed zones, states will gradually develop the right of capture of these resources by rushing and delivering licenses to operators to explore and exploit the latter while excluding or trespassing other potential parties' rights. However, the rejection of states' argument solely based on the presence of mineral resources as relevant circumstances in boundary delimitation is not absolute. It does depend on the way the claimant formulates its claim. Thus, ICJ's jurisprudence demonstrates that the claim built solely on the presence of mineral resources on the disputed zone may not be prosperous. Meanwhile, when the claim is built up by taking into consideration several factors including the presence of mineral resources on the disputed area, the Court seems to be convinced by the arguments of the claimant. This is the idea that the ICJ has opined in the North Sea Continental Shelf Cases when it asserts:

In fact, there is no legal limit to the considerations which States may take account of for the purpose of making sure that they apply equitable procedures, and more often than not it is the balancing-up of all such considerations that will produce this result rather than reliance on one to the exclusion of all others. The problem of the relative weight to be accorded to different considerations naturally varies with the circumstances of the case. Therefore, it could be contended that the main objective in boundary delimitation is to achieve an equitable delimitation and not an equal apportionment of maritime areas. Thus, when the sole consideration of the mineral resources' presence within the disputed area may entail an equal apportionment of this area, the relevance of the presence of mineral resources shouldn't be considered as such. Moreover, even the exploration or exploitation of mineral resources by concessionaires of states in dispute through could not serve as a solid relevant circumstance that could influence maritime boundary delimitation, except if these activities are carried on the basis of an express or a tacit agreement. The jurisprudence of ICJ provides evidence for this view in Nigeria/ Cameroon Case when it states that:

Overall, it follows from the jurisprudence that, although the existence of an express or tacit agreement between the parties on the siting of their respective oil concessions may indicate a consensus on the maritime areas to which they are entitled, oil concessions and oil wells are not in themselves to be considered as relevant circumstances justifying the adjustment or shifting of the provisional delimitation line. Only if they are based on express or tacit agreement between the parties may they be taken into account. In the present case there is no agreement between the Parties regarding oil concessions. The Court is therefore of the opinion that the oil practice of the Parties is not a factor to be taken into account in the maritime delimitation in the present case [28].

Again, remarkably, it should be noted that the exception of an express or a tacit agreement that is deemed to strengthen the presence of mineral resources on a disputed area as a relevant circumstance to be taken into account in maritime boundary delimitation seems not to be absolute. Indeed, in the phrasing of that exception as followed "Only if they are based on express or tacit agreement between the parties *may* they be taken into account", the Court used "may" rather than "should" which implies an obligation or a necessity whereas "may" is used to express permissibility of a fact.

Therefore, viewing the latitude left to the Court to consider this exception as relevant circumstance does explain the relativity of this exception. And, under this perspective, we could say that the Court does not take into consideration the conducts of states, or seemingly the Court is reluctant to consider states' conducts in maritime boundary delimitation. Moreover, in the case concerning the Delimitation of Maritime Areas between Canada and the French Republic (St. Pierre et Miquelon), the Tribunal concluded that it did not have a reason to consider the potential mineral resources as having bearing on the delimitation [29]. Nevertheless, it should be noted that ICJ's jurisprudence asserts that the presence of living resources in an area could be a determinant factor while tabling on a maritime boundary delimitation dispute. In fact, in the Greeland and Jan Mayen case, the ICJ did in fact take into account the presence of natural resources and divided the area in the way that the two Parties should enjoy equitable access to the fishing resources of this zone [30]. Hence, it should be noticed that the consideration of natural resources as a relevant circumstance in maritime boundary delimitation depends on the nature of maritime resources. Unlike to living resources, non-living resources, particularly mineral resources are subject to a relative appreciation of the Court before they could be considered as a relevant circumstance in maritime boundary delimitation.

Consequently, in order to reach an equitable delimitation, the presence of mineral resources which is one of the factors among others should be associated to other factors such as the physical and geological structure of the area in dispute, the general configuration of the coasts and the presence of any special or unusual features, and the element of a reasonable degree of proportionality.

5. Concluding Remarks

Just after independences, African states were facing instability due to the vagueness of their boundaries left to them by their former colonial powers. Thus, they wisely decided to keep these boundaries unchanged in order to restore peace and save it under the principle of inviolability of borders inherited from their colonial powers (hereafter called "principle of inviolability"). At the first glance, the principle of inviolability is closely intertwined with the manner in which the OAU decided to avoid armed conflicts over territorial claims and promote friendly relationship between African states, until a satisfactory and peaceful solution is found by the Parties to a territorial dispute in conformity with international law. This also encompasses their commitment to keep stability and unity among all African States through cooperation, particularly between neighboring states. In fact, under such a point of view, it could be rational to contend that this aim seems to fit with that of joint development on maritime resources through agreements in avoiding violence when negotiations between disputed states failed. Likewise, if African leaders intended to render boundaries less significant through the unification of the peoples of the continent in advocating the principle of inviolability, Joint Development Agreement (JDA) also focus more on maritime resources' development rather than maritime boundary dispute settlement. Furthermore, considering that 'the land dominates the sea' more than instructive, principle of inviolability can be imperious since the terrestrial delimitation could appear determinative in the delimitation of maritime boundary as we can notice in the aforementioned cases, notably those between Somalia and Kenya and Ghana- Ivory Coast. Even though the principle of inviolability is closely related to uti possidetis juris principle, the principle of inviolability does not provide a specific peaceful method or criterion to be used for ascertaining the pedigree of disputed boundaries. Intended to resolve disputes, the principle of inviolability does not specify any methods to be used for the peaceful settlement of boundary disputes. These disputes are left to be determined, on a case by case basis, by the states concerned. If this can be acceptable in terrestrial boundary determination, maritime boundary delimitation is regulated by UNCLOS. However, considering that the practice of ICJ and Tribunals demonstrate that not all relevant circumstances can be taken into account while delimiting maritime boundaries; one could assume that like in terrestrial boundary determination, maritime boundary delimitation seems to be done on a case by case basis. Nonetheless, what is clear in UNCLOS is that Articles 74 and 83 of UNCLOS call on states to effect delimitation through 'agreements'. Unfortunately, only few African states found out the wisdom to do so. Consequently, the application of the inviolability principle and the findings of agreements on boundary disputes result to the demarcation of only 30 percent of Africa's borders. And without being too much pessimist, I'd like to say that the rate of African maritime boundary issues will not increase if states do not stop displacing the symbols or monuments of borders' demarcation. This rate will remain unchanged as long as African states will continue challenging boundaries' delimitation principle that they have committed to respect. In fact, this is creating tension between countries seeking control of the region's natural resources, influencing political and international relations and triggering territorial disputes over maritime borders. Moreover, although the UNCLOS provides states the possibility to resolve a dispute peacefully through negotiation or other diplomatic measures, it is evident that these mechanisms mostly fail due to the existence of natural resources lying on the disputed zones. Lack of mutual trust between states could explain this failure in some extent. Reference to the ICJ or tribunals seems to be the last preference of African states while UNCLOS offers them a possibility to find out a practical arrangement to develop the zone while provisionally putting aside boundary delimitation issues. Therefore, African states would rather prevail 'practical arrangements' like JDA than the Court when we know that neither the implementation of the principle of inviolability nor the application of uti possidetis principle could help efficiently resolve a dispute on an area or a boundary full of natural resources. Although maritime claims must be consistent with international law, as well as the rights claimed by states within disputed maritime zones, JDA can be seen as one of the most suitable legal solutions that could help resolve conflicts on disputed maritime zones.

References

[1] Among these documents, we can cite: "Report of the United Nations Conference on Environment and Development, Rio de Janeiro, 3-14 June 1992, vol. I, Resolutions Adopted by the Conference (United Nations publication, Sales No. E.93. I. 8 and corrigendum), resolution 1, annex II"; "Report of the World Summit on Sustainable Development, Johannesburg, South Africa, 26August-4 September 2002(United Nations publication, Sales No. E.03. II. A. 1 and corrigendum), chap. I, resolution 2, annex" and "General Assembly resolution 66/288, annex".

[2] South Africa in 1930; Libya in 1951; Egypt, Morocco, Tunisia and Sudan in 1956 and Ghana in 1957. In addition it should be noted that Ethiopia and Liberia were free states and thus have never been colonized.

[3] Egypt, Sudan and Ethiopia.

[4] The Convention on the territorial sea and the contiguous zone, Apr.29, 1958, 516 U. N. T. S. 205; The Convention on the high seas, Apr. 29, 1958, 450 U. N. T. S.11; The Convention on the continental shelf, Apr. 29, 1958, 499 U. N. T. S. 311and the Convention on fishing and conservation of the living resources of the high seas, Apr.29, 1958, 599 U. N. T. S. 285.

[5] African delegations made important contributions to the interests of international community. These interests can be seen through Part XI (international seabed area and the institutional framework for deep seabed mining) and Part XV (dispute settlement) whereas those concerning the interests of individual state can be seen through the wordings of Part V (EEZ), Part X (rights of access to the sea and freedom of transit of landlocked states), Part XII (protection and preservation of the marine environment), Part XIII (marine scientific research) and Part XIV (development and transfer of marine technology).

[6] Cameroon, Equatorial Guinea, Ethiopia and Tanzania.

[7] Democratic Republic of Congo and Ethiopia.

[8] Benin, Sierra Leone, Congo and Liberia.

[9] Morocco, Ghana, Gambia, Senegal, Cote d'Ivoire, Nigeria, Togo, Equatorial Guinea, Cameroon, Cape Verde, Gabon, Guinea Bissau, Mauritania, Sao Tome and Principe, Somalia and South Africa.

[10] Land and Maritime Boundary between Cameroon and Nigeria (Cameroon v. Nigeria), 2002 I. C. J. p.443, para.295, recalling the North Sea Continental Shelf Cases, n.3, para. 91.

[11] Amongst 86 States that attended to 1958 Geneva Conference 49 were developing countries (20 States from Latin America and 29 States from Africa and Asia).

[12] A. D. Martinez, "The Third United Nations Conference on the law of the Sea: prospects, expectations and realities", Journal of maritime law and commerce, vol. 7, n° 1, October 1975, p. 261.

[13] The high seas being open to all nations, no State may validly purport to subject any part of them to its sovereignty. Freedom of the high seas is exercised under the conditions laid down by these articles and by the other rules of international law. It comprises, inter alia, both for coastal and non-coastal States: (1) Freedom of navigation; (2) Freedom of fishing; (3) Freedom to lay submarine cables and pipelines; (4) Freedom to fly over the high seas.

[14] North Sea Continental Shelf Cases, para. 85.

[15] Somalia institutes proceedings against Kenya with regard to "a dispute concerning maritime delimitation in the Indian Ocean", Paragraph 3 available at http://www.icj-cij.org/docket/files/161/18360.pdf accessed on 3/09/2015.

[16] http://www.worldbulletin.net/ghana/145272/ivory-coast-takes-maritime-dispute-with-ghana-to-arbitration.

[17] See Case No 23 on Dispute Concerning Delimitation of the Maritime Boundary between GHANA and CÔTE D'IVOIRE in the Atlantic Ocean, Request for the prescription of provisional measures, Special Chamber of the International Tribunal for the Law of the Sea, 25 April 2015, available at https://www.itlos.org/fileadmin/itlos/documents/cases/case_no .23_prov_meas/C23_Order_prov.measures_25.04.2015_orig_ Eng.pdf accessed on 25/08/2015.

[18] https://www.stratfor.com/analysis/angola-drc-dispute-maritime-boundary.

[19] http://menasborders.blogspot.com/2010/12/border-focus-angola-and-drc.html.

[20] Common interest zones (CIZs), which can be established when a deposit is located on the maritime borders of two or more states, consist of an ad hoc arrangement for joint administration of the maritime area in question. On this basis, Angola created another CIZ in June 2003 with Congo-Brazzaville, in which both countries agreed to share the revenues from the Lianzi oil field. "Champ pétrolier de Lianzi: plus d'un milliard de dollars pour l'exploitation", Journal de Brazza, 2 March 2012.

[21] "Zone pétrolière commune", Africa Energy Intelligence, no. 494, 22 September 2004.

[22] Lambert Mendé Omalanga, minister of hydrocarbon, approved it on 30 July 2007, and the National Assembly ratified it in November. Law 07/004 of 16 November 2007 authorising ratification of the agreement on the development and production of hydrocarbons in the maritime common interest zone signed by the DRC and Angola in Luanda, on 30 July 2007.

[23] http://www.irinnews.org/report/48239/gabon-equatorial-guinea-un-mediates-dispute-over-corisco-bay-islands.

[24] Anglo- German Demarcation Agreements of 11 March and 12 April 1913 are negotiated and contracted by the two colonial states (with virtually no knowledge of the peoples and no respect of their traditional boundaries) but nonetheless respected on principle of inviolability of African borders inherited from their former colonial powers, which is mostly referred to the international law principle of utipossidetis by ICJ.

[25] The Maroua Declaration (1 June 1975) is the outcome of the meeting held at Maroua from May 30 to June 1, 1975 between Cameroon and Nigeria on their borders' delimitation issues. So, unlike to the Anglo-German Demarcation Agreements, the Maroua Declaration is negotiated by the two countries as sovereign states, having regained their political independence from their respective colonial powers in the early 1960s.

[26] 'Annex to the Application Instituting Proceedings of the Government of the Republic of Guinea-Bissau', Arbitral Award of 31 July 1989 (Guinea-Bissau v Senegal) [1989] ICJ Pleadings 1. 152–3 [88] (MrBarberis).

[27] Continental Shelf (Libya/Malta), Chap. 3, para. 50.

[28] Land and Maritime Boundary between Cameroon and Nigeria (Cameroon v. Nigeria), supra note 10, in Chap. 3, paras. 303–304.

[29] Delimitation of Maritime Areas between Canada and the French Republic (St. Pierre et Miquelon), 10 June 1992, Para. 89.

[30] Maritime Delimitation in the Area between Greenland and Jan Mayen, Paras. 75-76.

22

Exploring the Socio-Ecological Characteristics of Gahkuch Marshland: A Unique Wetlands Ecosystem in Hindukush Mountain Ranges

Yawar Abbas[1, 2, *], Babar Khan[1], Farasat Ali[1], Garee Khan[1], Syed Naeem Abbas[3], Rizwan Karim[4], Saeed Abbas[1], Nawazish Ali[5], Ejaz Hussain[1]

[1]World Wide Fund for Nature-Pakistan, NLI Colony Jutiyal, Gilgit, Pakistan
[2]Department of Earth and Environmental Sciences, Bahria University, Islamabad, Pakistan
[3]Forest, Parks and Wildlife Department, Gilgit, Pakistan
[4]Department of Forestry and Range Management, PMAS-Arid Agriculture University, Rawalpindi, Pakistan
[5]Department of Agriculture and Food Sciences, Karakoram International University, Gilgit, Pakistan

Email address:

yawar_zaid@yahoo.com (Y. Abbas)

Abstract: Gahkuch marshland spreading over 133.54 hectare amidst Hindukush mountain ranges in northern Pakistan is characterized by typical wetlands ecosystem, comprising of small lakes, streams, peat lands, bogs, marshy areas and riverain forests. The area abodes largest resident population of waterfowl in Gilgit- Baltistan, in addition to providing wintering and staging ground for a large number of migratory birds and other aquatic life. A detailed socio ecological study conducted during August to September, 2011 revealed that the area is rich in biodiversity, harboring eight large and three small mammal species, 35 species of birds, seven species of fish, eight species of trees and 18 species of medicinal and economic plants and seventeen families of benthic-macro invertebrates. Moreover, six physical, nineteen chemical and three biological parameters of water bodies were also determined. In addition to its ecological significance the area also supports livelihoods of about 10000 people by providing timber, fuel wood, grazing ground and fish resources. Anthropogenic pressures includes solid waste, influent and illegal hunting were key threats to wetlands and its resources. Wetlands management planning in collaboration with key stakeholders would be effective approach to protect important biodiversity and wetlands resources of the area.

Keywords: Gahkuch, Marshland, Wetlands, WWF, Benthic, Hindukush

1. Introduction

Wetlands are areas of marsh, fen, peat land or water, whether natural or artificial, permanent or temporary, with water that is static or flowing, fresh, brackish or salt, including areas of marine water the depth of which at low tide does not exceed six meters. (Ramsar Convention, 1971). It is difficult to clearly demarcate boundaries between wetlands and adjacent deep-water systems, especially when water levels fluctuate greatly and frequently between extremes (seasonal wetlands) (RCS, 2006). Wetlands are sensitive ecosystems providing habitats for a variety of plants species, birds, small mammals and other aquatic biota. The term "wetland" is mostly used to explain diverse habitats where the land is wet for extensive period of the year or season but not essentially permanent waterlogged (Collins, 2005). 'Wetland' is a term used to define a variety of habitats in different climatic zones of the Globe. Ecologically, wetlands are transitional (ecotonal) systems between upland terrestrial and deep open water systems. They are characterized by aquatic vegetation (hydrophytes) (Mitsch and Gosselink, 2007). Wetlands are most important freshwater resources; fresh water forms the habitat of large number of species. These aquatic organisms and the ecosystem in which they live, represent a substantial sector of the Earth's biological diversity (UNEP, 1994). Around 50% of the world's wetlands have been lost in the past century alone due to urbanization, drainage for agriculture, and water system regulation (Shine and de Klemm, 1999).

High altitude wetlands, which include lakes, marshes, seeps, peat bogs etc, in the Himalayas have several characteristics that make them unique in terms of their biodiversity value. The plants and animals that occur in and around them are often endemic and highly adapted to their locations. (WWF, 2010). Alpine wetlands are critical ecosystems at risk from natural and manmade threats. In the Indian subcontinent, they are reservoirs of biodiversity, providing staging and breeding ground for migratory birds across Himalayas, and supporting many types of endemic wildlife. Human communities also inhabit the regions surrounding the high-altitude wetlands; they rely upon the ecosystem services (Srinivasan, 2002). High-altitude wetlands cover only approximately 3% of the total land area (Maltby and Immirzi, 1993). Alpine wetlands can store approximately 30% of the global terrestrial carbon (Gorham, 1991; Blodau, 2002). Pakistan supports an estimated 7,800,000 hectares (ha) of wetlands and in excess of 225 significant wetland resources are recorded by 2004, nineteen of these have been internationally recognized by the Ramsar Convention Bureau as being of global importance and about one-third of Pakistan bird species use wetlands for food, shelter, and (or) breeding (Ali, 2005). however, the birds that visit or breed in poorer quality habitats will not contribute to

a sustainable population through the years (Pulliam and Danielson 1991).

Gahkuch Marshland is one of such valued and critical wetland ecosystem, occurring at higher elevations but its socio ecological significance has never been documented. The current paper elaborates finding of a study conducted during August to September, 2011 to explore and record key socio ecological characteristics of Gahkuch Marshland.

2. Materials and Methods

2.1. Study Area

Gahkuch wetland complex located at 36° 10' 19.5" N, 73° 46' 25.3" E and at an elevation of 1899 m comprises of marshland, peats, riverain ecosystem along Ghizer River and more than a dozen lakes. The typical marshland spreading over 133.54 hectare is situated adjacent to Ghizer river in the district headquarter; Gahkuch and is owned by the Gilgit-Baltistan Forest & Wildlife Department and the local community, whereas, the upland cluster of three lake locally called Barosar, Chunosar and Photakesar covering some 427.24 sq km area are surrounded by lush green pastures abound in medicinal herbs.

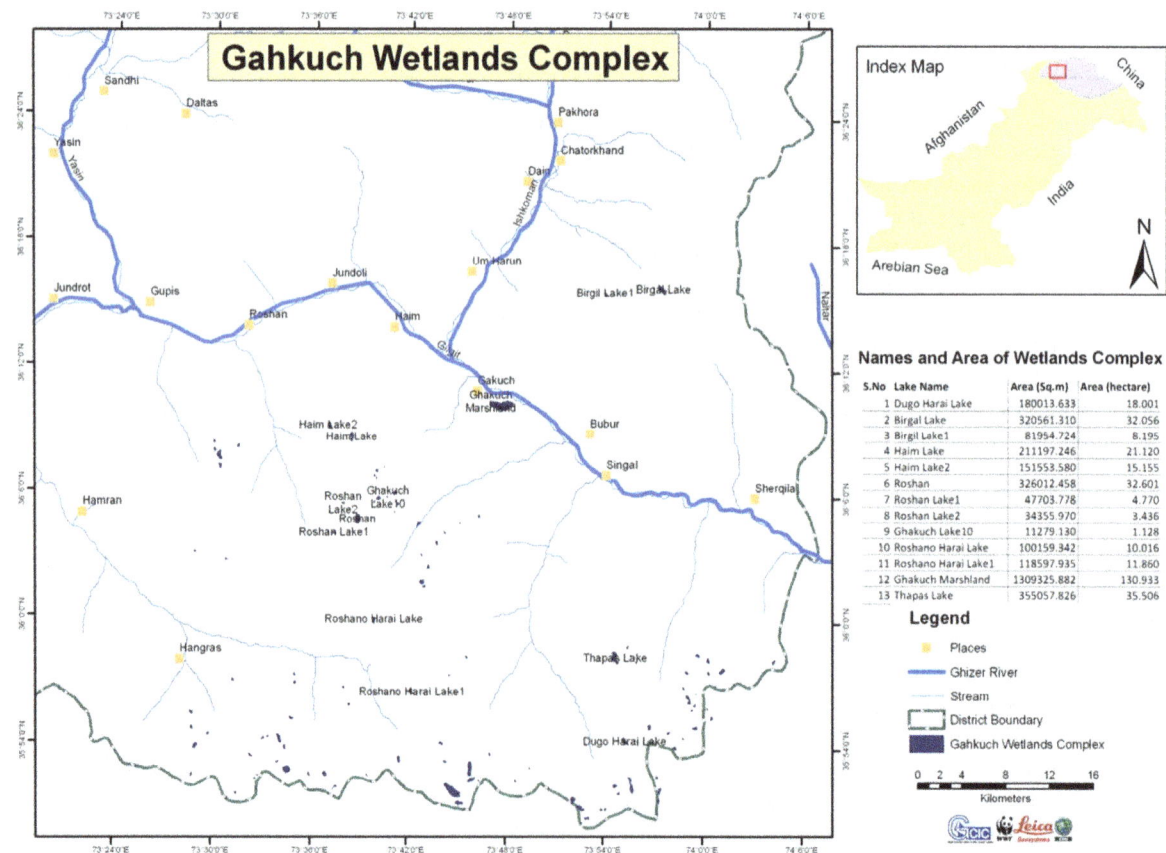

Figure 1. Map of the Study Area.

2.2. Methods

A random sampling procedure was followed to study the large mammals around the Gahkuch Marshland and in some

places transect methods were applied to estimate the population density of the large mammals while Transect method was also applied to study the small mammals of the areas. Techniques used during the studies are visual Sighting,

Odors, Spotlighting, Entry sites, Leftovers, Tracks and claw marks, Sherman Trap and also the information's gathered through interviews with local community resource persons.

Vegetation survey was under taken and collected plant specimens during their specific blooming season. Collected specimens were gathered with field information's and properly pressed with the help of plants presser. The ethno botanical information and traditional uses of plants were documented by interviewing local and filling of questionnaires from local experts both men and women. Species identified with help books and desk assessment was done to develop a check list of plants from Gahkuch Marshland. Equipment used during field survey were Garmin Global Positioning System (GPS), Columbia digital camera , plants pressers, blotting papers, note book, maps, pencils, plants pressers, polythene bags, Sesser and magnifying mirror, Bird's diversity was recorded through Direct and indirect counting methods. Direct observations include direct counts and specimen collection and indirect observations include information gathered from interviews and general discussions with the local community. Mist nets were used to capture the birds and nets were operated for all the 24 hours per day (beginning at local sunrise) during the study. All the birds caught were identified and blood samples were taken for DNA analysis. A 10x42 mm Olympus binocular, 20x45-60 mm Nikon telescope, Garmin Map 76 GPS receiver, note book, lead pencils, field guide, Birds of Pakistan by Grimmett et. al (2008). Pocket guide to the "Birds of Indian Subcontinent" by Richard Grimmettt (2001) and "a field guide to the Birds of the Indian Subcontinent" by KrysKazmierczak (2000) were used in the field.

Hydro chemical and physical properties were tested from a total of five sampling sites selected in Gahkuch Marshland on the basis of topography, surface geology, break and human made structures. During survey, each site was marked using a GPS. To study the affect of water quality sampling bottle of 450ml was filled completely from each sampling site mentioned above for the determination of water quality and heavy metals contents.All the physical and chemical parameters of water samples were determined referring to the American Public Health Association (APHA).

For fish surveys recognized methods used included bank-side counts, trapping, cast netting, seine netting, gill netting, and electro-fishing. Bank–side counts are preferred on the banks of clear shallow streams. Trapping used for specific species using specific baits. Gill netting and seine netting was mainly done in Marshland (Survey guidelines for Australia's threatened fish, 2004). All the specimen data and the relevant auxiliary information were recorded in the data sheet specially designed for these studies.

Baseline information on aquatic invertebrates (benthic macro-invertebrates and zooplanktons) of Gahkuch Marshland was studied by collecting three samples from different braided channels of Ghizer River. The samples were collected by two persons (first person hold D-frame dip net facing second person that help to disturb substrate by kicks). A total of 20 kicks were taken over the length of the stretch from all possible substrates. Four kicks were made at each transect. A total area of 3.1 m^2 was disturbed. After every 2 kicks, more often if necessary, the collected material was washed by running clean water through the net for two to three times. Then transferred to white enameled tray, thoroughly check all wooden debris for organism attachment and remove it from the sample. Then sample was transferred to preservation bottle having sufficient 5-10% formalin solution. External label was written with sampling locality, date of collection and collector name on external upper side of lid and lower bottom of sample container. An interior label written in pencil on waterproof paper was inserted as a backup. At last sampling bottle was placed inside an air tight ½ barrel hard plastic drum to avoid inhalation of formalin fumes and safe transportation during field operations.

To assess socio-economic conditions of local communities and understanding of extent of dependency on high altitude wetland, integrated management option for sustainable uses of natural resources a number of methodologies including group discussion, questionnaire survey and personal observation were employed during the study.

3. Results

3.1. Large Mammals and Small Mammals

A total of 8 species of mammals were recorded in the area. *Canis lupus* (Wolf), *Capra ibex* (Himalayan Ibex) and *Moschuschrysogaster* (Himalayan Musk Deer) were reported by the Wildlife staff and local peoples. Only one species of lagomorphs, *Lepus capensis* (Cape Hare) was seen. The fecal material of the *Vulpesvulpes* (Common Red Fox) was observed on different places of the area. The area is rich in large mammal species which adequate that the area is ecological rich in biodiversity. Three species of rodents, *Apodemousrusiges* (Himalayan Wood Mouse,), *Mus musculus* (House Mouse) and *Rattusturkestanicus* (Turkistan Rat) and one insectivore, *Crocidurapullata* (Asiatic White-toothed Shrew) were recorded in the study sites.

3.2. Records of Trapped Small Mammals in the Study Sites

Trapping results at Gahkuch Marshland shows that the site is productive as three species of rodents i.e., *Apodemousrusiges, Mus musculus, Rattusturkestanicus* and *Crocidurapullata* (Asiatic White-toothed Shrew) were trapped in the study area. The overall trapping success remains 51.76%. The 10.41% traps were tripped due to unknown reason while the 20.25% traps were dislocated from their original trapping stations, which can be ascribed about the fluent movement of some mammals at night. In terms of relative abundance, 45.30 %of the total trapped specimens were Himalayan Wood Mouse, 31.10% was *Mus musculus* (House Mouse), 15.11% *Rattusturkestanicus* (Turkestan Rat) while the 8.49% was *Crocidurapullata* (Asiatic White-toothed Shrew).

3.3. Taxonomic Records with Conservation Status

The conservation status of mammalian fauna in Gahkuch marshland is determined by referring to IUCN red list (2010), Out of 8 mammalian species of the area *Moschuschrysogaster* (Himalayan Musk Deer) are listed on CITES appendix 1 and are classified as Endangered , while the other 7 species are included as Least Concerned. Three species i.e., *Moschuschrysogaster* (Himalayan Musk Deer),

Canis lupus (Wolf), are rare in the area. *Vulpesvulpes* (Common Red Fox), *Lepus capensis* (Cape Hare), *Apodemousrusiges* (Himalayan Wood Mouse), *Mus musculus* (House Mouse) and *Rattusturkistanicus* (Turkistan Rat) are common in the area while the *Crocidurapullata* (Asiatic White-toothed Shrew) are less common as compared to the other small mammals of the area which are described in Table. 1.

Table 1. Taxonomic records and conservation status of the mammals in the study site.

Sr. #	Zoological Name	Order	Family	CITES	IUCN (Red list 2010)	Local status
1	*Canis lupus*	carnivora	Canidae	Least Concern	3.1	Rare
2	*Vulpesvulpes*	carnivora	Canidae	Least Concern	stable	Common
3	*Moschuschrysogaster*	Artiodactyla	Moschidae	Endangered	3.1	Endangered
4	*Lepus capensis*	Legomorpha	Leporidae	Least Concern	2.3	Common
5	*Apodemusrusiges*	Rodentia	Muridae	Least Concern	2.3	Common
6	*Rattusturkestanicus*	Rodentia	Muridae	Least Concern	2.3	Common
7	*Mus musculus*	Rodentia	Muridae	Least Concern	2.3	Common
8	*Crocidurapullata*	Insectivora	Soricidae	Least Concern	2.3	Common

3.4. Common Flora of the Area

The area falls in dry temperate zone, receiving no monsoon rains. It represents small patches of mixed, Blue pine, Spruce, Juniper, Salix and Birch forests. However, in community owned portion, Willow and Popular have been planted to meet timber and firewood needs. Major floral species identified were as follow in Table. 2.

Table 2. List of Tree identified during field survey.

S. #	Common Name	Botanical Name.	Family
1	Willow	*Salix tetrasperma*	Salicaceae
2	Juniper	Juniperusmacropoda	Coniferae
4	Poplar	*Populus alba*	salicaceae
5	Walnut	*Juglansregia*	Juglanceae
6	Russian olive	*Elgeanusangustifolea*	Elgeaniceae
7	Mulberry	*Morus alba*	Moraceae
8	Birch	*Betulautilis*	Cupuliferate

3.5. Avi Fauna

The recent study revealed 35 birds' species of 11 orders in the Gahkuch Marshland from which 11 were resident species 9 were summer migrants while 8 were winter visitor species. Migratory waterfowl were recorded in the study area were common pintail, Nothren shoveler, Common Teal, Eurasian Wigeon and Gadwal. Gahkuch Marshland, typical wetland in Gilgit-Baltistan is important in term of migratory waterfowl because it provides staging ground for migratory birds while

at the same time also harboring a sound population of resident. Mostly birds prefer this area because of food, shelter and nest. Identified birds in Gahkuch Marshland is given below in Table. 3.

3.6. Fish Fauna

Seven species of fishes have been recorded from the River Ghizer and in the marshland (Table 4). The fish *Salmo truttafario* is an exotic fish and introduced in the Gilgit-Baltistanin 1916 by the British representative. It is common in the water bodies of Gilgit-Baltistan, especially in the lakes and streams. The brown trout is confined to crystal clear cold waters of high altitude lakes and streams. They generally avoid entering into main rivers due to high turbidity generally prevailing in these rivers. It is one of the major source of protein for the masses and is mainly caught by rod and line but netting by cast net is also practiced which is a major threat for this species. Different teams from the area come with their tents and ration on specific trout spots and they catch fish for many days for commercial purposes.

The species *Schizothoraxplagiostomus* and *Racomalabiata* are also commercially important fishes and major part of the fish food for the local people. The population of *Racomalabiata* is, however, low in the area probably due to its food competition with the more successful fish *Schizothoraxplagiostomus*. The loaches *Triplophysamicrops, Triplophysatenuicauda,* and *Triplophysayasinensis* are small fish but an important component of the ecosystem as they are major food items for the brown trout. The species *Triplophysamicrops, Triplophysatenuicauda* are mainly found in the small waters of springs while the species *Triplophysayasinensis*is common in the river. All the loaches are endemic in the upper reaches of the Indus drainage. *Glyptosternumreticulatum* is the only cat fish in the area and is distributed up till Khalti Lake in the river Gilgit.

Table 3. Birds List.

Order	Families	Species	Common Name	Occurrence	IUCN 2012 Red list	Status
Anseriformes	Anatidae	*Anascrecca*	common teal	winter visitor	LC	Abundant
		Anasstrepera	Gadwall	winter visitor	LC	Common
		Anasacuta	NorhtrenPintatail	winter visitor	LC	Abundant
		Anasclypleata	NothernShoveler	winter visitor	LC	Abundant
		Anas Penelope	Eurasian Wigeon	winter visitor	LC	Abundant
Ciconiformes	Accipitridae	*Hieraaetuspennatus*	Booted Eagle	Resident	LC	Rare
		Accipiter nisus	Eurasian Sparrow Hawk	winter visitor	LC	Frequent
Ciconiformes	Falconidae	*Falco tinnunculus*	Eurasian Kestrel	Resident	LC	Common
		Falco subbuteo	Northern Hobby	summer breeding	LC	Frequent
Galliformes	Phasianidae	*AlectorisChakur*	Chakur	Resident	LC	Common
Gruiformes	Rallidae	*Fulicaatra*	Common coot	Resident	LC	Abundant
		Gallinulachloropus	common Morhen	Resident	LC	Abundant
Gruiformes	Gruidae	*Anthropoidesvirgo*	Demoiselle Crane	Migrant	LC	Frequent
Columbiformes	Columbidae	*Streptopeliaorientalis*	Oriental Turtle Dove	Summer Visiter	LC	Common
		Columba livia	Rock Pigeon	Resident	LC	Abundant
Apodiformes	Apodidae	*Apus apus*	Common swift	summer breeding	LC	Abundant
Upupuiformes	Upupidae	*Upupaepops*	Hoopoe	summer breeding	LC	Common
Piciformes	Picidae	*Picussquamatus*	Scaly-bellied Woodpecker	Resident	LC	Common
passeriformes	Motacilidae	*Motacillaflava*	Yellow Wagtail	Migrant	LC	Common
		Motacilla alba personata	White Wagtail	summer breeding	LC	Common
passeriformes	Cinlidae	*Cincluspallasii*	Brown Dipper	Resident	LC	Common
passeriformes	Troglodytidae	*Troglodytes troglodytes*	Winter wren	Resident	LC	Common
passeriformes	Alaudidae	*Alaudaarvensis*	Common Skylark	Migrant	LC	Common
		Galeridacristata	Horned Lark	Resident	LC	Common
		Alaudagulgula	Oriental skylark	summer breeding	LC	Common
		Calandrellabrachydactyla	Greater Short toed Lark	Migrant	LC	Abundant
passeriformes	Prunellidae	*Prunellastrophiata*	Rufous-breasted Accentor	summer breeding	LC	Frequent
		Prunellafulvescens	Brown Accentor	Resident	LC	Common
		Prunellarubeculoides	Robin Accentor	summer breeding	LC	Frequent
passeriformes	Turdidae	*Phoenicurusphoenicurus*	Black Redstart	Migrant	LC	Common
		Phoenicurusphoenicurus	Common Redstart	Migrant	LC	Common
		Luscinia pectoralis	Himalayan Ruby throat	summer breeding	LC	Common
		Lusciniasvecica	Blue Throat	winter visitor	LC	Frequent
		Tarsigercyanurus	Orange-flanked Bush Robin	summer breeding	LC	Frequent
		Phoenicurus frontalis	Blue fronted Redstart	summer breeding	LC	Frequent

Table 4. Fish Fauna of Ghizer River in the Gahkuch Marshland Areas.

S. No.	Fish Species	Common Name	Food Habit	Status in the area	IUCN Status	Distribution status
1	*Salmo truttafario*	Brown trout	Carnivore	Common	LC	Exotic
2	*Racomalabiata*	KunarSnowtrout	Omnivore	Rare	Data deficient	Wide
3	*Schizothoraxplagiostomus*	Golden snow trout	Omnivore	Common	Vulnerable	Wide
4	*Triplophysamicrops*	Leh loach	Carnivore	Rare	LC	Endemic to upper Indus drainage
5	*Triplophysatenuicauda*	Short tailed loach	Carnivore	Rare	Data deficient	Endemic to upper Indus drainage
6	*Triplophysayasinensis*	Yasin loach	Carnivore	Common	Data deficient	Endemic to upper Indus drainage
7	*Glyptosternumreticulatum*	Turkestan catfish	Carnivore	Common	LC	Wide

3.7. Aquatic Invertebrates

There is no published information regarding the invertebrates of Gahkuch Marshland. Present baseline survey was conducted to investigate the first hand knowledge of Benthic Macro-invertebrates fauna. Twenty taxa (Seventeen families and eleven generas) of Benthic-macro invertebrates were identified from 1198 individuals at Gahkuch Wetland Complex (see Table 5 and Figure 2). The number of individuals from S_1, S_2 and S_3 were 585, 208 and 405 respectively. The most abundant taxa at S_1 were Tanypodinae (Chironomidae) 210 individuals followed by *Baetis* sp. 158 individuals and Chironomidae 123 individuals. Nine taxa including *Mesonemoura* sp., *Hydropsyche* sp., Psychomyiidae, *Omphiogomphus* sp., *Probezzia*sp.,

Tabanidae, Hydrophilidae, Naididae and Tubificidaewere not recorded from S_1. Sampling station S_2 was found abundant with Chironomidae 52 individuals followed by *Baetis* sp. 36 individuals and Tanypodinae (Chironomidae) 31 individuals. Following six taxa were not recorded at S_2, *Epeorus* sp., *Lepidostoma* sp., Simuliidae, *Culex* sp., Elmidae and Hydracarina (Arachnida). Sampling station S_3 was again dominated by Chironomidae 147 individuals followed by Tanypodinae 94 individuals and *Baetis* sp. 84 individuals. Five taxa including, *Mesonemoura* sp., *Epeorus* sp., *Probezzia*sp., Elmidae and Psychomyiidae were not recorded from S_3. Organism's pollution tolerance was taken from HKHbios (Hindukush Himalayan score Bioassessment) scoring list. Extremely pollution sensitive organisms richness were maximum at sampling station S_2, *e.g.,* *Mesonemoura* sp. and *Rhithrogena* sp. Extremely pollution tolerant taxa were *Bezzia* sp. and *Probezzia* sp. among them first one was

found at all sampling sites while later in S_2 only.

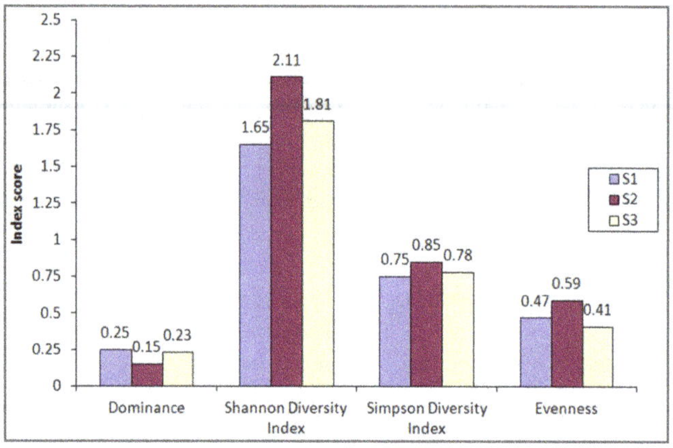

Figure 2. *Diversity indices of Benthic Macro-invertebrates at Gahkuch marsh land.*

Table 5. *Benthic.*

S. No.	Taxa (Family/ Class)	Genus	S_1	S_2	S_3	Functional feeding groups	Pollution Tolerance (Based on HKH bios scoring list)
1	Nemouridae	*Mesonemoura* sp.	-	4	-	Shredders	9
2	Heptageniidae	*Rhithrogena* sp.	14	23	8	Scrapers	9
3		*Epeorus* sp.	8	-	-	Scrapers	8
4	Baetidae	*Baetis*sp.	158	36	84	Collector gatherers	-
5	Hydropsychidae	*Hydropsyche* sp.	-	6	3	Collector filterers	7
6	Psychomyiidae	-	-	1	-	Collector gatherers/ Scrappers	7
7	Gomphidae	*Omphiogomphus* sp.	-	3	5	Predators	-
8	Lepidostomatidae	*Lepidostoma* sp.	5	-	8	Shredders	8
9	Chironomidae	-	123	52	147	Unknown	-
10		Tanypodinae	210	31	94	Predators	-
11	Ceratopogonidae	*Bezzia* sp.	16	5	10	Predators	2
12		*Probezzia*sp.	-	1	-	Predators	2
13	Simuliidae	-	27	-	15	Collector filterers	7
14	Tabanidae	-	-	2	2	Collector gatherers / Predators	6
15	Culicidae	*Culex* sp.	8	-	3	Collector filterers	-
16	Elmidae	-	-	4	-	Scrappers / Collector gatherers	8
17	Hydrophilidae	-	-	3	2	Unknown	6
18	Naididae	-	-	25	10	Collector gatherers	3
19	Tubificidae	-	-	16	9	Collector gatherers	1
20	Hydracarina (Arachnida)	-	12	-	5	Predators	-
	No. of Individuals		585	208	405	Total No. of Individuals= 1198	
	Number of taxa		11	14	15		

The diversity indices of benthic macro-invertebrates (BMIs) of Gahkuch Wetland Complex were displayed. The dominance index score 0.25 was detected higher at S_1. It was attained due to the increase of relative abundance of single taxonTanypodinae 35.9% of the total count, consequently other taxon were poorly represented. For that reason, the diversity indices including Shannon Weiner and Simpson score were lower due to higher dominance score of sampling station S_1. These results were a sign of less equitable distribution of BMIs taxa at S_1 and S_3. At sampling station S_2 the dominance index score 0.15 was two times less than the other sampling stations. The taxa relative abundance is relatively evenly distributed at S_2, therefore, diversity indices Shannon Weiner, Simpson and Evenness score of S_2 were

higher than the S_1 and S_3. It showed that sampling station S_2 was more equitable in distribution of taxa as compared to S_2 and S_3.

Macro-invertebrates, number of individuals, relative abundance (values inside parenthesis) functional feeding groups and pollution tolerance values at Gahkuch Wetland Complex. (Very sensitive taxa are those taxa given HKHbios score of 10 & 9 extremely tolerant taxa are those taxa given a HKH bios score of 1 & 2)

Many species of benthic macro-invertebrates are diagnostic of certain kinds of habitats and their water quality, commonly known as indicator organisms. These organisms become numerically dominant only under a specific set of environmental conditions. The Ephemeroptera (mayflies),

Nemouridae (stoneflies), Trichoptera (caddisflies) and Diptera (true flies) are commonly, or perhaps always, the four orders used as indicator organisms in environmental impact assessments. For this reason, more emphasis is placed on these orders than on other orders of insects. The highly pollution sensitive taxa *Mesonemoura* sp. and *Rhithrogena* sp. were found at S_2. Therefore, above mentioned organisms would respond by first disappearing after further environmental degradation or perturbation at sampling station S_2.

The overall conclusions revealed that

- Among benthic macro-invertebrates Tanypodinae (35.9% of total count) was dominant at sampling station S_1. Chironomidae were dominant at S_2 and S_3, while share 25% and 36.3% of total count respectively
- Organism's pollution tolerance was taken from HKHbios (Hindukush Himalayan score Bioassessment)

scoring list. Extremely *sensitive organisms include Mesonemoura sp. (Nemouridae: Plecoptera) and Rhithrogena sp. (Heptageniidae: Ephemeroptera).* Extremely pollution tolerant taxa were belonging to family *Ceratopogonidae* including *Bezzia sp.* and *Probezzia sp.*

3.8. Water Quality

3.8.1. Physical Properties

The values of all physical properties have been shown in Table 6. All physical properties of marshland water at five different points were found to be normal in normal range suggested by USEPA, WHO and Pak EPA. Difference in values of physical properties due to distance, geographic location, biological and chemical activities in and around marshland.

Table 6. Physical parameters of water at Gahkuch Marsh Land Lake.

	Physical Parameters	Units	Site-I	Site-II	Site-III	Site IV	Site V	Standard Value
1	Temp Air	°C	17.5	16.8	17.2	16.6	16.5	40.0
2	Temp Water	°C	12.0	11.4	12.3	12.9	10.0	35.0
3	pH		07	07	07	07	07	6.2-7.5
4	Color	TCU	36	22	32	28	35	50.0
5	Odor		No	No	No	No	No	Un-objectionable
6	Turbidity	NTU	14.0	13.2	14.3	13.2	14.2	25.0

3.8.2. Chemical Parameters

The values of all chemical properties were found to be normal as suggested by USEPA, WHO and Pak EPA. There were no Zn and Manganese was found. Differences in values and chemical properties of Marshland was due to location. Results are shown in Table. 7.

Table 7. Chemical Properties of five Different Sites.

S.No	Parameters	Units	Site-I	Site-II	Site-III	Site IV	Site V	Standard Value
1.	DO	mg/l	4.12	4.38	4.21	4.78	4.23	3.000
2.	Alkalinity	mg/l	276	354	245	231	329	-
3.	Carbonates	mg/l	53	43	64	52	64	-
4.	Bicarbonates	mg/l	312	276	265	312	218	-
5.	Hardness	mg/l	321	307	343	210	286	500.0
6.	Ca	mg/l	123	118	154	132	128	200.0
7.	Mg	mg/l	165	189	127	165	194	500.0
8.	Chloride	mg/l	176	189	212	354	276	600.0
9.	Chlorine	mg/l	5.6	3.9	5.8	5.0	5.3	1.0
10	Electrical Conductivity	µs/cm	298	287	265	276	215	-
11	Flouride	mg/l	2.12	1,87	1.43	1.98	1.32	1.5
12	Manganese	mg/l	ND	ND	ND	ND	ND	0.5
13	Nitrate	mg/l	14	13	15	17	13	45.0
14	Nitrite	mg/l	0.03	0.06	0.04	0.03	0.02	0.010
15	Salinity	mg/l	ND	ND	ND	ND	ND	1000.0
16	Sulphate	mg/l	132	187	232	165	243	400.0
17	Sulphide	mg/l	3.32	3.11	2.43	3.21	2.54	-
18	TDS	mg/l	123	132	143	132	127	1500.0
19	Zn	mg/l	ND	ND	ND	ND	ND	15.0

3.8.3. Biological Parameters

The value of biological parameters has been shown in table 8. The higher value of ammonia was found at site IV i.e. 0.007mg/l followed by site III i.e. 0.005mg/l. The lower value was found at site V i.e. 0.002mg/l. The higher value of BOD was found at site III i.e. 3.231mg/l followed by site I i.e. 3.213mg/l. The higher value of COD were found at sites V i.e. 3.413mg/l followed by site III i.e. 3.321mg/l. The significantly lower value was found at site II i.e. 2.213mg/l. The values were found to be in normal range as suggested by US EPA, WHO and Pak EPA. According to Pakistan NEQS, the maximum limit of COD for municipal and industrial effluents is 150mg/l for inland waters.

Table 8. Biological parameters of water at Gahkuch Marsh Land.

S.No	Parameters	Units	Site-I	Site-II	Site-III	Site IV	Site V	Standard value
1	Ammonia	mg/l	0.004	0.003	0.005	0.007	0.002	0.500
2	BOD	mg/l	3.213	3.141	3.231	3.222	2.132	6.0
3	COD	mg/l	2.654	2.213	3.312	3.221	3.413	10.0

3.8.4. Socio Economic

There are total 600 households in Gahkuch Bala with a population of 5400 individuals, while in Gahkuch Paeen is comprised of 500 households with population of 4500 individuals. Natural forest of Gahkuch Nallah is government protected, but some concession on fuel wood collection and grazing right are granted to local community to fulfill their domestic requirement.

Major land use in the area is subsistence farming and

livestock rearing. It is clear that farming is major form of local land use. But still traditional agriculture is being techniques practiced in upper Gahkuch and adjust areas. The alluvium traces and fens mostly used growing for fruits trees and fodder cultivation, but during study it was revealed that average crop production per household wheat is 456 kg, maize 36 kg and 804 kg potatoes respectively. Gahkuch Paeen falls under double cropping zone while Gahkuch Bala is single cropping zone. Mostly the villager's prefer potatoes and wheat than maize and barley.

Livestock is integral part of the livelihoods in the study area, livestock contributing inputs such as farm and manure. In Gahkuch valley 65% people have livestock, of whom 35% sale livestock products in markets and rest utilized for domestic purposes. Average livestock population per household was 11.4 including cows, goats and sheep respectively. Livestock population in upper Gahkuch is higher than lower Gahkuch. Goats and sheep mostly grazed in Gahkuch Nullah without a proper grazing system.

Socio-economic study in the area revealed that community is dependent on natural resources *e.g.* forest, wildlife, agriculture, water, wetlands and pastures because lack of opportunities for jobs, and alternate source of energy.

4. Conclusion

The baseline socio ecological studies revealed that Gahkuch Marshland is one of the significant wetland providing habitats for small and large mammals, migratory birds, fish species and other aquatic life and floral species. But the area is facing serious natural and manmade threats including illegal hunting, fishing and poaching of wildlife, solid waste and water pollution, deforestation, lack of awareness, poverty, institutional and policy, Climate change impacts (floods, sedimentation, turbidity, loss of habitat) and ill managed tourism are common threats for the marshland. Dependency of agro pastoral community on wetland resources is high due to lack of alternate source of energy, poverty and awareness. The conservation and protection of the area needs wetlands conservation planning and management through collaborative approaches

Acknowledgement

Socio ecological studies conducted under World Wide for Natural Pakistan and its Saving Wetlands Sky High Programme and Pakistan Wetlands Programme. The authors wish to thank World Wide Fund for Nature Pakistan for supporting the field studies.

References

[1]   Ali Z. 2005. Ecology, distribution and conservation of migratory birds at Uchalli Wetlands Complex, Punjab, Pakistan. A thesis submitted to the University of the Punjab in partial fulfillment of the requirements for the degree of Doctor of philosophy in Zoology (unpublished).

[2]   Blodau C. 2002. Carbon cycling in peatlands. A review of processes and controls; *Enviorn. Rev.*; 10(2): 111–134.

[3]   Collins NB. 2005. Wetlands: *The basics and some more. Free State Department of Tourism, Environmental and Econimic Affairs.* ISBN 0-86886-708-X, Bloemfontein.

[4]   Gimmet R, Inskipp C, Inskipp T. 2001. *Birds of the Indian Subcontinent.* (Revised reprint 2001), Christopher Helm, London. Pp. 384.

[5]   Gimmet R, Tom R, Inskipp T. 2008. Birds of the Pakistan. Christopher Helm, London. Pp. 1-255.

[6]   Gorham E. 1991.*Role in the global carbon cycle and probable response to climate warming*; *Ecol.* Appl.; 1: 182–193.

[7]   http://www.ramsar.org/cda/en/ramsar-activities-cepa-classification-system/main/ramsar/1-63-69%5E21235_4000_0__Accessed 9.46 am hrs, 02-05-2012

[8]   Kazmierczak K. 2000.*A field guide to the birds of the Indian Subcontinent.* Pica press Sussex. Pp. 1-352.

[9]   Maltby E, Immirzi P. 1993. Carbon dynamics in peatlands and other wetland soils, regional and global perspectives; *Chemosphere;* 27: 999–1023.

[10] Mitsch WJ, Gosselink JG. 2007. *Wetlands* (4th edn). New York: John Wiley.

[11] Kazmierczak K. 2000.*A field guide to the birds of the Indian Subcontinent.* Pica press Sussex. Pp. 1-352.

[12] Pulliam HR, Danielson BJ. 1991. Sources, sinks and habitat selection. A landscape perspective on population dynamics: *The American Naturalist.* 137: 850-866.

[13] RCS. 2006. The Ramsar Convention manual: A *guide to the convention on wetlands (Ramsar, Iran, 1971) (4th edn). Gland* (Switzerland): Ramsar Convention Secretariat.

[14] Shafiq. 1998. *Status of large mammal species in Khunjerab National Park.* The Pakistan journal of forestry vol 48(1-4) 1998 p 91-96.

[15] Shine C, de Klemm C. 1999. Wetlands, water, and the law: Using law to advance wetland conservation and wise use (IUCN environmental policy and law paper no. 38).Gland, Switzerland: IUCN.

[16] Srinivasan S. 2002. Mapping the Tso Kar basin in Ladakh Gathering Spatial Information from a Nomadic Community, The Journal of Community Informatics. ISSN: 1712-4441.

[17] Survey guidelines for Australia's threatened fish. 2004. Guidelines for detecting fish listed as threatened under the Environment Protection and Biodiversity Conservation Act 1999.

[18] UNEP. 1996. Groundwater: A threatened resource, UNEP Environment Library no. 15, UNEP, Nairobi, Kenya.

[19] WWF - Pakistan. 2010. Project document Saving Wetlands Sky High Programme Extension phase, un published report Pp 1.

23

Water Velocity, Vorticity and Bed Deviation Modeling for a Reach from Damietta Branch Using K-ε Turbulence Model Solved by Cubic Interpolated Pseudo (CIP) Method

Mohammed Ibrahiem Ibrahiem Mohammed, Mohamed Ahmed Abdel Hady Eid

Irrigation and Hydraulics Department, Faculty of Engineering, Mansoura University, Mansoura, Egypt

Email address:

mimi_elgamal@yahoo.com (M. I. I. Mohammed), ma_abdelhady@yahoo.com (M. A. A. H. Eid)

Abstract: A multi-meander reach with length of 20 km located between km 130.0 and km 150.0 downstream of Delta barrages - Damietta branch was selected and numerically studied. This study aimed to simulate and study numerically the water velocity, vorticity and bed deviation of the curved zones for the reach under study and illustrating the relationship between them. Consequently, the vulnerable zones subjected to maximum velocities were accurately determined. Field data were collected and analyzed for the modeling process. A 3-D model called iRIC (International River Interface Corporative) based on an explicit finite difference method (Abbott-Ionescu scheme) was applied. Therefore, in order to fulfill such objective, standard K-ε turbulence model was employed using Cubic Interpolated Pseudo (CIP) method for solving the advection terms. For illustrating obviously the variation of water velocity with vorticity and deviation of bed elevation, two important zones consisted of several meanders were selected, the first zone located from km: 132.00 to km: 137.33 while the second located between km: 137.33 and km: 142.67. Through the modeling process, it was assumed that the sediment particles move in the bed layer zone only. From this study, it was found that for both selected zones, the velocity value was ranged between 0.13 m/sec and 0.24 m/sec, and it could be considered as a small range to make scouring process. It was observed also, that there was a noticeable relationship between water velocity, vorticity and deviation of bed elevation.

Keywords: Flow Characteristics, Modeling, Turbulence Model, Sediment, K-ε model, CIP Method

1. Introduction

Rivers have been widely considered as an interesting and attractive study subject by engineers and scientists who have been fascinated by the self-formed river geometric shapes and their responses to changes in nature and human interferences. In addition, understanding river behavior and providing flood protection is also essential to environmental enhancement.

Flow in curved river reaches is usually under the influence of centrifugal acceleration, which induces transverse velocity component (helical flow currents) and super elevation in water surface. Although, these curved reaches are sometimes stable, there are general tendency of bank failure and bed scour at the outer bend followed by sedimentation at the inner bend. Therefore, lateral migration of the reach planform occurs, consequently several morphological and navigational problems take place. Due to

these dynamic interactions, the transverse velocity profile, shear stress on channel bed, lateral bed slope, sediment size distribution, and energy expenditure will be changed (Grade, 1995).

A meander is a bend in a sinuous watercourse or river. It is formed when the moving water in a stream erodes the outer banks and widens its valley. A stream of any size could be assumed a meandering course, alternately eroding sediment from the outside of a bend and depositing them on the inside. The result is a snaking pattern as the stream meanders back and forth across its down-valley axis. When a meander gets cut off from the main stream, an oxbow lake is formed. (Hickin, 2003).

Stable alluvial river in natural state tends to maintain water conveyance of a specific magnitude by managing its sediment movement and deposition. Variation in local sedimentation, valley slope, geologic properties, and hydrograph influence its geometry as well as the arrangement of the sediment. Therefore, river meandering, lateral migration, deterioration of local navigation depths and flood conveyance are the result of the movement and deposition of bed sediment.

Attia and El-Saied (2004) investigated the statistical nature of river bends along Damietta branch. In this study, three bend types were defined as: free, limited, and forced; which were classified according to the physical and morphological characteristics and degree of freedom to attain the lateral shifting. They concluded that Damietta branch is changing in its planform several times down its course. Also, they summarized meander dimensions of many investigators such as given by Inglis (1938), Leoplod and Wolman (1960, 1964) and Zeller (1967). Based on the analytical regression of the non-linear relationships, there study derived many formulas for Damietta branch concerning the three mentioned types of bends. These formulas linked different parameters of meander geometrical sizes (Ahmed, 2010).

As the combined transport of water and sediment in rivers is a complex process, on-site investigations, evaluation of experience and large scale prototype tests are needed for verifying the results obtained from any mathematical or physical models.

For computation of the bed formation in river bends or near bifurcations, it is important to develop one dimensional models to be two dimensional models. In such models, the two dimensional flow equations in the x and y directions are used.

Three dimensional models need long time and large cost in computations (Wang, 1988; Wang et al., 1989; and Shimizu et al., 1990). In these models the state of turbulence is characterized by turbulence models such as standard k-ε, RNG k- ε and zero equation models.

Ibrahiem (2015) used iRIC 3-D model for modeling a reach from Damietta branch using upwind scheme for studying numerically the navigation way through this reach. Through his study a comparison between k-ε turbulence model and zero equation turbulence model was illustrated. It was important to illustrate the relationship between flow characteristics and associate between each other. So, in this paper the relationship between water velocity, vorticity and bed deviation was determined accurately through a long reach from Damietta branch of 20 km length using CIP method.

2. Site Description

As Damietta branch is very well concerned by Ministry of Water Resources and Irrigation and Ministry of Transport, Egypt, for passing maximum required discharges as well as to develop such safe navigation waterway, the study reach under consideration was selected. This reach is approximately 20 km long which locates from km 130 to km 150 downstream of Delta barrages, Damietta branch, Fig. (1).

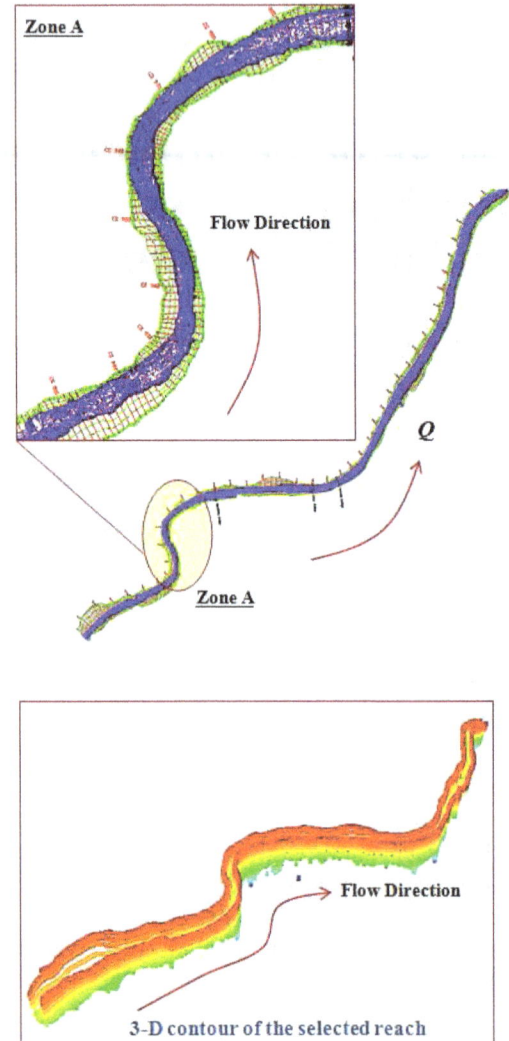

Fig. (1). Description of the reach under study.

3. Data Collection

The hydrographic survey of the study reach was carried out by Hydraulics Research Institute "HRI" of the National Water Research Center, Ministry of water resources and Irrigation, Egypt. Using the provided echo-sounder light boat, riverbed bathymetric survey was carried out along the branch following zigzag pathway trans-sections between the two river sides which are roughly spaced at 50 m intervals in stream wise direction. Moreover, in order to cover the study reach area, three longitudinal sections located at the left, middle, and right sides of the river reach were acquired.

Consequently, the provided differential Global Positioning System (GPS) were employed to record each data set point consisting of X and Y positions as well as the flow depth at an interval of one second on the equipped data logger.

Due to the significance of the acquired measurements, differential GPS system was utilized to provide a global accuracy of nearly 1.0 m in the plan direction with a relative depth accuracy of +/-10 cm. While the applied echo-sounder system permits flow depth measurements and consequently determining bed elevation with a relative accuracy of +/- 5 cm.

For shallow areas, where the flow depths are less than 0.75 m, another total station system was used which was launched on a light rubber boat (Zodiac). Then, the file of these coordinates (X, Y and Z coordinates) was prepared in the form of (XYZ.tpo) file that would be required for the 3D simulation. This file consisted of 102230 points, each point had X, Y and Z coordinates.

The velocity measurements were carried out at locations of 1.200 km, 9.000 km and 11.839 km from the upstream boundary of the reach under study which is located downstream km 130.

The grab sediment sampler was used to collect 48 bed material samples at different locations to prepare (d_{50}.anc) file used for the 3-D modeling process. The bed sample locations were selected to cover the entire features of the study reach and to represent the difference in the value of the Manning roughness. The samples were analyzed for grain size distribution, according to the relevant specifications, in the Hydraulics Research Institute.

Eight years data between 2005 and 2012 were collected downstream of Delta barrages, for estimating the maximum and minimum discharges from by Hydraulics Research Institute "HRI".

The resulted discharges from analysis of the obtained data could be given as:

- The measured discharge (35.50 M. m³/day): is the measured flow discharge during field measurements which equals to 35.50 million m³/day.
- The minimum flow discharge (9.90 M. m³/day): is the minimum recorded value downstream of Delta barrages throughout the studied years;
- The maximum flow discharge (62.10 M. m³/day): is the maximum recorded value downstream of Delta barrages throughout the investigated years; and
- The future discharge (80 M. m³/day): which are stated by the Nile Research Institute as future peak discharges for Damietta branch rehabilitation.

4. Model Set Up

- Governing Equations

Momentum equation in X-direction

$$\frac{\partial (uh)}{\partial t} + \frac{\partial (u^2 h)}{\partial x} + \frac{\partial (uvh)}{\partial y} = -gh\frac{\partial H}{\partial x} - \frac{\tau_x}{\rho} + D_x \qquad (1)$$

Momentum equation in Y-direction

$$\frac{\partial (vh)}{\partial t} + \frac{\partial (v^2 h)}{\partial y} + \frac{\partial (uvh)}{\partial x} = -gh\frac{\partial H}{\partial y} - \frac{\tau_y}{\rho} + D_y \qquad (2)$$

Continuity Equation

$$\frac{\partial (h)}{\partial t} + \frac{\partial (vh)}{\partial y} + \frac{\partial (uh)}{\partial x} = 0.0 \qquad (3)$$

where:

v : velocity in Y-direction;
u : velocity in X-direction;
τ_x : shear stress at X-direction;
τ_y : shear stress at Y-direction;
t : time;
h : water depth at any point;
ρ : water density; and
g : gravitational acceleration.

The governing equations were converted from co-orthogonal coordinates (X and Y coordinates) to represent the local stream lines into river coordinates, non-orthogonal coordinate system, (general coordinates or ξ and η coordinates). The non-orthogonal coordinate system allows more precise fitting of the coordinate system to suit arbitrary channel curvature and variable width. More importantly, the more detailed treatment of turbulence and large eddies allow predictions of time-variable behavior even for steady discharges.

- Standard K-ε Turbulence Model

Turbulent flow is dissipative, which means that kinetic energy in the small dissipative eddies are transformed into internal energy. The small eddies receive the kinetic energy from slightly larger eddies. The slightly larger eddies receive their energy from even larger eddies and so on. The largest eddies extract their energy from the mean flow. This process of transferred energy from the largest turbulent scales (eddies) to the smallest is called cascade process.

This model is represented by the following equations:

$$\upsilon = C_\mu \frac{k^2}{\varepsilon} \qquad (4)$$

$$\frac{\partial k}{\partial t} + u\frac{\partial k}{\partial x} + v\frac{\partial k}{\partial y} = \frac{\partial}{\partial x}\left(\frac{\upsilon}{\sigma k}\frac{\partial k}{\partial x}\right) + \frac{\partial}{\partial y}\left(\frac{\upsilon}{\sigma k}\frac{\partial k}{\partial y}\right)$$
$$+ P_h + P_{kv} - \varepsilon \qquad (5)$$

$$\frac{\partial \varepsilon}{\partial t} + u\frac{\partial \varepsilon}{\partial x} + v\frac{\partial \varepsilon}{\partial y} = \frac{\partial}{\partial x}\left(\frac{\upsilon}{\sigma \varepsilon}\frac{\partial \varepsilon}{\partial x}\right) + \frac{\partial}{\partial y}\left(\frac{\upsilon}{\sigma \varepsilon}\frac{\partial \varepsilon}{\partial y}\right)$$
$$+ C_{1\varepsilon}\frac{\varepsilon}{k}P_h + P_{\varepsilon v} - C_{2\varepsilon}\frac{\varepsilon^2}{k} \qquad (6)$$

where:

υ : eddy viscosity;
u : water velocity in X-direction;
v : water velocity in Y-direction;
k : turbulence kinetic energy;
ε : turbulence dissipation; and
t : time.
and,

$$P_h = \upsilon\left[2\left(\frac{\partial u}{\partial x}\right)^2 + 2\left(\frac{\partial v}{\partial y}\right)^2 + \left(\frac{\partial u}{\partial y} + \frac{\partial v}{\partial x}\right)^2\right]$$

$$P_{kv} = C_k \frac{u_*^3}{h}, \quad P_{\varepsilon v} = C_\varepsilon \frac{u_*^4}{h^2}$$

$$u_* = \sqrt{C_f \left(u^2 + v^2\right)}$$

$$C_k = \frac{1}{\sqrt{C_f}}, C_\varepsilon = 3.6 \frac{C_{2\varepsilon}}{C_f^{\frac{3}{4}}} \sqrt{C\mu}$$

where,

Cμ	C₁ε	C₂ε	σₖ	σε
0.09	1.44	1.92	1.0	1.30

- Sediment Transport Model

In general, it is assumed that the direction of sediment transport is the same as direction of flow and the sediment particles move in the bed layer zone only, Fig. (2).

The general modeling equation is:

$$\frac{\partial z_b}{\partial t} + \frac{1}{1-\lambda}\left(\frac{\partial q_{bx}}{\partial x} + \frac{\partial q_{by}}{\partial y}\right) = 0.0$$

where:
Z_b : elevation of bed;
t : time;
λ : porosity of sediment mixture;
q_{bx} : sediment rate in X-direction; and
q_{by} : sediment rate in Y-direction.

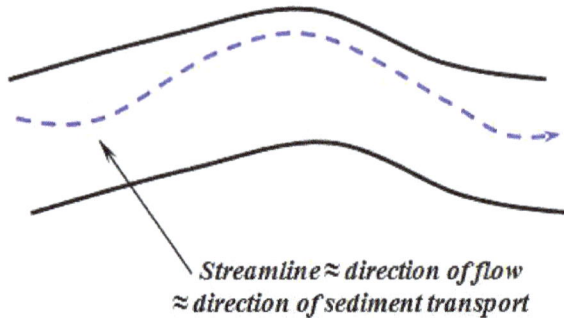

Streamline ≈ direction of flow
≈ direction of sediment transport

Fig. (2). *Definition sketch shows the direction of sediment transport.*

Watanabe gave the following equations for the sediment transport rates in X and Y directions as (Shimizu, 2012):

$$q_{bx} = q_b \left[\frac{u_b}{V_b} - \gamma\left(\frac{\partial z_b}{\partial x} + \cos\theta s \frac{\partial z_b}{\partial y}\right)\right]$$

$$q_{by} = q_b \left[\frac{v_b}{V_b} - \gamma\left(\frac{\partial z_b}{\partial y} + \cos\theta s \frac{\partial z_b}{\partial x}\right)\right]$$

according to Hasegwa's formula,

$$\gamma = \sqrt{\frac{1}{\mu_s \, \mu_k}}$$

where:
μ_s : static friction factor = 1.0
μ_k : kinematic friction factor = 0.45
- Vegetation Model

For modeling the effect of vegetation, the following equation is included in the model,

$$\frac{F_v}{\rho} = \frac{1}{2}C_{dv}\lambda_v \left(u^2 + v^2\right) hv$$

where:
F_v : drag force due to vegetation;
C_{dv} : drag coefficient;
λ_v : vegetation density; and
h_v : water depth or vegetation height, Fig. (3).

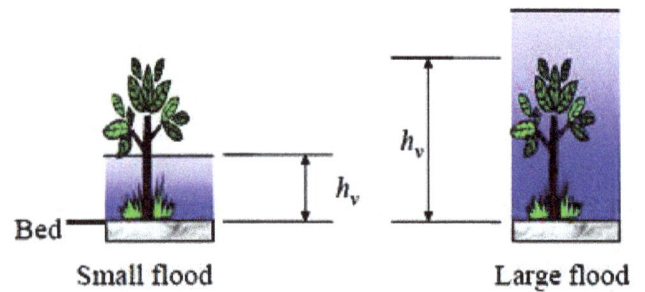

Fig. (3). *Indication of term h_v used in vegetation model.*

where,

$$\lambda_v = \frac{nd}{s^2}$$

s^2 is the area of grid cell, n is the number of stems of vegetation in the cell, d is the averaged diameter of each stem, Fig. (4).

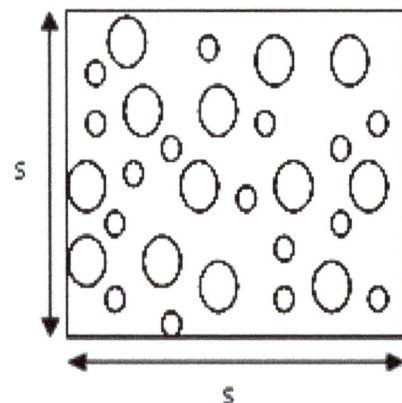

Fig. (4). *Plan view of vegetation points.*

- Grid Generation

The three dimensional model called IRIC (International River Interface Corporative), is a model used for supporting the numerical modeling of river morphodynamics using different turbulence models based on an explicit finite difference method (Abbott-Ionescu scheme) with CIP method for solving the advection terms.

The description of grid is given as:
- number of streamwise points = 61;
- number of cross-stream points = 11;
- number of iterations = 25; and
- centerline tensions = 10;
- standard relaxation coefficient = 0.2.
- stream wise increment = 347.63;
- cross stream wise increment = 30.00;
- number of sub segments = 41;
- number of triangles = 191298;
- number of vertices = 102230;
- number of incircle tests = 804607; and
- number of 2D orientation tests = 1072079.

The basic goal of mesh design is creating a representation of the water body that provides an adequate approximation of the true solution of the governing equations.

The stage of network design is finished when the contour of the whole reach can be plotted by the program. The different grid elements for the whole reach could be plotted as given in Fig. (5).

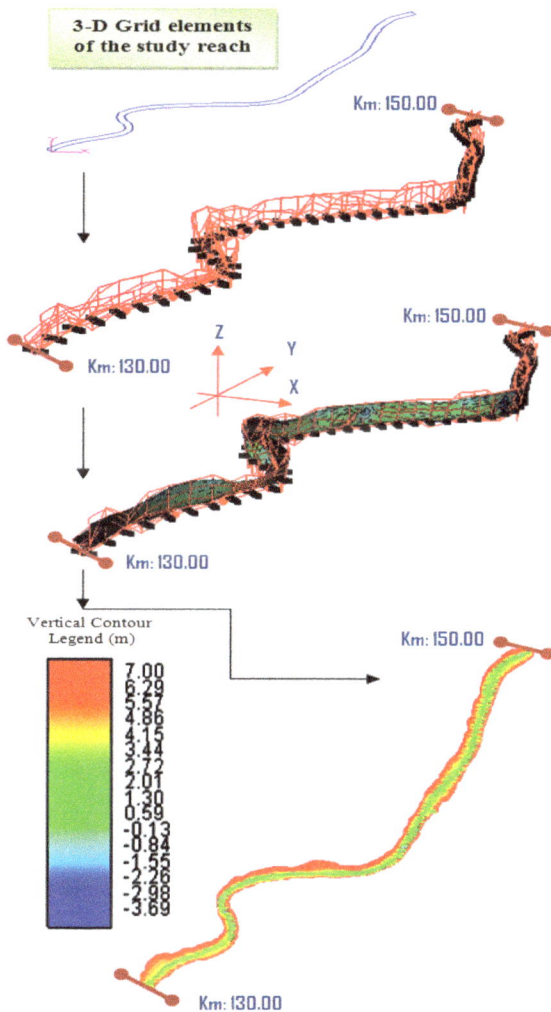

Fig. (5). *3-D contour map for the reach under study.*

For all runs the following items have to be adjusted:
- output time interval = 1 sec;
- calculation time step = 0.001 sec (minimum time step);
- start time for output = 0.0 sec;
- end of time steps calculations= 50 sec; and
- start time for bed deformation = 0.0 sec.

The boundary conditions can be specified for all runs as:
-no periodic boundary conditions;
-upstream velocity will be calculated according to uniform flow principles; and
-boundary condition slope will be estimated from geometric data.
-downstream water levels for measured, minimum, maximum and future discharges are 9.00 m, 8.10 m, 9.21 m and 10.00 m, respectively.

According to initial conditions; the initial water surface should be calculated according to the principles of non-uniform The median diameter of bed material (d_{50}) can be entered as a file with the extent of (.anc), and the standard value of critical angle of repose (φ) for bed material is used and equal to 0.3 (Shimizu, 2012).

In this model; both the length and width of the reach are divided into 60 and 10 units, respectively.

5. Modeling Process

In this research, the modeling process was carried out using the measured flow discharge of 410.87 m^3/sec.

To illustrate obviously the variation of water velocity with vorticity and deviation of bed elevation, two important zones consisted of several meanders were selected, Fig. (6).

Zone (1): [5.33 km length] lies from km: 132.00 (unit 6) to km: 137.33 (unit 22)

Zone (2) [5.34 km length] locates between km: 137.33 and km: 142.67 (unit 38).

For each zone, three longitudinal sections at 30%Bi (unit 3), 50% Bi (unit 5) and 70%Bi (unit 7) were selected. Where, Bi is the variable width of the reach at any cross section (i).

Fig. (7) shows the contours of water velocity at the two selected zones.

Fig. (8) and Fig. (9) exhibit the water velocity through the three longitudinal sections (at 30% Bi, 50% Bi and 70% Bi) for both zone (1) and zone (2), respectively.

Figs. (10) through (15) illustrate the water velocity, vorticity and bed deviation for the three selected longitudinal sections for zone (1) and zone (2), respectively.

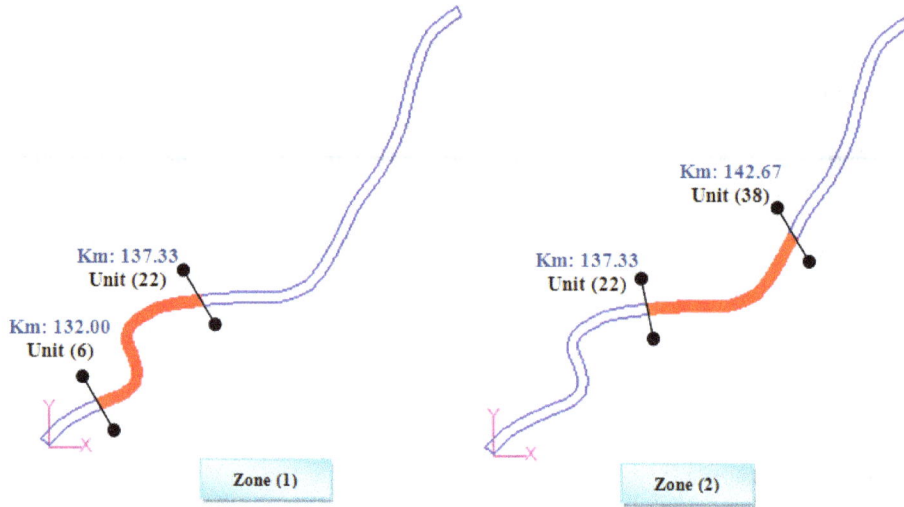

Fig. (6). *Selected two zones.*

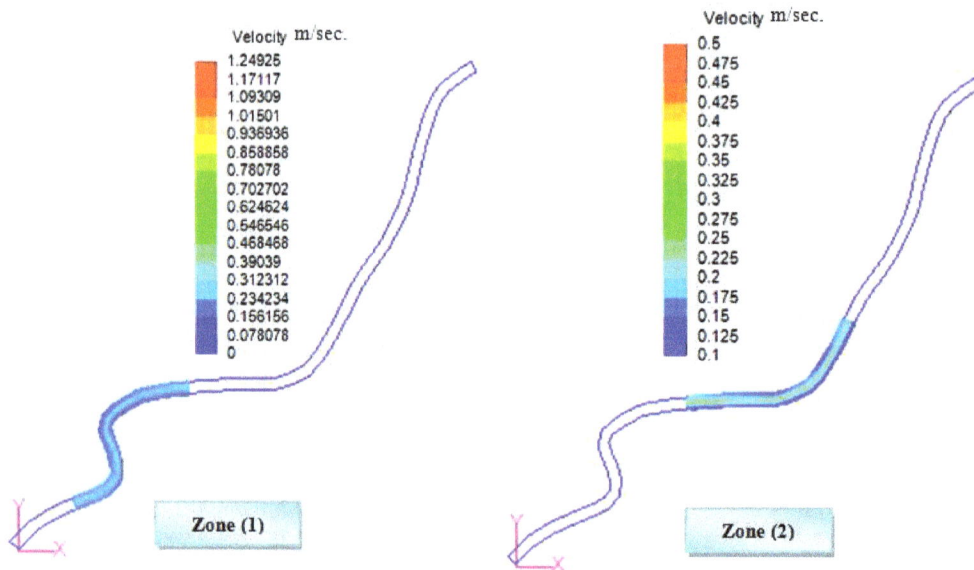

Fig. (7). *Filled velocity contours for the selected two zones.*

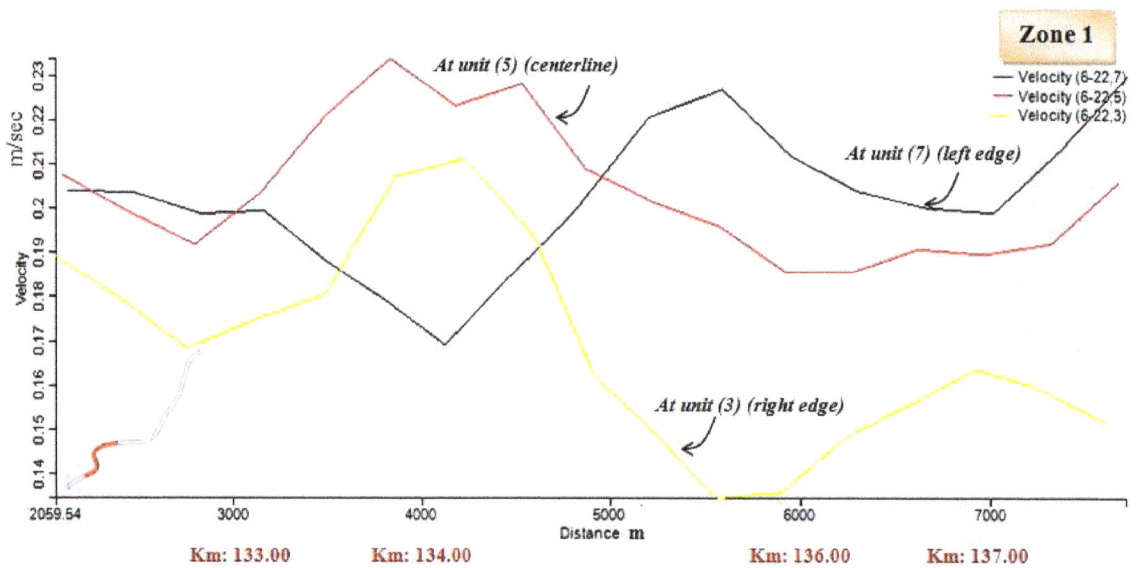

Fig. (8). *Water velocity for the three selected longitudinal sections at zone (1).*

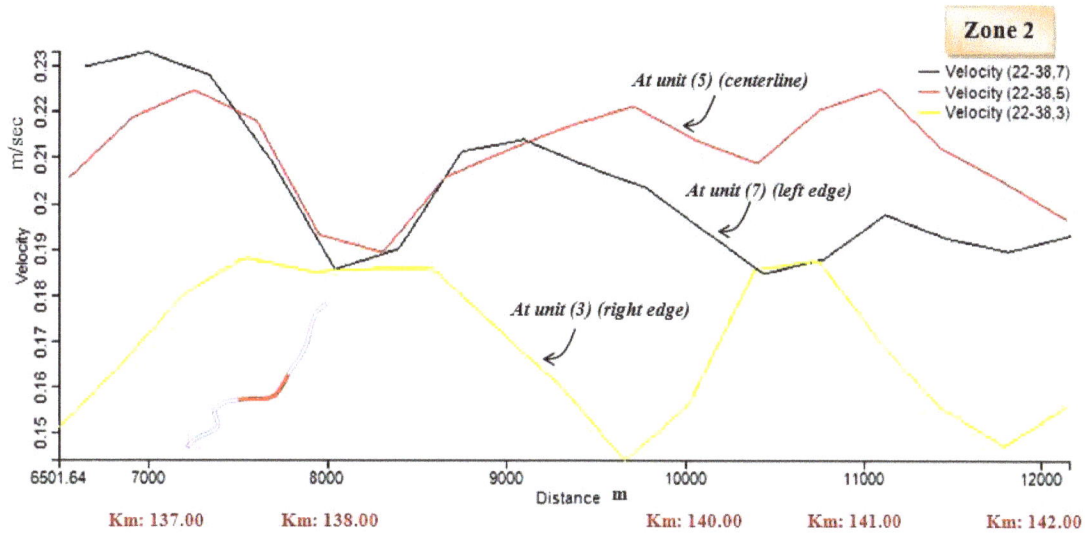

Fig. (9). *Water velocity for the three selected longitudinal sections at zone (2).*

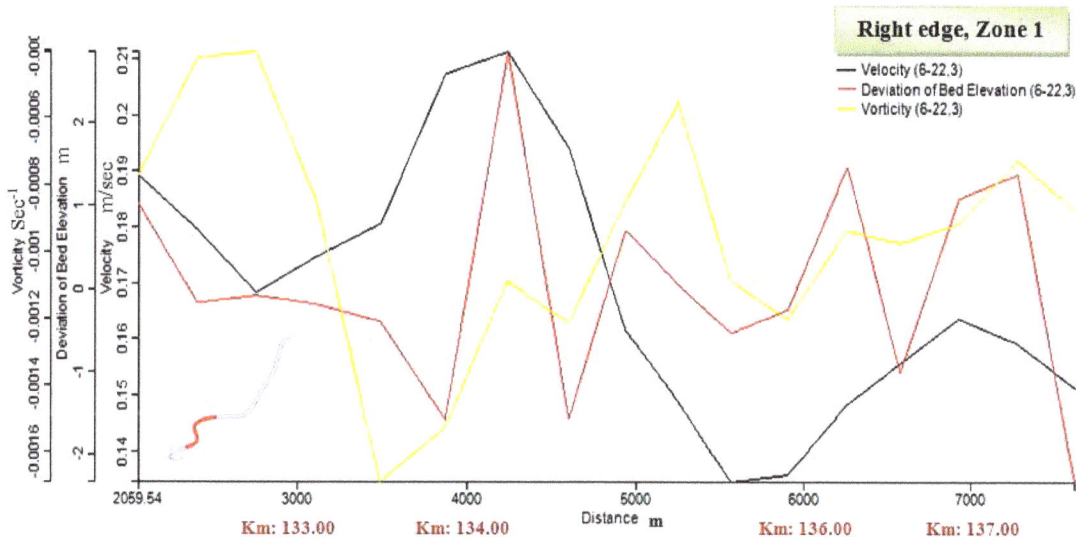

Fig. (10). *Water velocity, bed deviation and vorticity for zone (1) at longitudinal section (1) [30%Bi, right edge].*

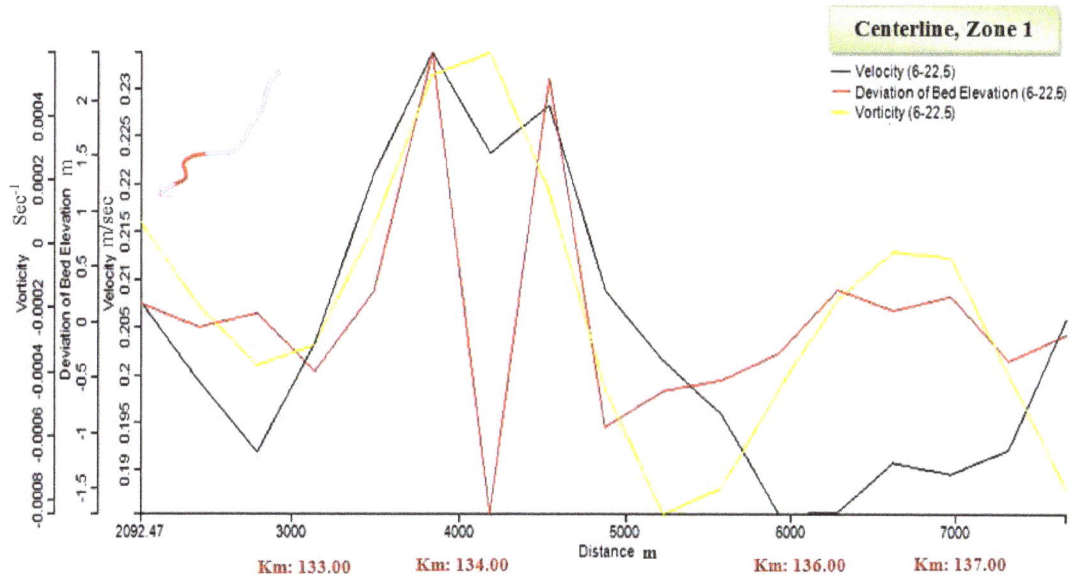

Fig. (11). *Water velocity, bed deviation and vorticity for zone (1) at longitudinal section (2) [50%Bi, centerline].*

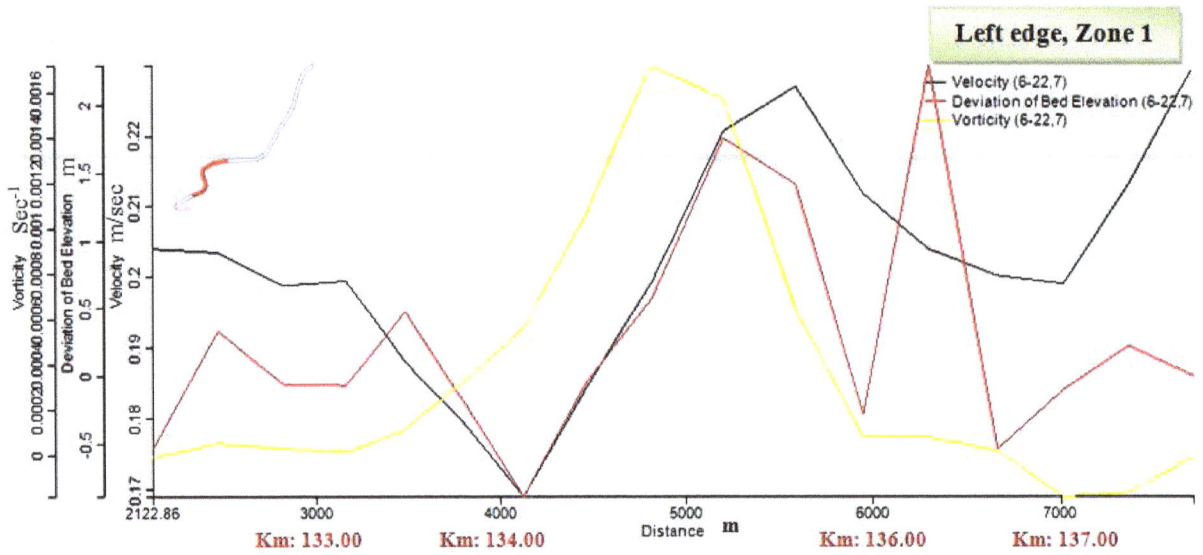

Fig. (12). *Water velocity, bed deviation and vorticity for zone (1) at longitudinal section (3) [70%Bi, left edge].*

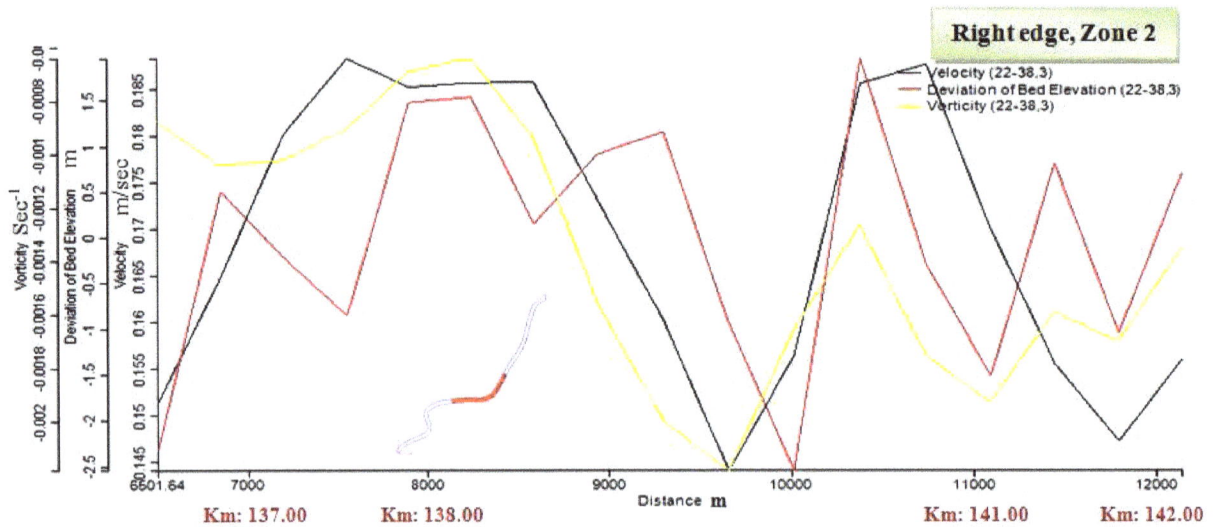

Fig. (13). *Water velocity, bed deviation and vorticity for zone (2) at longitudinal section (1) [30%Bi, right edge].*

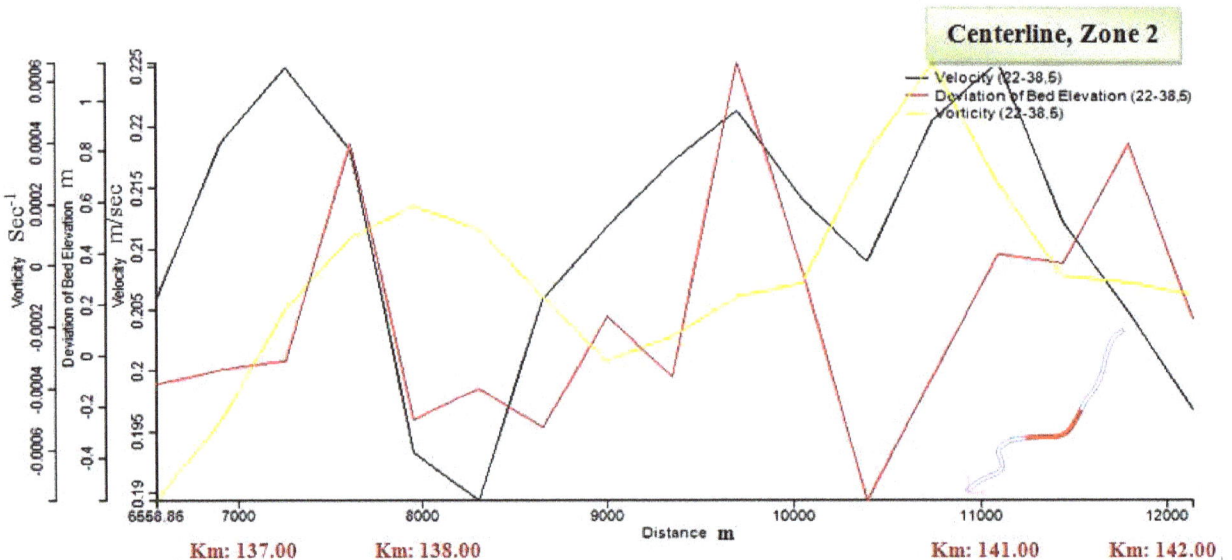

Fig. (14). *Water velocity, bed deviation and vorticity for zone (2) at longitudinal section (2) [50%Bi, centerline].*

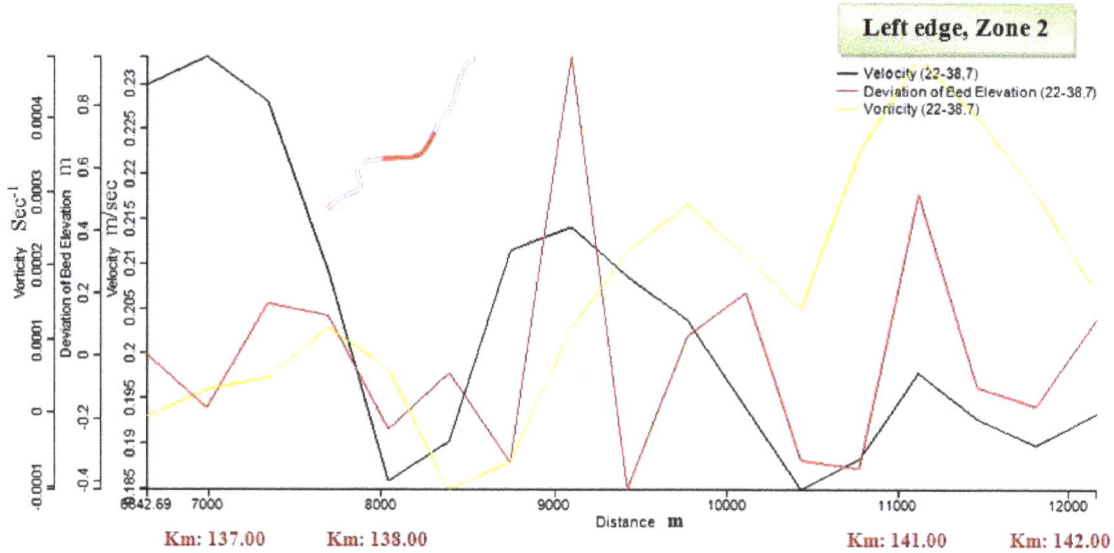

Fig. (15). Water velocity, bed deviation and vorticity for zone (2) at longitudinal section (3) [70%Bi, left edge].

From Figs. (8) and (9), it was observed that:-

- For zone (1), the water velocity ranged from 0.13 m/sec to 0.21 m/sec at right edge, from 0.17 m/sec to 0.23 m/sec for left edge and from 0.19 m/sec to 0.24 m/sec at the centerline.

- For zone (2), the water velocity ranged from 0.14 m/sec to 0.19 m/sec at right edge, from 0.19 m/sec to 0.23 m/sec for left edge and from 0.19 m/sec to 0.23 m/sec at the centerline.

- For zone (1), the maximum value of velocity at the right edge, left edge and centerline was found at km: 134.5, km: 135.8 and km: 133.85, respectively.

- For zone (2), the maximum value of velocity at the right edge, left edge and centerline was found at km: 137.5, km: 137.2 and km: 141.3, respectively.

It was noticed from these figures also, that the water velocity at the left edge and centerline was higher than the corresponding one at the right edge of the reach for both zones. This means that the effect of centrifugal force for bends here is very small and its effect could be neglected. It was also found that for both zones the values of velocities were ranged between 0.13 m/sec and 0.24 m/sec, and this range could be considered small to make scouring process.

It is important also to explain that these velocity values could arrive to the sedimentation range and may cause a serious danger for formation islands in this reach.

From analysis of Figs. (10) through (15), it was showed that there was a noticeable relationship between water velocity and vorticity for both zones, because when the water velocity increased the corresponding vorticity decreased and the corresponding bed deviation increased.

This action occurs when the X and Y components of velocity increases in the longitudinal direction and then, the corresponding vorticity of water in the vertical direction decreases and the deviation of bed elevation could increase.

Also, it was determined that the values of vorticity at the center line of the reach ranged between - 0.0008 sec^{-1} and + 0.0004 sec^{-1} for zone (1) and ranged from -0.0006 sec^{-1} to + 0.0006 sec^{-1} for zone (2), while the vorticity values at the right

edge were higher than the corresponding ones at the left edge.

The vorticity values at the centerline of the reach were less than the corresponding ones for both right and left edges.

Fig. (16) show the resulted deviation of bed elevation for the whole reach starting from km: 130.00 to km: 150.00. It is resulted from this figure that there was a noticeable sedimentation process covered the most of the reach.

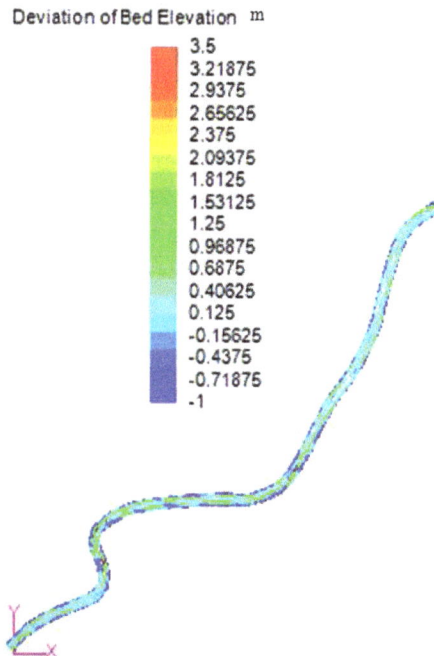

Fig. (16). Filled contours for resulted bed deviation.

6. Conclusions

A multi-meandering reach with length of around 20 km located between km 130 and km 150 downstream of Delta barrages - Damietta branch was selected and numerically studied. The aims of this study were simulating and modeling numerically the water velocity, vorticity and bed deviation for the curved zones of the reach under study and then, to determine accurately the vulnerable zones subjected to maximum velocities. Field data were collected and analyzed for the modeling process. A 3-D model called iRIC (International River Interface Corporative) based on an explicit finite difference method (Abbott-Ionescu scheme) was used. A standard K-ε turbulence model was employed using Cubic Interpolated Pseudo (CIP) method for solving the advection terms. The effect of vegetation and interactions between sediment motion and vegetation were added in the modeling process.

From this study, it was observed that the water velocity at the left edge and centerline was higher than the corresponding one at the right edge of the reach for the selected curved zones. This means that the effect of centrifugal force for bends here is very small and its effect could be neglected.

It was also found that for both selected curves, the values of velocities were ranged between 0.13 m/sec and 0.24 m/sec, and this range could be considered small to make scouring process, and then it could arrive to the sedimentation range and cause a serious danger for formation islands in this reach.

It was showed that there was a noticeable relationship between water velocity, vorticity and deviation of bed elevation for curved zones.

Acknowledgments

Grateful thanks for all staff of Hydraulics Research Institute (HRI), National Water Research Center, for their help in obtaining data.

References

[1] Ahmed A. F. (2010). "Improving navigation in river bends by bottom vanes", Scientific Bull., Faculty of Engineering, Ain Shams University, ISSN 1110-1385, Vol. 32, No. 1.

[2] Attia K. M. and El-Saied N.A. (2004). "Plan form geometry of river meander at Damietta branch", Scientific Bulletin, Faculty of Engineering, Ain Shams University, ISSN 1110-1385, Vol. 39, No. 1, pp.359-379.

[3] Grade R. J. (1995). "History of fluvial hydraulics", New Age Publishers. p. 14.and p.19, ISBN 812240815X. OCLC 34628134.

[4] Hickin J.S. (2003). "Meandering channels", International Journal of Recent Trends in Engineering, India, Vol. 1, No. 6, pp. 430–434.

[5] Ibrahiem, M.I., Zidan, A.A., El-Alfy, K.S. and Abdalla, M. G. (2015). "Numerical modeling of flow conditions and bed deviations of a reach from the river Nile", PhD thesis, Irrigation and Hydraulics Dept., El-Mansoura University.

[6] Inglis C.D. (1938). "Relationship between meander belts", Central Irrigation Hydrodynamic, India, Tech. Note No. 12.

[7] Leopold L. B. and Wolman M. G. (1960). "River meanders", Geological Society, Bull., Vol. 71, pp. 769-794.

[8] Leopold L. B., Wolman M. G., and Miller J. P. (1964). "Fluvial processes in geomorphology", W. H. Freeman, San Francisco, California.

[9] Shimizu Y.R., Itakura T.G., and Yamaguchi H.K. (1990). "Three dimensional computation of flow and bed deformation", Journal of Hydraulics, ASCE, Vol.116, No. 9, pp. 1090-1108.

[10] Shimizu Y.R. (2012). Lecture Notes, "3-D river hydraulics modeling", Hokkaido University, Japan.

[11] Wang S.S. (1988). "Three dimensional models for fluvial hydraulic simulation", Proceeding of the International Conference on Fluvial Hydraulics, Budapest, Hungary, 88-30.

[12] Wang S.S., Combs P.X., and Hu K.K. (1989). "New developments in modeling 3D sedimentation phenomena", New Orleans, Louisiana, Vol.14, No.18, pp. 33-38

[13] Zeller J.D. (1967). "Meander channels in Switzerland", Cranfield University Report, Britannia.

24

Pesticide Contamination Monitored by Passive Sampling in Environmental Water of Japanese Coral Island

Yutaka Tashiro[1], Yutaka Kameda[2]

[1]Faculty of International studies, Meio University, Nago, Okinawa, Japan
[2]Faculty of Engineering, Chiba Institute of Technology, Narashino, Chiba, Japan

Email address:

tashiro@meio-u.ac.jp (Y. Tashiro)

Abstract: The pesticide contaminations in the water of rivers and an estuary of a Japanese coral island with unique ecosystems in enclosed moats and on fringing reefs were analyzed by means of passive sampling. Samplers were deployed in the rivers and estuary for each 2 weeks of August through December of 2013 and of February of 2014. One to 12 kinds of pesticides were detected from all the samplers at the river sites with the maximum amount of 260 ng day^{-1} per sampler for EPN. The detected compounds and their amounts fluctuated widely at each sampling occasion. The analyses of the grab water samples detected much less compounds in comparison with the passive samples from the same sites. The time weighted average (TWA) concentrations of these pesticides are estimated as several micrograms per litter in the river waters. Further, 0.1 ng day^{-1} of procymidone was also detected from a sampler in the middle of the estuary where the river water is largely diluted with seawater. This amount of pesticide corresponds to a TWA concentration in the estuary water with the order of several ng L^{-1}. Considerable discharges of chemicals into coastal water by intensive agricultural practices such as flower cultivations on the island is concerned.

Keywords: Coastal Environment, Coral Reef, Agriculture, Pesticide, Water Pollution, Passive Sampling

1. Introduction

Coral reefs are one of the most biologically productive and diverse ecosystems in the world. Although they are important as the basis of marine fisheries and tourism in tropical and subtropical regions, they have been damaged by human activities, such as reclamations and water pollutions. Japanese Southwest Islands, which are composed of about 200 islands between Japanese main islands and Taiwan including the largest Okinawa Island (1,204 km^2), are surrounded by fringing reefs with coral ecosystems of prominent biodiversity [1]. As agriculture is one of the important industries on these islands, farms occupy a significant part of the land-uses there. Annual rainfall on these islands is approximately 2,000 mm, and a considerable part of the rain water is drained through the surface soil of the farms into coastal seawater as surface water and shallow groundwater. Although sugar cane has been traditionally cultivated on the majority of these farms, vegetable, fruit and flower cultivations are also increasing in recent years. As residual pesticides are generally controlled only in edible products, farmers tend to use a large amount of pesticides especially for the cultivation of flowers to avoid any pests which may damage their appearance. Previous researches on one of these islands revealed that more than ten times larger amount of pesticides are used for flowers in comparison with those for vegetables [2]. Although these backgrounds have been arising concerns about the impacts of pesticides discharged into coastal seawater on the coral ecosystems, information about the pollution level of water environment by chemicals in these islands is limited [3-6].

As the quantity and quality of agricultural drainages often change drastically with rainfall events and those of coastal seawaters with tide, grab sampling at river and/or sea is not always efficient to figure out the state of pollution. Also meaningful comparison between samples with the concentrations near the detection limit tends to need intensive laboratory works. The passive samplers such as Chemcatchers® [7] are based on the diffusion of target compounds through a membrane and the subsequent accumulation in a solid phase extraction disk. As the amount of the accumulated compounds reflect their time weighted average (TWA) concentrations in the environmental water where a sampler was deployed, these samplers are widely used for the assessments of pollutants with low and/or fluctuating concentrations in aquatic environment [8-11]. The present study revealed actual occurrences of pesticides in the rivers and coastal seawater of the Okinawa Island, the largest and the most populated one of these islands, by means of passive sampling.

2. Materials and Methods

2.1. Sampling Sites

Six sampling sites (A-F in Fig.1) were established on the Okinawa Island. Sites A and B were in the rivers of the northern area of the island, while sites C and D were in the central area. These sites were selected to represent the rivers which are located in the typical agricultural areas of the island and which receive the drainages from flower (chrysanthemums) farms. The distance along the river from each site to the coast is also shown in Fig.1. As these sites were on a small island, all of them were close to the coasts and thus to the fringing reef ecosystems around the island. Sites E and F were located at the inner part and at the middle of an estuary which is downstream of site B, respectively.

2.2. Target Compounds

Table 1 shows the 28 pesticide which were selected as the targets of the monitoring in the present study. They were consumed with large quantities in Okinawa Prefecture, Japan [12], and are the compounds which can be analyzed in aquatic samples by means of solid phase extractions and quantification with a gas chromatograph mass spectrometer [13].

2.3. Sampling Procedures and Chemical Analyses

The passive samplers which were consisted of a Empore® styrenedivinylbenzene (SDB-XC) disk (47 mm diameter, 3M) as a receiving phase fitted with a diffusion limiting membrane of low density polyethylene (Omnipore® 0.2μm pore size, 47 mm diameter, Merck) were used for sampling. Chemcatchers® were purchased from 3M Japan Ltd. (Tokyo, Japan), and the samplers with more compact design (Fig.2), which were similar to those used by Vermeirssen et al. [14] in the sense that they held the disk and membrane between the plates, were fabricated for the use in shallow rivers.

Samplers were deployed at sites A-D for each 2 weeks of August through December of 2013, and at sites E and F for 2 weeks of February of 2014. Afterwards they were brought back to the laboratory for disassembly. The SDB-XC disk was taken off the sampler body, dried under vacuum using a vacuum manifold for about 20 min, and subsequently eluted

twice with 5 mL acetone with 5 min ultrasonic treatment. The eluate was evaporated to 0.2 ml under a gentle stream of nitrogen and analyzed for pesticides using a GC-17A gas chromatograph (Shimadzu Corp., Kyoto, Japan) coupled to a QP5050 mass spectrometer (Shimadzu Corp., Kyoto, Japan) [15]. Samples were introduced with a 1 μL splitless injection and separation was achieved using a HP-5 capillary column (30 m x 0.32 mm; 0.25 μm film thickness) (J&W Scientific, Folsom, USA). Amounts of eluted pesticides from each sampler were calculated from detected pesticide peaks, and divided by the number of days of deployment to give TWA amounts of absorbed pesticides by the sampler per day.

One litter of grab samples of river water were also collected at sites A-C on the retrieval of the passive samplers in December 2013. They were extracted by solid phase extraction with SDB-XC disks, eluted with acetone, and analyzed in the same way as for the passive samples.

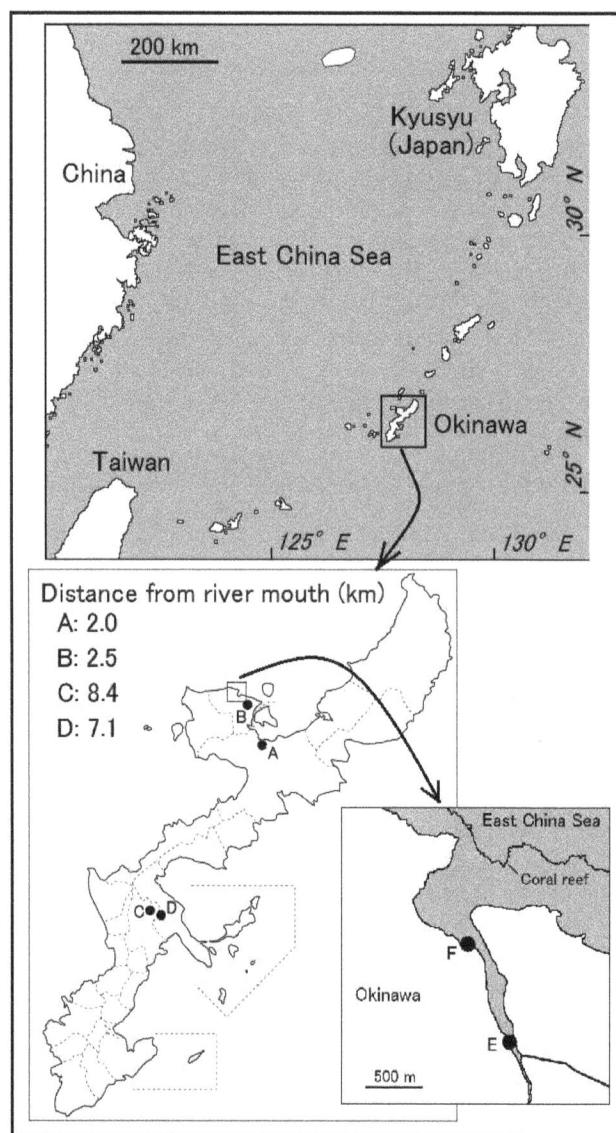

Figure 1. Location of the sampling sites.

Table 1. *Pesticides analyzed in the present study, and their consumptions [12] in Okinawa Prefecture in 2011 fiscal year.*

Insecticides	No. in Fig. 2	Consumption (ton)	Fungicides	No. in Fig.2	Consumption (ton)	Herbicides	No. in Fig.2	Consumption (ton)
MEP	1	19	Metalaxyl	16	26	Pendimethalin	23	1.5
Benfuracarb	2	6.1	TPN	17	5.9	Metribuzin	24	0.90
BPMC	3	5.9	Tolclofos-methyl	18	5.1	Esprocarb	25	0.31
Prothiofos	4	5.0	Captan	19	3.1	Mefenacet	26	0.12
Carbosulfan	5	4.5	Triflumizol	20	1.5	Pretalachlor	27	0.09
Carbofuran*	6		IBP	21	1.4	Dimethametryn	28	0.01
Ethylthiometon	7	4.3	Procymidone	22	1.3			
Isoxathion	8	2.4						
Diazinon	9	2.1						
Malathion	10	1.9						
DMTP	11	1.3						
EPN	12	1.3						
DEP	13	1.2						
MPP	14	1.0						
PAP	15	0.67						

*Carbofuran is a metabolite of carbosulfan.

3. Results and Discussion

One to 12 kinds of pesticides were detected from all the samplers deployed at sites A-D with the maximum amount of 260 ng day^{-1} for EPN from a sampler at site C (Fig.3). Only 8 among the all analyzed compounds were not detected from any of the samplers. The detected compounds and their amounts fluctuated widely at each sampling occasion. The analyses of the grab water samples detected much less compounds in comparison with the passive samples from the same sites. These results suggest that passive sampling could figure out the pollutions of this island more in detail than grab sampling.

The samplers at all these sites absorbed carbosulfan, carbofuran, and tolclofos-methyl in November, and procymidone in December. On the other hand, EPN and PAP were detected in November only at sites C and D in the northern area of the island, whereas BPMC was detected in September only at sites A and B in the central area, possibly reflecting different pesticide applications in each region. These features of pollution suggest that the fluctuation of the amounts should be attributed to the temporal discharges of pesticides from the farms caused by the spraying practices of the pesticides by the farmers and/or flushing by rainfall events during the deployment terms.

According to the calibration by Gunold et al. [16], the sampling rate of Chemcatcher® with SDB-XC disk without diffusion-limiting membrane for semi-polar pesticides is 0.12-0.44 L day^{-1}. It is also reported that the use of a membrane decreased the sampling rate for atrazine by a factor of 10 [15]. Based on these data, the TWA concentrations of the pesticides detected by the samplers in the present study are estimated as several micrograms per litter in the river waters. Thus, it is probable that the concentration of EPN at site C in November was exceeding the guideline value of 6 µg L^{-1} for EPN in water environments by Japanese Ministry of the Environment.

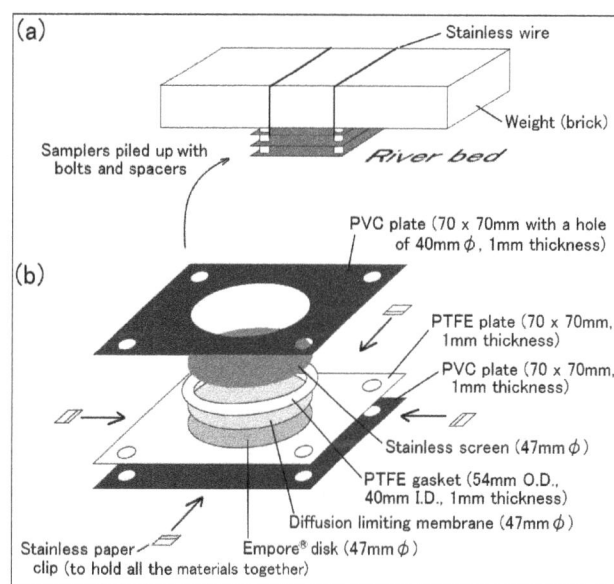

Figure 2. *Installation (a) and structure (b) of passive samplers.*

Sheikh et al. [5] detected 753 ng L^{-1} as maximum concentration of diuron, a herbicide commonly used in Japan, in a grab sample of river water from Ishigaki, another island of the Southwest Islands. They also detected 50 ng L^{-1} of diuron at another site about 4 km downstream of the same river near the coast. This reduction for about one order in concentration may be attributed to the dilution and/or decomposition of the compound in river water. In the present study, three pesticides, including procymidone with the largest amount of 2.8 ng day^{-1}, were detected at site E which is located at the inner part of an estuary of a river. As many of the pesticide amounts absorbed by the samplers at the river sites on this island were some dozens of ng day^{-1}, the reduced concentrations detected in the estuary were in harmony with the results in Ishigaki. Further, 0.1 ng day^{-1} of procymidone was also detected at site F in the middle of the same estuary where the river water is largely diluted with sea water. This amount of pesticide corresponds to a TWA concentration in water with the order of several ng

L^{-1}, which is in the same order of the concentrations of herbicides detected in inshore reef and river mouth sites of the Great Barrier Reef [17]. As the chemicals applied on the farms and discharged into fresh water run off to the coast within a relatively short time on the small islands, they potentially

approach the coral reefs without enough time and contacts with river beds necessary for degradation and/or absorption. Also the small water catchments of the rivers on the islands can provide a limited amount of water for dilution before the pesticides are discharged into the coastal seawater.

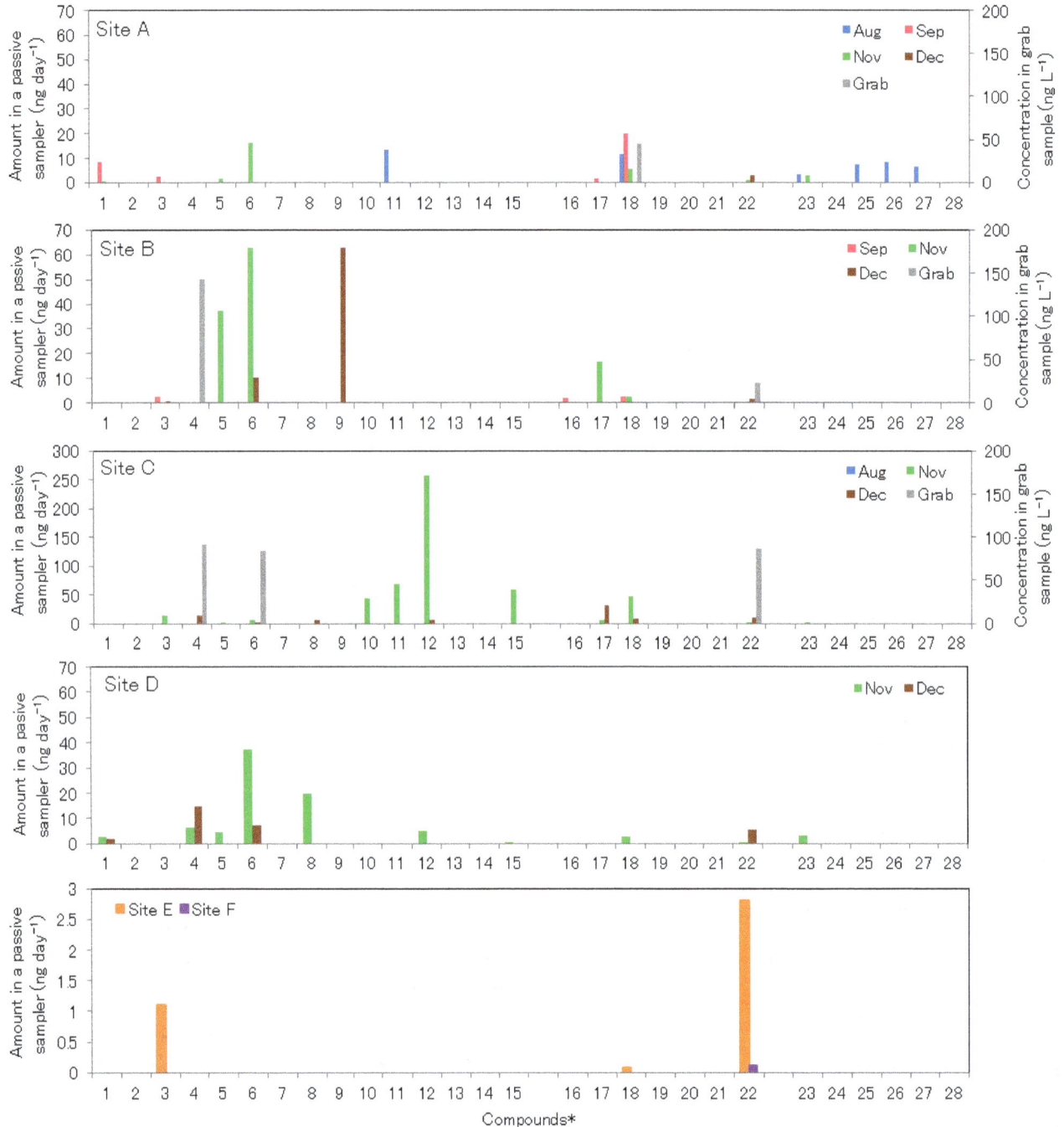

Figure 3. Amounts of pesticides absorbed by passive samplers and concentrations in grab samples (: compounds are indicated with the numbers corresponding to Table 1).*

4. Conclusions

The results of passive samplings revealed that a variety of pesticides are discharged through rivers into coastal seawater around the Okinawa Island, where the conservation of aquatic environments is important to maintain the unique ecosystems

in the enclosed moats and on the fringing reefs. Intensive agricultural practices such as flower cultivations may lead to considerable discharges of chemicals into coastal water even in the case of small islands. Although the concentrations of pesticides detected from the estuary in the present study were low, chronic environmental impacts and higher levels of temporal pollutions by occasional discharges are concerned. The importance of continuous efforts to improve the understandings on the sources and the distribution of seawater pollutions around these islands is emphasized.

Acknowledgements

This study was supported by Grant-in-Aid for Scientific Research of the Japan Society for the Promotion of Science (Grant No. 25340062).

References

[1] C. M. Roberts, C. J. McClean, J. E. N. Veron, J. P. Hawkins, G. R. Allen, D. E. McAllister, C. G. Mittermeier, F. W. Schueler, M. Spalding, F. Wells, C. Vynne and T. B. Werner, Marine biodiversity hotspots and conservation priorities for tropical reefs, Science, 295(5558), 2002, 1280-1284.

[2] Y. Tashiro and T. Taniyama, Pesticide contamination in groundwater on Okinoerabu Island, an intensive agricultural district in Japan (in Japanese), Jpn. J. Crop. Sci., 65, 1996, 77-86.

[3] S. T. Imo, M. A. Sheikh, K. Sawano, H. Fujimura and T. Oomori, Distribution and possible impacts of toxic organic pollutants on coral reef ecosystems around Okinawa Island, Japan, Pacific Science, 62(3), 2008, 317-326.

[4] Y. Kitada, H. Kawahata, A. Suzuki and T. Oomori, Distribution of pesticides and bisphenol A in sediments collected from rivers adjacent to coral reefs, Chemosphere, 71, 2008, 2082-2090.

[5] M.A. Sheikh, T. Oomori, H. Fujimura, T. Higuchi, T. Imo, A. Akamatsu, T. Miyagi, T. Yokota and S. Yasumura, Distribution and potential effects of novel antifouling herbicide diuron on coral reefs, in Herbicides - Environmental Impact Studies and Management Approaches, Dr. Ruben Alvarez-Fernandez (Ed.), ISBN: 978-953-307-892-2, 2012, InTech,

[6] Y. Tashiro and Y. Kameda, Concentration of organic sun-blocking agents in seawater of beaches and coral reefs of Okinawa Island, Japan, Marine Pollution Bulletin, 77, 2013, 333–340.

[7] J. Kingston, R. Greenwood, G. Mills, G. Morrison and L. Pesson, Development of a novel passive sampling system for the time averaged measurement of a range of organic pollutants in aquatic environments, Journal of Environmental Monitoring, 2, 2000, 487–495.

[8] R. B. Schäfer, A. Paschke, B. Vrana, R. Mueller and M. Liess, Performance of the Chemcatchers passive sampler when used to monitor 10 polar and semi-polar pesticides in 16 Central European streams, and comparison with two other sampling methods, Water Research, 42, 2008, 2707 – 2717.

[9] M. Shaw, M. J. Furnas, K. Fabricius, D. Haynes, S. Carter, G. Eaglesham and J. F. Mueller, Monitoring pesticides in the Great Barrier Reef, Marine Pollution Bulletin, 60, 2010, 113–122.

[10] F. Sánchez-Bayo, R. V. Hyne, G. Kibria and P. Doble, Calibration and field application of Chemcatcher passive samplers for detecting amitrole residues in agricultural drain waters, Bull. Environ. Contam. Toxicol., 90, 2013, 635–639.

[11] F. Sánchez-Bayo and R. V. Hyne, Detection and analysis of neonicotinoids in river waters – Development of a passive sampler for three commonly used insecticides, Chemosphere, 99, 2014, 143–151.

[12] Japan Plant Protection Association, "Noyaku Yoran (Pesticide Handbook) 2013" (in Japanese), 2013, JPPA.

[13] Japanese Ministry of the Environment, "Interim manual for investigations on endocrine disrupting compounds" (in Japanese), 1998, JME.

[14] E. L. M. Vermeirssen, C. Dietschweiler, B. I. Escher, J. van der Voet and J. Hollender, Uptake and release kinetics of 22 polar organic chemicals in the Chemcatcher passive sampler, Anal. Bioanal. Chem., 405, 2013, 5225–5236.

[15] A. T. K. Tran, R. V. Hyne and P Doble, Calibration of a passive sampling device for time-integrated sampling of hydrophilic herbicides in aquatic environments, Environ. Toxicol. Chem., 26, 2007, 435-443.

[16] R. Gunold, R. B. Schäfer, A. Paschke, G. Schüürmann and M. Liess, Calibration of the Chemcatcher passive sampler for monitoring selected polar and semi-polar pesticides in surface water, Environmental Pollution, 155, 2008, 52-60.

[17] M. Shaw and J. F. Mueller, Preliminary evaluation of the occurrence of herbicides and PAHs in the Wet Tropics region of the Great Barrier Reef, Australia, using passive samplers, Marine Pollution Bulletin, 51, 2005, 876–881.

PERMISSIONS

LIST OF CONTRIBUTORS

Dike Henry Ogbuagu
Department of Environmental Technology, Federal University of Technology, PMB 1526, Owerri, Nigeria

Ebenezer Temitope Adebayo
Department of Fisheries and Aquaculture Technology, Federal University of Technology, Owerri, Nigeria

Emmanuel Ikechukwu Iwuchukwu
Department of Agricultural Engineering, Federal Polytechnic, Nekede, Owerri, Nigeria

El Nadi M. H. A.
Public Works Dept., Faculty of Eng., ASU, Cairo, Egypt

El Sergany F. A. G. H.
Civil Eng. Dept., HIT, 10th of Ramadan City, Egypt

El Hosseiny O. M.
Civil Eng. Dept., Future Univ., Cairo, Egypt

Abed El Rahman Hassoun, and Elissar Gemayel
National Council for Scientific Research, National Center for Marine Sciences, Batroun, Lebanon
IMAGES_ESPACE-DEV, Universite de Perpignan Via Domitia, Perpignan Cedex, France
ESPACE-DEV, UG UA UR UM IRD, Maison de la teledetection, Montpellier Cedex, France

Veronique Guglielmi, Catherine Goyet and Franck Touratier
IMAGES_ESPACE-DEV, Universite de Perpignan Via Domitia, Perpignan Cedex, France
ESPACE-DEV, UG UA UR UM IRD, Maison de la teledetection, Montpellier Cedex, France

Marie Abboud-Abi Saab
National Council for Scientific Research, National Center for Marine Sciences, Batroun, Lebanon

Michele Giani4, Gianmarco Ingrosso
OGS (Istituto Nazionale di Oceanografia e di Geofisica Sperimentale), Oceanography Section, Via A. Piccard, Trieste, Italy

Patrizia Ziveri
ICREA - Institute of Environmental Science and Technology (ICTA), Universitat Autonoma de Barcelona, Ed. Z, ICTA-ICP, Carrer de les columnes, E- 08193 Bellaterra, Barcelona, Spain

Mohamed Gharbi and Amel Soualmia
Laboratory of Water Sciences and Technology, National Institute of Agronomy of Tunisia, University of Carthage, Tunis, Tunisia

Denis Dartus and Lucien Masbernat
Institute of Fluid Mechanics of Toulouse, National Polytechnic Institute of Toulouse, University of Toulouse, Toulouse, France

Inyang, Aniefiok I, Antai, Ekpo E, Dan and Monica U
Department of Marine Biology, Akwa Ibom State University, Ikot Akpaden Mkpat Enin, Nigeria

Sayed Sharaf El-Din
University of Alexandria, Faculty of Science, Department of Oceanography, Alexandria, Egypt

Mohamed Elsharkawy
National Institute of Oceanography and Fisheries, Physical Laboratory, Alexandria, Egypt

Ayelech Belayneh, Worku Batu
Department of Chemistry, Adigrat University, Adigrat, Tigray, Ethiopia

Josephat Alexander Saria
The Faculty of Science, Technology and Environmental Studies, The Open University of Tanzania, Dar es Salaam, Tanzania

Matobola Joel Mihale
Department of Physical Sciences, Open University of Tanzania, Dar es Salaam, Tanzania

Isabela Thomas Mkude
Dept. of Environmental Studies, Open University of Tanzania, Dar es Salaam, Tanzania

Alexei Popov
Mechanical Department, National University of Shipbuilding, Nikolaev, Ukraine

George Nerubenko
NERMAR Limited, Toronto, Canada, National University of Shipbuilding, Nikolaev, Ukraine

Solomon Melaku Melese
Department of Forestry, College of Agriculture, Wollo University, Dessie, Ethiopia

Giorgi Metreveli
Institute of Applied Ecology, FENS, Ivane Javakhishvili Tbilisi State University, Tbilisi, Georgia

Lia Matchavariani
Department of Geography, Faculty of Exact & Natural Sciences, Ivane Javakhishvili Tbilisi State University, Tbilisi, Georgia

George Nerubenko
NER*MAR Limited, Toronto, Canada – National University of Shipbuilding, Nikolaev, Ukraine

Ogeleka Doris Fovwe and Okpako Solomon
Department of Chemistry, Federal University of Petroleum Resources, Effurun, Delta State

Okieimen Felix Ebhodaghe
University of Benin, Geo-Environmental and Climate Change Adaptation Research Centre, Benin City

Masafumi Sasaki and Noboru Endoh
Department of Mechanical Engineering, Kitami Institute of Technology, Kitami, Japan

Samira Fuladavand
Faculty of Agriculture, Shahid Chamran University of Ahvaz, Ahvaz, Iran

Gholam Abbas Sayyad
Department of Soil Science, Faculty of Agriculture, Shahid Chamran University of Ahvaz, Ahvaz, Iran

Lingling Liu, Lixin Yi and Xiaoqing Cheng
College of Environmental Science and Engineering, Nankai University, Tianjin, China

Djibril Moudachirou
Institute of International Law, School of Law, Wuhan University, Wuhan, China

Yawar Abbas
World Wide Fund for Nature-Pakistan, NLI Colony Jutiyal, Gilgit, Pakistan
Department of Earth and Environmental Sciences, Bahria University, Islamabad, Pakistan

Babar Khan, Farasat Ali, Garee Khan, Ejaz Hussain and Saeed Abbas
World Wide Fund for Nature-Pakistan, NLI Colony Jutiyal, Gilgit, Pakistan

Syed Naeem Abbas
Forest, Parks and Wildlife Department, Gilgit, Pakistan

Rizwan Karim
Department of Forestry and Range Management, PMAS-Arid Agriculture University, Rawalpindi, Pakistan

Nawazish Ali
Department of Agriculture and Food Sciences, Karakoram International University, Gilgit, Pakistan

Mohammed Ibrahiem Ibrahiem Mohammed and Mohamed Ahmed Abdel Hady Eid
Irrigation and Hydraulics Department, Faculty of Engineering, Mansoura University, Mansoura, Egypt

Yutaka Tashiro
Faculty of International studies, Meio University, Nago, Okinawa, Japan

Yutaka Kameda
Faculty of Engineering, Chiba Institute of Technology, Narashino, Chiba, Japan

Index